LAND USE AND LAND COVER SEMANTICS

Principles, Best Practices, and Prospects

LAND USE AND LAND COVER SEMANTICS

Principles, Best Practices, and Prospects

Edited by

Ola Ahlqvist
Dalia Varanka
Steffen Fritz
Krzysztof Janowicz

CRC Press
Taylor & Francis Group
Boca Raton London New York

CRC Press is an imprint of the
Taylor & Francis Group, an **informa** business

CRC Press
Taylor & Francis Group
6000 Broken Sound Parkway NW, Suite 300
Boca Raton, FL 33487-2742

First issued in paperback 2017

© 2016 by Taylor & Francis Group, LLC
CRC Press is an imprint of Taylor & Francis Group, an Informa business

No claim to original U.S. Government works

ISBN-13: 978-1-4822-3739-9 (hbk)
ISBN-13: 978-1-138-74799-9 (pbk)

Visit the Taylor & Francis Web site at
http://www.taylorandfrancis.com

and the CRC Press Web site at
http://www.crcpress.com

To Pete

Contents

Preface

After decades of accomplishments and faced with new technological and scientific insights, the field of land use and land cover (LULC) is seemingly at a crossroads for effective and open uses of data. The use of categorical LULC data in computer-based land analysis poses a significant challenge because it usually leads to a binary treatment of the information in subsequent analysis. Still, LULC data offer a rich and generic resource and are often used for purposes other than just finding out what the land cover is at a location; examples include climate modeling, integrated assessment studies, land use modeling, monitoring of habitats and biodiversity, and simulation of urban expansion. As a result, the objectives for LULC semantics have moved from those of increasingly accurate technological representations of spatially explicit change to representation of the integrated roles LULC play within a broader environmental context. Many of the uses call for deeper understanding of the categories in order for the data to be repurposed. As more and more land cover data sets have been developed, there is also an increased recognition that variation in nomenclature and class definitions poses significant hurdles to effective and synergistic use of LULC resources.

The idea to compile a full book on information semantics specific to the area of LULC studies is not entirely intuitive, yet the result we have in front of us speaks to the rich and diverse aspects of this problem domain. It was clear from the outset that this book project was not something that one author could possibly cover, so it was decided that this had to be an edited book. But even as an edited book, it would have been hard for only one editor to span the disciplinary diversity needed for this project. As a result, we formed a team of coeditors lead by Ola Ahlqvist, with expertise and professional networks that spanned many of the pertinent areas related to this subject: LULC studies, ontology, semantic uncertainty, information science, earth observation, and more.

We knew from the start that there really was no one resource where a concentrated and diverse treatment of this problem area existed. We also felt that this area is gaining increased attention from multiple stakeholders in both academic and professional geoscience communities. We are therefore grateful that CRC Press encouraged us to develop this resource in the first place. Our intent was to offer a platform for scholars to use as a state of the art resource, reassess the field, affirm successful practical approaches, and point to future possibilities in advancing LULC semantics.

The thirteen chapters are the result of an open call and a peer-review process that was followed by revisions and editing. It is always hard to predict the outcome of such a process, but we are pleased to see the quality and variety in perspectives represented in these chapters. Although hard to subdivide into sections, the book does progress from introductory and foundational chapters, through contributions that focus on current best practice, to the later chapters, with their emphasis on ongoing developments and future prospects. We hope that this organization will help the readers to find their way, whether their interest is to review local to global practical contexts where LULC semantics are playing an important role or look further into aspects of semantic data modeling, or maybe learn about current and emerging approaches to manage LULC semantics.

By asking authors to provide fairly complete introductions to their specific issue, we hope that the book will be accessible to a broad range of users in the public and private sectors, including researchers and students as well as practitioners, such as LULC data producers, local and national authorities, and managers. In fact, one of the key concerns from the early review stage of this book was to ensure that the practical aspects would be as present as the theoretical ones. As a result, we hope that this will allow more stakeholders to adopt current best practices in their own work with land cover and land use information.

Finally, we thank all of the reviewers listed as follows for their constructive feedback on chapter manuscripts during the peer-review process: Elżbieta Bielecka, Gary Berg-Cross, Boyan Brodaric, Lex Comber, Robert Czerniak, Curdin Derungs, Wim Devos, Giles Foody, Anders Glimskär, Torsten Hahmann, Louisa Jansen, Marinos Kavouras, Margarita Kokla, Małgorzata Luc, Gaurav Sinha, Helle Skånes, Geoff Smith, Lynn Usery, Dalia Varanka, Jan Oliver Wallgrün, and Nancy Wiegand.

Ola Ahlqvist
Ohio State University

Dalia Varanka
U.S. Geological Survey

Steffen Fritz
*International Institute for
Applied Systems Analysis*

Krzysztof Janowicz
University of California

Editors

Ola Ahlqvist, PhD, is associate professor in geography at Ohio State University, Columbus, Ohio. He has worked professionally with local and regional environmental planning in Finland and Sweden in the early 1990s. After completing a PhD in geography at Stockholm University, Stockholm, Sweden, in 2001 followed by postdoctoral training at Pennsylvania State University, State College, Pennsylvania, he joined the Geography Department at Ohio State University in 2005. His research interest spans three areas: (1) semantic uncertainty and formal ontology in analysis of land cover change, landscape history, and visualization; (2) how online maps, social media, and games form a nexus for spatial collaboration, socioenvironmental simulation, and decision making; and, (3) the scholarship of engagement in geographic information system and service learning.

Dalia Varanka, PhD, is research geographer at the U.S. Geological Survey and adjunct professor at Johns Hopkins University, Baltimore, Maryland. After working at the Field Museum of Natural History and the Newberry Library in Chicago, Illinois, she received her PhD in geography from the University of Wisconsin–Milwaukee, Wisconsin, in 1994. She began her federal career as a Physical Science Technician with the U.S. Bureau of Land Management in Milwaukee, Wisconsin. She joined the Mid-Continent Mapping Center, Rolla, Missouri, in 1997 and became a research scientist in 1999. She has worked in a professional capacity with a wide range of geospatial information technologies and applications, including GIS, remote sensing, cartography, geospatial statistics, and geographical texts. She conducts research in geospatial semantics and ontology and teaches a graduate-level course on that subject.

Steffen Fritz, PhD, is a researcher at the International Institute for Applied Systems Analysis, Austria. He has a master of science degree in geographical information for development from the University of Durham, Durham, United Kingdom, and a PhD from the School of Geography at the University of Leeds, Leeds, United Kingdom. As a postdoctoral fellow at the Joint Research Centre (JRC Ispra), Italy, his main focus was to mosaic, harmonize, and produce the Global Land Cover

GLC2000 database, but he has also studied the uncertainties in global land cover. His research interests include crowdsourcing, volunteered geographical information, semantics, land-use science, earth observation, citizen science, global and regional vegetation monitoring, wild land mapping, crop yield and crop acreage estimations, serious gaming, gamification, in situ data collection of land use and land cover via crowdsourcing, and the potential of using crowdsourcing in developing countries. He is the initiator and driving force behind Geo-Wiki.org and Geo-Wiki mobile.

Krzysztof Janowicz, PhD, is an assistant professor for geographic information science at the Geography Department of the University of California, Santa Barbara (UCSB), California. He is the chair of UCSB's Cognitive Science Program and one of two editors-in-chief of the *Semantic Web* journal. Before moving to Santa Barbara, he was an assistant professor at the Pennsylvania State University, State College, Pennsylvania, and a postdoctoral researcher at the University of Muenster, Muenster, Germany. He is studying the role of space and time for knowledge organization and is especially interested in geosemantics, geo-ontologies, and geographic information retrieval.

Contributors

Helbert Arenas
Checksem Team, Laboratory Le2i
University of Burgundy
Dijon, France

Stephan Arnold
Federal Statistical Office
Wiesbaden, Germany

Alkyoni Baglatzi
School of Rural and Surveying
    Engineering
National Technical University of
    Athens
Athens, Greece

Gebhard Banko
Environment Agency Austria
Wien, Austria

Gary Berg-Cross
Spatial Ontology Community
    of Practice
Potomac, Maryland

Elżbieta Bielecka
Military University of Technology
Warsaw, Poland

Alexandra Björk
Flen municipality
Flen, Sweden

Michael Bock
German Aerospace Center
Bonn, Germany

Alexis Comber
Department of Geography
University of Leicester
Leicester, United Kingdom

Christophe Cruz
Checksem Team, Laboratory Le2i
University of Burgundy
Dijon, France

Robert J. Czerniak
Department of Geography
New Mexico State University
Las Cruces, New Mexico

Wim Devos
Monitoring Agricultural Resources
    Unit
Institute for Environment and
    Sustainability
European Commission, Joint
    Research Centre
Ispra, Italy

Peter Fisher
Department of Geography
University of Leicester
Leicester, United Kingdom

Anders Glimskär
Department of Ecology
Swedish University of Agricultural
    Sciences
Uppsala, Sweden

William J. Gribb
Geography Department
University of Wyoming
Laramie, Wyoming

Benjamin Harbelot
Checksem Team, Laboratory Le2i
University of Burgundy
Dijon, France

Gerard Hazeu
Alterra—Wageningen UR
Wageningen, The Netherlands

Louisa J. M. Jansen
Climate, Energy and Tenure Division
Food and Agriculture Organization
    of the United Nations
Rome, Italy

Marinos Kavouras
School of Rural and Surveying
    Engineering
National Technical University of
    Athens
Athens, Greece

Alexander Klippel
The Pennsylvania State University
State College, Pennsylvania

Margarita Kokla
School of Rural and Surveying
    Engineering
National Technical University of
    Athens
Athens, Greece

Barbara Kosztra
Institute of Geodesy, Cartography
    and Remote Sensing (FÖMI)
Budapest, Hungary

Małgorzata Luc
Institute of Geography and Spatial
    Management
Jagiellonian University in Krakow
Krakow, Poland

David Mark
NCGIA and Department of
    Geography
University at Buffalo
Buffalo, New York

Pavel Milenov
Remote Sensing Application Center
and
Agency of Sustainable Development
    and Eurointegration
Sofia, Bulgaria

Christoph Perger
International Institute for Applied
    Systems Analysis
Laxenburg, Austria

Nuria Valcarcel Sanz
National Geographic Institute
Madrid, Spain

Helle Skånes
Department of Physical Geography
    and Quaternary Geology
Stockholm University
Stockholm, Sweden

Geoffrey Smith
Specto Natura Ltd.
Cambridge, United Kingdom

Tomas Soukup
GISAT
Prague, Czech Republic

Kevin Sparks
The Pennsylvania State University
State College, Pennsylvania

Geir-Harald Strand
Norwegian Forest and Landscape
    Institute
Ås, Norway

E. Lynn Usery
U.S. Geological Survey
Rolla, Missouri

Dalia E. Varanka
U.S. Geological Survey
Rolla, Missouri

Richard Wadsworth
Department of Biological Sciences
Njala University
Freetown, Sierra Leone

Jan Oliver Wallgrün
The Pennsylvania State University
State College, Pennsylvania

Nancy Wiegand
University of Wisconsin–Madison
Madison, Wisconsin

Naijun Zhou
University of Maryland
College Park, Maryland

1

Land Use/Land Cover Classification Systems and Their Relationship to Land Planning

William J. Gribb and Robert J. Czerniak

CONTENTS

ABSTRACT Current and accurate information about the natural and human landscape is critical to assist governments in planning for the future needs of their citizens. To assist in the planning effort, governments identify the goals and objectives that the citizens want them to use as the direction for decision making. Land use/land cover analysis assists policy makers and planners to obtain a working knowledge of land activity occurring in their community, the region, state, or nation. The linkage between planning goals, objectives, and policy and land use classification systems is the focus of this chapter. Integrating the current land use and land cover information with indices and models of landscape change provides a means by which decision makers can establish policies that sustain the population in the future. The knowledge of land utilization and consumption is essential to effective and efficient land planning and management. The expanding spatial dimension of the urban-metropolitan area can be three times larger than the actual population increase. The land consumption rate is often evidenced as urban sprawl and the demand for land can increase faster than population growth would dictate. However, the different land use/land cover classifications were often designed to describe the land use and cover of Earth's surface, not necessarily to inform decisions about land use planning. A land classification system should address three criteria: (1) describe the nature of existing land uses accurately and in adequate detail; (2) fit consistently with the logic and classes of future land use plans; and (3) be compatible with the typology of uses in the development-management policies, regulations, and ordinances. The importance of the classification system is not in its level of detail or the method of collecting, compiling, and calculating the data. It is whether or not and to what level the data support the decision-making process and the policies that are created for a sustainable community that are important. Three examples are used to demonstrate the need for land use/land cover information at different planning scales. The hierarchical structure of most land use/land cover systems provides a means by which broad categories can become more specific as the need for detail increases. The main issue is that the land use/land cover classification system should meet the needs of the overall land use plan and be flexible enough to be incorporated into the policy analysis. This requires that the data collection techniques, land use/land cover categories, analytical techniques, and policy decisions are linked so that the appropriate information can be utilized by the different levels of government, including decision makers, planning staff, and stakeholders.

KEY WORDS: *Land use planning, land use/land cover classification system, decision making.*

1.1 Introduction

Local and state governments undertake land planning to meet current and future needs of their citizens regarding the environmental, economic, and social dimensions of their lives. This includes protecting and maintaining the natural landscape and determining ways and means by which the natural landscape can be used to sustain the population. Human needs for the essentials of life (water, food, housing, and energy) have to meet current and future population needs at a sustainable rate. The built environment can be effectively planned to satisfy the changing needs of the population, if the resources and land are available.

Current and accurate information about the natural and human landscape is critical to assist governments in planning for the future needs of their citizens. Understanding the nature and amount of urban growth or decline and identifying goals and objectives to maximize the benefit from these changes requires a direct and clear linkage to land cover/land use classifications. Many land cover and land use schemes have been proposed. Some classifications help policy makers and planners obtain a working knowledge of land activity occurring in their community and region. The need for this knowledge can also be applied to larger areas such as a state or nation. The linkage between planning goals, objectives, and policy and land use classifications is the focus of this chapter. The linkage is discussed in relation to a sustainable land planning process and how this process should strongly influence the way a land information system, including land use and land cover, is created and used.

Integrating the current land use and land cover information with indices and models of landscape change provides a means by which decision makers can establish policies that may sustain the population in the future. There are a multitude of ways by which current landscape information can be collected: field observation, field questionnaires, public surveys, mapping, and remote sensing. The different forms of data collection can be integrated in providing essential information about the landscape. Incorporating data from the past can provide a temporal sequence that can offer insight into landscape change; an important element in developing land use policy.

1.2 Population Demands and Change to the Landscape

Human-induced change to the landscape is a function of demand. The 2012 United Nations (UN) World Population Prospects estimated the world population to be 7.3 billion in mid-2013. Using a medium-variant projection, the population could increase to 8.1 billion by 2025, an increase of almost 11% in the next decade. This rate of change is not universal and regions of the world will experience higher or lower rates of change. No matter which country or region of the world in which the population is changing, there are continual demands on world resources, in some areas more intense than others. The Population Institute (2007) identified three elements in the environment that are critical to human life: water, food, and energy. The amount and type of land that is consumed is an important contributor to the provision of these elements. Knowledge of land utilization and consumption is essential to effective and efficient land planning and management.

According to the UN report on the state of cities, the largest growth in the world population will occur in a developing country's cities (UN-Habitat 2013). It is estimated that by mid-twenty-first century, 7 out of 10 world inhabitants will be living in cities (UN-Habitat 2013, p. 25). Historically, migration from the rural to urban areas was the largest contributor to urban population expansion. However, in the past decades, it is estimated that natural increase is contributing almost 60% to urban population growth. And, that rural settlements are becoming larger and are being "reclassified" as urban areas, which accounts for another 20% of the urban growth rate (UN-Habitat 2013, p. 25). The urban growth, however, is occurring in two different realms: changing population distribution and changing density.

There is an overall decline in the population of central cities and an overwhelming increase in the periphery of the city-urban decline and suburban growth. The expanding spatial dimension of the urban-metropolitan area can be three times larger than the actual population increase. The land consumption rate is often evidenced as urban sprawl, and the demand for land increases faster than the population. As an example, over the past 30 years, Mexico's urban expansion averaged 7.4% per year, four times the urban population growth over the same period (UN-Habitat 2013, p. 32). At the same time, even though the urban population is growing, the density in the urbanized region is declining. According to the UN report, in a survey of 88 cities, 77 cities recorded declines in population density, from an average of 174 persons/ha to 137 persons/ha, whereas in some Asian cities, the decreases were more than 25% and in periphery regions more than 50% (UN-Habitat 2013, pp. 32–33). The land consumption of urbanizing areas will continue as megacities increase and the proliferation of small- and medium-sized cities grow.

Governments need to provide a means by which decision makers and land managers have access to information about the lands under their jurisdiction. This should include the location of parcels, detailed information on land use, zoning, infrastructure, land value, land and building conditions, and the physical and environmental attributes of the parcel. Overall, there are four main functions of a land management system: land tenure, land value, land use, and land development (Williamson et al. 2010). Local and state governments should have the institutional capability to enforce legislation; develop and maintain land records; and provide the land information for management, taxation, land markets, and land planning (UN-Habitat 2012).

1.2.1 Planning Environments

The global landscape can be divided into four general environments: natural, rural, urban, and transition (Botequilha Leitao et al. 2006) (Figure 1.1).

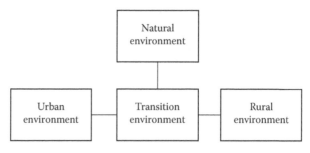

FIGURE 1.1
Planning environments.

Their size and structure, and how they are interconnected, are important to global, national, regional, and local decision making. The natural environment includes those areas that have had minimal or no visible impact by human activity. Obviously, the urban environment drastically alters the natural environment and exhibits maximum impact from human activity. The rural environment alters the natural environment but does not reach the level of development required for urban activity. Both the scale and intensity of rural development is relatively at a low level. Rural environments could include agricultural areas or forested areas that have a majority of the area in replacement and harvestable species. The transition environment is the space in between the other three levels of development. For instance, an urban area that undergoes transitions into the peri-urban fringe into a rural environment, which then transitions to a national forest. Although each of the environments carries its own significance, the rural environment, because of its relatively lighter effect on the landscape, is not described in this chapter. The other three environments are each discussed. The urban environment is of the highest concern because of its enduring effects on the natural landscape, the high demand for detailed land use and land cover information, and the need to carefully plan for future demands.

1.2.2 Natural Environment

The natural environment is identified as areas that have been minimally impacted by human activities (Hendee, Stankey, and Lucas 1990). These areas serve as the most valuable areas for natural ecosystem services (Daily 1997; deGoot, Wilson, and Boumans 2002), a function that has to be maintained if the human species is to survive into the future. Natural science can identify what services are needed for human survival; what is not known is the minimum spatial and species extent needed to provide the services. Thus, in land use decision making, identifying, protecting, maintaining, and rehabilitating the natural landscape are essential elements in conserving all elements of an ecosystem.

Ecosystem management requires the balancing of four different needs required by the human population (Gregg 1994):

- Materials and energy needs: The production of materials, goods, and energy that is needed for the world population to survive or used for the development of new resources or technologies. This incorporates not only energy resources but also timber and food production, water resources, and minerals.
- Social needs: The ecosystems in which settlement and human interactions take place. The environmental factors that influence where people settle or enjoy interacting with other people and with the natural environment through recreation.

- Spiritual needs: The places in the ecosystem that have cultural, historic, or aesthetic significance to the population. Sites that represent religious, symbolic, or metaphysical connections to the natural environment.

- Informational needs: The areas in the natural environment and eco-system that provide the opportunity to expand our knowledge of the biotic and abiotic world.

Each of the activities identified above requires specific land information if we are capitalizing the benefits from them. For example, the identification of high-quality land for timber, food, and/or mineral production is essential.

1.2.3 Urban Environment

As mentioned earlier, the world population is becoming more urbanized. In 2010, for the first time over 50% of the world population lived in urban areas. Future urbanization trends will have definite landscape effects. According to the UN, three urban growth trends are emerging: mega-cities, urban corridors, and city regions (UN-Habitat 2009). Megacities are the places exceeding 20 million people or more. For example, the Hong Kong–Shenzhen–Guangzhou area has a combined population of over 120 million people. Similarly, the Tokyo–Nagoyo–Osaka–Kyoto–Kobe region is expected to have 60 million residents by 2015 (UN-Habitat 2009, p. 7). Their size and complexity require the collection of detailed land information in support of land planning and decision making.

Urban corridors are configured in a different spatial pattern linking urban employment along transportation routes, which may include megacities. An elongated urban corridor is developing in the industrialized section of India, stretching over 1500 km from Mumbai to Delhi. In West Africa, the urban corridor connects four countries between Ibandan–Lagos–Accra, a stretch of over 600 km, and is the heart of this portion of Africa's economic develop-ment. An urban corridor is a linkage between economic, development, land, and employment in select regions of the world that have the resources to pro-duce goods but, more importantly, can transport them to markets efficiently and inexpensively. An urban corridor's land information and transport information must be brought together to support travel demand modeling, which has the potential to reduce congestion and air pollution while maxi-mizing the efficient movement of people and goods.

The third trend is development of the city region, which is a combination of expanding urban areas and accompanying suburbanization. City regions are large areas of urbanization that, as the population expands horizontally, consume adjacent cities and towns. The population density decreases on the edges of urban sprawl, yet continues to expand with increases in transporta-tion networks and commercial/industrial development. The Bangkok urban region is expected to extend its urban edge 200 km from the city center by 2020. Metropolitan Sao Paulo is over 8000 km² with a population of almost 16.4 million people, both an area and population larger than some countries (UN-Habitat 2009, p. 10). In city regions, land information is needed to anticipate urban expansion through the use of modeling, which requires parcel-level information (Waddell 2003).

1.2.4 Transition Environment

The spectrum of human activities from natural areas to urban areas can be visualized as areas that are untrammeled to urban scenes that are completely built environments. In most cases this includes the rural environment, with sparse settlement, utilizing the natural resources or land for agriculture: either grazing, cropping, or a combination of both. Encroachment into the natural environment can take the form of recreation activities, trails, and campsites. Or, it can be the intrusion for natural resource extraction, from mining to lumbering. Agricultural activities in natural areas are mainly grazing. More intense interface activities can include cropping the newly cleared forest regions along with reforesting for commercial timber yields. In rural areas, agriculture can be disrupted by increased intensity of housing, transportation networks, and industrial development. In 2006, the Food and Agricultural Organization (FAO) calculated that approximately 13 million hectares of forest/woodland are cut down every year and converted to agricultural uses. In addition, the UN estimates that between 1950 and 1990 over 22% of all cropland, forested areas, and pastureland was degraded (UN-DESA 2008), thus requiring more land for production.

1.3 Land Planning

In a survey, the UN found that urban planning was the most crucial element in urban prosperity (UN-Habitat 2013). As stated in the report, "Against a background of rapid urbanization, urban planning is a necessity not a luxury, as demonstrated, in the many cities in which it is lacking" (p. xvi). Differing opinions state that the purpose of land planning is to manage the built environment or that it is to provide direction for market sector development (Fainstein and Campbell 2003). In a broader context, Berke et al. believe that land use planning is an advocacy for a community "on protecting the environment, advocating equity, promoting livable cities, and supporting economic development" (2006 p. 35).

Land planning and management are the keys to the future successful development of the local or regional population, especially as it becomes more urbanized and there are mounting pressures on natural and resource areas. A UN report on building a land framework identified five criteria for developing a useful land information system (UN-Habitat 2012, p. 15) as follows:

1. Level of governance.
2. Embedding land information in a stable land institution.
3. Identifying the essential elements of a land information system.
4. Involving stakeholders.
5. Access and use of land information.

The level of governance requires a transparent, efficient, and effective governing body that has the techniques and procedures in place for land planning and the political will to institute land management. The decision makers, however, need to rely on a stable institution that can collect, manage, maintain, and have the technical expertise to analyze land data. Consistency in data capture, database management, analysis, and display are some of the essential elements of the land information system. It is critical that stakeholders are an integral part of the process as members of advisory boards, technical committees, and review panels associated with the land information system and most importantly decision making. The stakeholders are an assortment of concerned citizens, professionals, associations, and agencies involved with land resources, land development, conservation, and economic development. A high level of transparency in the access and use of land information provides a democratic means by which citizens, agencies, and organizations can use information about their land (Figure 1.2).

The transparency and use of information become relevant when discussing land use dynamics in any region. Understanding change, however, is possible only if there is a baseline on which to measure it. In the realm of land planning that baseline is the land use and land cover of the landscape at a specific point in time. Researchers, planners, and land managers use a wide range of land use/land cover classification systems based on their purpose and needs. DiGregorio and Jansen (2000) provide a synopsis of the different classification systems and their purposes (http://www.fao.gov).

A classification system that was developed by Anderson et al. (1976) served some of the initial purposes and needs for land cover/land use studies of the natural landscape and agricultural lands. Other systems have been developed by a number of organizations, for instance, the FAO (LCCS), the European Union (CORINE), the American Planning Association's Land-Based Classification Standards (LBCS), and a revised Anderson (National Land Cover Data and the National Vegetation Classification Standard). These systems provide a standard by which human activity on the land and natural land cover can be categorized. It is from these categories with spatial and temporal dimensions that land use dynamics can be identified and analyzed for land policy decisions. The different land use/land cover classifications,

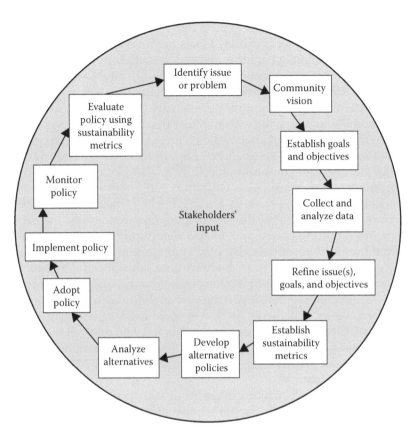

FIGURE 1.2
The planning process for a sustainable community.

however, were often designed to describe the land use and cover of Earth's surface, not necessarily to inform decisions about land use planning. The categories of land represent human activities or the plant, soil, water, or geology interpreted from the surface. Jeer (2001) identified other aspects that must be considered in any land use classification system that is used for land planning, including land activity, function, structure type, development characteristic, and ownership. Others have also begun to examine the semantics of the land use/land cover classifications to determine if the name of the class category actually represents activity or cover (Bishr 1998; Feng and Flewelling 2004; Comber, Fisher, and Wadsworth 2005; Ahlqvist 2008; Jansen and Veldkamp 2012).

The utilization of land use/land cover data is important to the successful implementation of a sustainable planning process (Figure 1.2). Community vision statements are generally the foundation on which a community or organization establishes their mission. For example, the U.S. Forest Service has their mission: "Sustain the health, diversity, and productivity of the

Nation's forests and grasslands to meet the needs of present and future generations" (U.S. Forest Service 2014). Goals are broad-based statements that provide a more specific set of ideas and concepts that are focused on aspects of the vision or mission statement(s). Objectives are measureable statements that assist in directing actions or policies to obtain the goals. A typical objective may state that the community will increase affordable housing units by a given number and size to reach the goal that every citizen has access to housing. The collected and analyzed data are a key element in the planning process (Figure 1.2), once the goals and objectives have been determined.

In this process, the goals and objectives are developed (at least in draft form) prior to land information collection and analysis. In fact, the data that are employed in the analysis are specified by the identified goals and objectives. In addition, it is the use of this data that assists the decision makers in examining policy options and determining sustainable metrics. The land use/land cover data also play an important role in evaluating the policy decisions that have been made. A time sequence of data can reveal the changes that have taken place on the landscape, and that information can be examined in relationship to the sustainability metrics that have been established. Policy statements are the specific ways to reach an objective and/or goal. Clearly, the land cover/use data must be relevant to the vision statement and goals and objectives to be useful and meaningful to decisions makers and stakeholders who are either influencing or making and implementing policy.

Overall, there are six major categories of land planning policy: regulations, incentives, acquisition, capital improvements to infrastructure, financing, and education (Platt 2004). Each policy type, however, has a number of different options. For instance, within regulations the most frequently used options are zoning, subdivision regulations, building codes, and architectural standards. Lesser-employed policy options may include incentives such as tax relief; density bonuses; and infrastructure development or education, providing maps, tables, and figures to the general public about current and future conditions or needs.

In land planning, there are several types of land data that need to be available within the policy-making process. Meck (2002) identified six types of studies for this purpose:

1. An inventory of the amount, type, and intensity of existing land uses
2. Land areas served by public utilities, including water and sewer line capacity and their location
3. An analysis of existing land use patterns and trends for land development along with open space and recreation resources
4. An analysis of infrastructure carrying capacity and identification of areas of need and projected demand

5. An identification of areas for potential land redevelopment and rehabilitation

6. Projections of land use 20 years in the future based on population, employment, and transportation

The fundamental study is a well-defined inventory of existing land uses. According to Berke et al. (2006, p. 206), a land classification system should include three criteria: (1) describe the nature of existing land uses accurately and in adequate detail; (2) fit consistently with the logic and classes of future land use plans; and (3) be compatible with the typology of uses in the development–management policies, regulations, and ordinances. The three criteria set the standard that is most compatible with land planning. The importance of the classification system is not in its level of detail or the method of collecting, compiling, and calculating the data. It is whether or not and to what level the data support the decision-making process and the policy or policies that are created for a sustainable community that is important.

1.3.1 Examples of Land Planning and the Use of Land Information

The following examples demonstrate the need for land use/land cover information at different planning scales. The hierarchical structure of most land use/land cover systems provides a means by which broad categories can become more specific at the different categorical levels. In most governance situations, there is a correlation between the land use/land cover categorical level and the spatial resolution of the land use goals and objectives (Figure 1.3). At the national level, the broader land use goals may require only general land use/land cover classifications and a coarse spatial resolution such as 1 km². The next level of government—the state, province, or canton—requires more detailed land use/land cover goals and objectives and thus more specific categories and finer spatial resolution. The local government will provide the most specific land use goals and objectives and the highest level of use/land cover classification and spatial resolution.

1.3.2 National Scale Classification

Great Britain has long been a leader in land use/cover classification and land use planning for urban and rural settings. Beginning in the early 1990s, the Office of the Deputy Prime Minister and the Planning Land Use Statistics Division began the process of developing a standardized land cover/land use classification system for the country (Harrison 2006). Although a number of land use classification schemes had been previously developed, such as the National Land Use Classification in the 1970s, none of them provided the broad coverage or detail necessary for land planning. In addition the predecessor schemes had a number of deficiencies, including uncoordinated

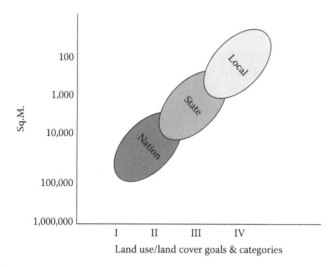

FIGURE 1.3
Land use goals in relation to land use/land cover and spatial resolution.

approaches to classification and collection, the need to rework data from study to study, and data that were incompatible across community boundaries and out-of-date or inappropriate for a given type of study. The National Land Use Database (NLUD), intended to cover the entire country, possesses a comprehensive classification system and addresses identified deficiencies of previous classifications.

To address the issue of land use versus land cover, the report recognizes them as two distinct dimensions of the land. Land use is defined as "the activity or socio-economic function for which land is used," and land cover is defined as "the physical nature or form of the land surface" (Harrison, p. 16). Figure 1.4 demonstrates how the same area can be viewed differently when land use or land cover is classified.

Similar to the American Planning Association's LBCS (Jeer 2001), the NLUD is a multidimensional schema. It covers a broad range of uses/cover and can classify land for different types of studies. It is a clear attempt to modernize and rationalize the classification of land at a national level in the context of local land-planning initiatives.

The Harrison report (2006) identifies the need to link land use and land cover data to the policies of government.

Reliable and up-to-date geo-referenced information on land use is required to provide a basis for the sustainable development of land resources in both urban and rural contexts and to inform the development of policies across all areas of human activity at national, regional, and local levels, including planning and regeneration, housing, employment, transport, agriculture, environment, and recreation. Within government the need for information on land use is evident through published policy documents and through the

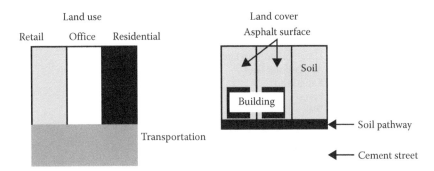

FIGURE 1.4
Comparison of land use and land cover.

large number of surveys sponsored by government and other bodies to collect such information since the mid-1970s. (Harrison, p. 8)

It is difficult to imagine, given the comprehensiveness, the level of pretesting (almost 10 years), and the ability to revise and adapt the system, that it will not support the planning process. In their concluding remarks about the NLUD, the authors state the following about the review of the classification system: "The classification has been the subject of two rounds of consultation with potential user organisations and experts and has been evaluated as part of three trial data collection exercises. These consultations and data trials have extended across England, Scotland and Wales" (Harrison, p. 29). It is clear that that the NLUD can fit the needs of local communities for a variety of purposes.

1.3.3 State/Province Scale Classification

The state of Oregon in the United States northwest passed legislation in 1973 that listed 19 statewide planning goals regarding land use (ORS-197). The local governments can incorporate these goals into their comprehensive plans and establish local policies that will begin to reach the specified goal. To meet the goals, there is a need to inventory the land use/land cover at the highest spatial resolution required (usually the parcel level) and with enough specificity to know if a goal has been reached or not. For instance, Goal 4—"To conserve forest lands by maintaining the forest land base and to protect the state's forest economy by making possible economically efficient forest practices that assure the continuous growing and harvesting of forest tree species as the leading use of forest land consistent with sound management of soil, air, water, and fish and wildlife resources and to provide for recreational opportunities and agriculture" (Oregon Department of Land Conservation and Development 2010)—relates to the conservation of forest lands. To accomplish this goal, an inventory of forested lands needs to be completed and policies implemented to preserve forest lands and insure that

TABLE 1.1

Level I of the Anderson Classification System

1. Urban or built-up land
2. Agricultural land
3. Rangeland
4. Forest
5. Water
6. Wetland
7. Barren
8. Tundra
9. Perennial snow

forest development should not exceed carrying capacity (Oregon Department of Land Conservation and Development 2010, p. 4-2). Thus, the land use/land cover classification system needs to identify forest lands through use and cover. The Anderson classification system could be employed. It includes four levels of specificity (level I is shown in Table 1.1), with increasing levels of specificity as one moves from levels I to IV (Anderson). The level I, category 4—Forest, is sufficient to meet the inventory goal.

However, to meet the other aspects of the goal—sound management of soil, air, water, fish, and wildlife resources, and recreational opportunities—requires more detail, both categorically and spatially, land use/land cover data, at a level II, III, or IV. As Berke et al. (2006) stated, the land use/land cover categories need to provide the needed information, thematically, spatially, and temporally, for analysis and decision making.

1.3.4 County/City Scale Classification

Clackamas County, Oregon, is a local jurisdiction that is required to follow the State of Oregon's land use goals. This county surrounds Portland, Oregon, to the south and east, providing both urban and rural environments, and it has a large percentage of its land in forests. In the county's comprehensive plan of 2001, it recognizes the same goals as the State of Oregon; however, the different elements have been updated separately since 2001. For instance, Chapter Four—Land Use was updated in 2014, with other elements modified within the past 8 years.

Fourteen major issues were identified in the land use chapter:

1. Supply and location of land for urban uses
2. Density of residential uses
3. Intensity of commercial and industrial uses
4. Proximity of mutually supporting land uses
5. The cost impacts of various land uses

6. Compatibility or conflict between land uses
7. Competing demands for land having certain characteristics
8. Compatibility of city and county plans
9. Supply and location of land for rural uses
10. Preservation of land for agricultural and forestry uses
11. The character and appearance of neighborhoods
12. Compatibility of land use with supportive systems such as transportation and sewerage
13. Protection of natural features and waterways from the impact of development
14. Provision of open spaces within the urban environment

From this list of issues, there are several aspects of land use planning that are important beyond the change in land use types. Three major themes can be identified: the relationship between land use types, the connection to infrastructure services, and the demands for land and governmental services. For this analysis of land use/land cover classification, three major land uses within the plan will be explored: residential, agriculture, and forestry. The residential portion of the land use chapter identified five major categories of residential: low density (residential lots ranging from 2,500 to 30,000 sq. ft.), medium density (up to 12 units/acre), medium high density (up to 18 units/acre), high density (up to 25 units/acre), and special high density for high-rise multifamily housing (up to 60 units/acre). Overall, there are six residential goals as follows:

1. Protect the character of existing low density neighborhoods.
2. Provide a variety of living environments.
3. Provide for development within the carrying capacity of hillsides and environmentally sensitive areas.
4. Provide opportunities for those who want alternatives to the single family house and yard.
5. Provide for lower cost, energy-efficient housing.
6. Provide for efficient use of land and public facilities, including greater use of public transit (pp. 4-20–4-21).

To identify the land use/land cover categories and to present their distribution to accomplish these goals, the county created a land use map (Figure 1.5). However, the land uses listed on the map are a mix of land use categories, land ownership, land jurisdictions, and zoning. Surprisingly, the categories on the map were not as specific as the categories identified in the plan, thus creating a disconnect between the goals, the land use categories,

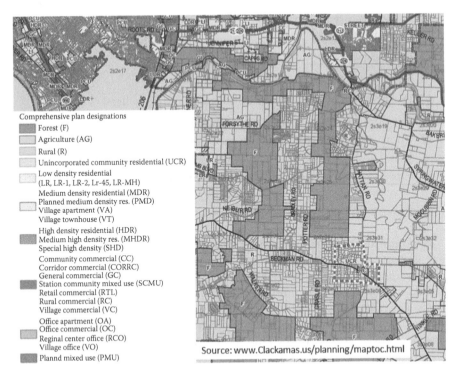

Comprehensive plan designations

Forest (F)
Agriculture (AG)
Rural (R)
Unincorporated community residential (UCR)
Low density residential
(LR, LR-1, LR-2, Lr-45, LR-MH)
Medium density residential (MDR)
Planned medium density res. (PMD)
Village apartment (VA)
Village townhouse (VT)
High density residential (HDR)
Medium high density res. (MHDR)
Special high density (SHD)
Community commercial (CC)
Corridor commercial (CORRC)
General commercial (GC)
Station community mixed use (SCMU)
Retail commercial (RTL)
Rural commercial (RC)
Village commercial (VC)
Office apartment (OA)
Office commercial (OC)
Reginal center office (RCO)
Village office (VO)
Plannd mixed use (PMU)

Source: www.Clackamas.us/planning/maptoc.html

FIGURE 1.5 (See color insert.)
Clackamas County, Oregon: A portion of land use plan map.

and the map. This is a clear example of a land use classification not supporting the goals of a land use plan.

Agricultural land is a major component of the rural environment in
Clackamas County (Figure 1.4). There is no distinction between the different
types of agricultural lands on the map. The plan lists seven goals for agricultural lands:

1. Preserve agricultural use of agricultural land.
2. Protect agricultural land from conflicting uses, high taxation, and
   the cost of public facilities unnecessary for agriculture.
3. Maintain the agricultural economic base of the county and increase
   the county's share of the agricultural market.
4. Increase agricultural income and employment by creating conditions that further the growth and expansion of agriculture and
   attract agriculturally related industries.
5. Maintain and improve the quality of air, water, and land resources.

6. Conserve scenic and open space.

7. Protect wildlife habitats (p. 4-57).

To reach the goals, a number of different types of information are needed: specifics on agricultural lands, changes in agricultural lands, quality of agricultural lands, identification of scenic and open space, and the range of wildlife habitats. Ten specific policies are listed in the plan for agriculture, with the major policies designed as regulations to preserve, protect, and maintain agricultural lands. Analysis of the impacts on agricultural land are needed with any change in land use or improvements to the land through the development of infrastructure (roads or sewer systems). Similarly, land quality relative to water, air, and soil needs to be analyzed along with any changes to wildlife habitat, all requiring information on land cover and changes to that cover. For example, in a 2012 study of the potential to expand agricultural productivity, 13 categories of agricultural activity were identified (Ecotrust 2012).

Forests in Clackamas County are a mix of natural landscape and second, third, or fourth growth harvested forests. Forest lands in Clackamas County make up approximately 54% of the county (1097 sq. mi.). Similar to the category for agriculture, there is only one broad category for forest lands on the land use map (Figure 1.5); however, in the more detailed "Northwest Oregon Forest Management Plan" (2010) there are five categories of forest lands (Regeneration, Closed Single Canopy, Understory, Layered, and Older Forest Structure). Thus, as we have observed with residential and agricultural goals, the map does not provide the detail needed to employ and evaluate the policies identified to reach the forestry goals.

1.4 Summary and Conclusion

Land planning and management require knowledge of the current and possible land use/land cover changes to the landscape. Population needs and growth along with local and global markets are major factors that influence the utilization of the landscape. The global population is growing at a substantial rate, over 1% per annum, and this population will primarily locate in urban areas. In some portions of Africa, the urban growth rate is double that of the rest of the world. To be successful, these places will necessitate land planning and linked land-information systems.

The three planning examples demonstrate that knowledge of the existing land use/land cover patterns are important to attain the goals, objectives, and the effective implementation of policies to gain maximum benefit from the creation of future land uses. The land use/land cover classes mapped must provide sufficient thematic and spatial resolution of the description and distribution of the categories, so they are specific enough for substantial analysis of land utilization. For example, the single agricultural class for Clackamas County would not be sufficient to meet the needs for an analysis of the county's rural economy. A separate study had to analyze the agricultural data in more detail, 13 categories, to examine the potential for agricultural development.

Overall, Berke et al. (2006) identified the issue that the land use/land cover classification system should meet the needs of the overall plan and be flexible enough to be incorporated into the policy analysis and eventually the monitoring/evaluation of the policy implementation. This requires that the data collection techniques, the land use/land cover categories, the analytical techniques, and the policy decisions are linked so that the appropriate information can be used by different layers of government. Examining land use/land cover categories, the semantics and linkages are as critical as the categories in making the data and information operable for planning and decision making.

References

Ahlqvist, O. 2008. "Extending post-classification change detection using semantic similarity metrics to overcome class heterogeneity: A study of 1992 and 2001 U.S. National Land Cover Database changes," *Remote Sensing of Environment*, 12: 1226–1241.

Anderson, J. R., E. E. Hardy, J. T. Roach, and R. E. Witmer. 1976. *A Land Use and Land Cover Classification System for Use with Remote Sensor Data*, U.S. Geological Survey Professional Paper No. 964, Washington, DC: Government Printing Office.

Berke, P. R., D. R. Godschalk, E. J. Kaiser, and D. A. Rodriguez. 2006. *Urban Land Use Planning*, 5th Ed., Urbana, IL: University of Illinois Press.

Bishr, Y. 1998. "Overcoming the semantic and other barriers to GIS interoperability," *International Journal of Geographical Information Sciences*, 12: 299–314.

Botequilha Leitao, A., J. Miller, J. Ahern, and K. McGarigal. 2006. *Measuring Landscapes: A Planner's Handbook*, Washington, DC: Island Press.

Clackamas County. 2001. "Map IV-6 Urban North Clackamas County," *Clackamas County Comprehensive Plan 2001*, Oregon City, Oregon.

Comber, A., P. Fisher, and R. Wadsworth. 2005. "You know what land cover is but does anyone else?… an investigation into semantic and ontological confusion," *International Journal of Remote Sensing*, 26: 223–228.

Daily, G. C. 1997. *Nature's Services: Societal Dependence on Natural Ecosystems*, Washington, DC: Island Press.

deGoot, R. S., M. A. Wilson, and R. M. J. Boumans. 2002. "A typology for the classification, description and valuation of ecosystem functions, goods and services," *Ecological Economics*, 41(3): 393–408.

Ecotrust. 2012. *GIS Agricultural Land Analysis for Clackamas County*, Agricultural Investment Plan, Oregon City, OR.

Fainstein, S. S. and S. Campbell, eds. 2003. *Readings in Planning Theory*, 2nd Ed., Malden, MA: Blackwell Press.

FAO. 2014. *State of the World's Forests, 2014: Enhancing the Socioeconomic Benefits from Forests*, Rome, Italy.

Feng, C. C. and D. M. Flewelling. 2004. "Assessment of semantic similarity between land use/land cover classification systems," *Computers, Environment and Urban Systems*, 28: 229–246.

Harrison, A. R. 2006. *National Land Use Database: Land Use and Land Cover Classification Version 4.4*, London, Queen's Printer and Controller of Her Majesty's Stationery Office.

Hendee, J. C., G. H. Stankey, and R. C. Lucas. 1990. *Wilderness Management*, 2nd Ed., Golden, CO: North American Press.

Jansen, L. J. and T. Veldkamp. 2012. "Evaluation of the variation in sematic contents of class sets on modelling dynamics of land use changes," *International Journal of Geographical Information Science*, 26: 1–30.

Jeer, S. 2001. *Land-Based Classification Standards*, Chicago, IL: American Planning Association.

Meck, S., ed. 2002. *Growing Smart Legislative Guidebook: Model Statutes for Planning and the Management of Change*, Chicago, IL: American Planning Association.

Oregon Department of Forestry. 2010. *Northwest Oregon: State Forests Management Plan*, Revised Plan, April, Salem, OR.

Oregon Department of Land Conservation and Development. 2010. *Oregon's Statewide Planning Goals & Guidelines*, Salem, OR.

Platt, R. 2004. *Land Use and Society, Revised Edition: Geography, Law and Public Policy*, Washington, DC: Island Press.

Population Institute. 2007. *The Population Challenge: Key to Global Survival*, The 21st Century Papers, No. 2, Washington, DC.

UN-Habitat. 2008. *State of the World Cities, 2010/2011: Bridging the Urban Divide*, London, United Kingdom: Earthscan.

UN-Habitat. 2012. *Sustaining Urban Land Information: A Framework Based on Experiences in Post-Conflict and Developing Countries*, Global Land Tool Network, Guide No. 2, Nairobi, Kenya.

UN-Habitat. 2013. *State of the World Cities, 2012/2013: Prosperity of Cities*, London, United Kingdom: Earthscan.

U.S. Forest Service. 2014. *The U.S. Forest Service—An Overview*, Washington, DC: Government Printing Office.

Waddell, P., A. Borning, M. Noth, N. Freier, M. Becke, and G. Ulfarsson. 2003. "UrbanSim: A simulation system for land use and transportation," *Networks and Spatial Economics*, 3: 43–67.

Williamson, I., S. Enemark, J. Wallace, and A. Rajabifard. 2010. *Land Administration for Sustainable Development*, Redlands, CA: ESRI Press.

2

Ontology for National Land Use/Land Cover Map: Poland Case Study

Małgorzata Luc and Elżbieta Bielecka

CONTENTS

ABSTRACT The history of land use mapping in Poland is almost 100 years old. Since that time, numerous attempts have been made to represent forms of land use/cover across the country on maps. They were characterized not only by the variable scale, but also the different number of classes used, that is, from 5 to 25, which shows how difficult it was to reach a consensus on the methodology. Over time, the methodology of land use and land cover mapping has changed. Many European countries, including Poland, entered projects in which databases such as CORINE LC, PELCOM, LUCAS, GTOS, VMapL2, and BDOT were created. Some of them have been widely applied in spatial planning and crisis management. Currently, CORINE and BDOT are the databases most commonly used in Poland for the purposes of research and spatial planning. In 2007, Poland became a signatory to the infrastructure for spatial information in Europe (INSPIRE) directive, whose primary objective is to facilitate access to and use of spatial data related to, among others, land use and land cover. Using the example of maps and databases, this chapter draws attention to the problems of semantic integration of land use/cover data in two transborder projects in the Oder River Basin and in the Carpathians. Also, harmonization and interoperability of spatial data in Poland are discussed in a context of semantic plasticity or relations between theory and reality in its pragmatic dimension. We emphasize how the projects coped with problems caused by the terminology.

KEY WORDS: *Land use/land cover map, land use/land cover database, Poland.*

2.1 Introduction

The term "ontology" derives from two Greek words: *onto*, which means "being; that which is" and *logia*—"science, study, theory." It, therefore, means "a theory concerning the kinds of entities" (Goove 1993, p. 1577) and describes the relations between entities (Mark et al. 2004), but most importantly it lays the theoretical foundations of every discipline of science. In the case of land use and land cover (LULC), ontology is crucial for classification and for defining land use although distinguishing it from land cover. The understanding of classification terms varies, which is inconvenient for the purpose of retrospective analysis and comparison, and acts as a barrier for international, transborder, and interdisciplinary cooperation. Therefore, one of the main tasks for European Union (EU) countries is to simplify access to spatial data associated with the environment and to use it in broadly understood spatial planning, crisis management, and sustainable development. Maps and databases of LULC often comprise a basic source of information for many projects undertaken in this field, including in Poland.

The approach to LULC has evolved greatly in the past 100 years. Currently, land cover is understood as the physical and biological features of the Earth's surface as "observed" from the perspective of man as well as satellite. Included are areas of vegetation (trees, bushes, fields, and lawns), bare soil (even if there is a lack of cover), hard surfaces (rocks, buildings), and wet areas and water bodies (sheets of water and watercourses and wetlands) (EUROSTAT 2001). Land use, on the other hand, is considered in terms of functionality in a socioeconomic dimension, which leads to the identification of areas used for residential, industrial, or commercial purposes; for farming or forestry; and for recreational or conservation purposes, among others. As such, land use has close links with land cover. Mücher et al. (1993 *vide* EUROSTAT 2001) distinguishes the so-called "sequential" approach, which has found use particularly in agriculture. They associate the term "land use" with a series of operations on land, carried out by humans, with the intention to obtain products and/or benefits, for example, plowing, seeding, weeding, fertilizing, and harvesting. A similar twofold approach to defining land use is suggested by Comber (2008), who maintains that built-up and agricultural areas comprise two basic classes. Land use cannot simply be observed; additional information is required. In other words, one can accept the formula derived by Burley (1961): land cover + land utilization = land use. A slightly different approach is used by Jankowski (1975), according to whom maps of land use show spatial distribution of various forms of land cover and utilization by man (including areas of low anthropogenic pressure), as well as spatial relations between them.

In this chapter, we use historical maps of the entire area of Poland to retrospectively analyze LULC changes in transborder terrains. Characterized resources vary thematically as well as in the minimal mapping unit (MMU) size and classification model. However, they all show LULC without any possibility of unambiguously distinguishing land forms into either LU or LC classes. On the basis of two case studies, we describe the issues surrounding combining differently classified data and data of numerous origins, with differing scales and created using varying methods, also for classification. These Polish examples demonstrate the use of semantic plasticity when faced with LULC ontological ambiguity.

2.2 Land Use and Land Cover Mapping in Poland

The past 100 years in Poland have a long history of cartographical attempts to present different forms of LULC for the entire country. Primarily, such maps were created based on topographic maps. Later on satellite images were used for cartographic representations, and finally numeric databases were developed using aerial photographs and satellite data as a primary source. Besides varying source data, all of these resources also vary in MMU, and scale and number of LULC classes. An overview is provided in Table 2.1 together with the CORINE Land Cover (CLC) data resources for comparison purposes.

2.2.1 Poland's General Land Utilization Map in 1:1,000,000 Scale by Uhorczak

It was not until 1957 that the first map of land use of all of Poland in a small scale was published (Uhorczak 1969). A very interesting aspect of it lies in a unique methodology developed specifically for this purpose. The main source of data was military topographic maps in 1:100,000 scale from the

TABLE 2.1

Specification of General Information about Polish LULC Data Sets

	Poland's General Land Utilization Map by Uhorczak	Satellite Map of Poland	Topographic Land Use/Land Cover Database (BDOT)	Polish Carpathian Mountains	CORINE Land Cover
MMU (ha)	10	25	1	5	25
Map scale	1:1,000,000	1:500,000	1:10,000	1:100,000	1:100,000
Number of LULC classes	5	11	24	11	32 (out of 44 classes)
Source data	Topographic maps 1:100,000	Landsat MSS Salut-6 images	Orthophotomaps	IRS-6	Landsat TM, SPOT, IRS
Conformance with the CLC data level	Level 1	Level 2	Level 3	Level 1	n/a

MSS, multispectral scanner; TM, thematic mapper; SPOT, satellite probatoire d'Observation de la terre; IRS, Indian remote sensing satellites

1930s, classified into five main groups: (1) waters (rivers and lakes), (2) meadows and pastures, (3) woodland, (4) arable land, and (5) habitation. The result was five maps with a resolution of 1 ha (1 mm² on maps), each representing data from one class. The delimitation of classes did not pose any problems, as cartographic symbols used for their representation were clearly distinguished and followed the methodology published in the technical part of the topographic instruction (Military Geographical Institute 1925). For the reason of technical restrictions, the preparation of a ten fold photographic scale reduction was impossible, so a threefold photographic generalization for each separate element was applied twice. As a result, land use is presented in a scale of 1:1,000,000; however, the accuracy is as high as on 1:100,000 maps, and even after double rescaling, individual symbols from original maps are clearly seen.

The title of the Uhorczak's map suggests the classification of land in terms of its utilization. However, semantic analyses of distinguished classes clearly show that only class (2)—meadows and pastures, presents land from the functionality point of view. The other classes are based on physical and biological features of the Earth's surface, so they belong to a land cover classification system. Still, if we follow Burley's (1961) aforementioned definition of land use, this map is the only Polish example of land use representation. To clarify the ontological relations and possibility for data harmonization for a variety of Polish maps and databases, Table 2.2 presents LULC classifications mapped to level 3 of the CLC database.

TABLE 2.2

Mapping Polish Maps and LULC Databases to CLC Level 3

CORINE Land Cover Level 3	Poland's General Land Utilization Map by Uhorczak	Satellite Map of Poland	Topographic Database Land Use/Land Cover (BDOT)	Polish Carpathian Mountains
111 Continuous urban fabric surfaces	Habitation	Built-up areas	Urban fabric	Settlement areas
112 Discontinuous urban fabric				
121 Industrial or commercial units		Industrial and storage areas	Industrial or commercial	Communication and industrial areas
122 Road and rail networks and associated land			Road and rail networks and associated land	
123 Port areas			Port areas	
124 Airports			Airports	
131 Mineral extraction sites			Mineral extraction sites	Mineral extraction sites
132 Dump sites			Dump sites	Fallow land
133 Construction sites			Construction areas	Fallow land
141 Green urban areas		Built-up areas	Parks and green areas	Grassland
142 Sport and leisure facilities			Sport and leisure facilities	
211 Nonirrigated arable land	Arable lands	Arable land or arable land with the dominance of large-scale land management	Arable lands	Arable land
222 Fruit trees and berry plantations		Arable land	Plantations	Fruit trees and shrubs
231 Pastures	Meadows and pastures	Meadows and pastures	Meadows and pastures	Grassland

(Continued)

TABLE 2.2 *(Continued)*

Mapping Polish Maps and LULC Databases to CLC Level 3

CORINE Land Cover Level 3	Poland's General Land Utilization Map by Uhorczak	Satellite Map of Poland	Topographic Database Land Use/Land Cover (BDOT)	Polish Carpathian Mountains
242 Complex cultivation patterns	Arable lands	Arable land	Not distinguished due to 1 ha MMU	Arable land
243 Land principally occupied by agriculture, with significant areas of natural vegetation				Fallow land
311 Broad-leaved forest	Woodlands	Deciduous forest	Forest and shrubs	Forests
312 Coniferous forest areas		Coniferous forest		
313 Mixed forest		Mixed forest	Mixed forest	
321 Natural grasslands vegetation associations		Wasteland	Natural grassland	Grassland
322 Moors and heathland			Forest and shrubs	Grassland
324 Transitional woodland shrub				Forests
331 Beaches, dunes, sands			Sands	Other areas
332 Bare rocks			Bare rocks	
333 Sparsely vegetated areas			Sparsely vegetated areas	Grassland
334 Burnt areas			Not distinguished	Not distinguished
411 Inland marshes			Marshes	Grassland
412 Peat bogs			Peat bogs	
511 Water courses	Waters	Water courses	Water courses	Water
512 Water bodies		Water bodies	Water bodies	
521 Coastal lagoons			Not distinguished	Not distinguished
523 Sea and ocean		Not distinguished	Sea	Not distinguished

2.2.2 Satellite Map of Poland

Nearly, a quarter of a century following the printing of the Uhorczak's map (in 1975), a new cartographic edition emerged in Poland in 1:500,000 scale. It was also the first in Poland Land Cover Database based on satellite images (from Landsat 1975–1976 and the Soviet orbit station Salut-6 1977–1978). The MMU was 25 ha and 80 m for linear objects. On the basis of the satellite images, 11 types of LULC were distinguished (Table 2.2). It was relatively easy to identify arable lands by the high spectral differentiation (mosaic of crops). Classification difficulties involved mostly woodland, habitation, and river networks. The mixed woodland class was finally differentiated based on an equally proportionate space occupied by spectral signatures attributed to deciduous and coniferous forest within a spatial unit (1 cm^2 in map scale). Built-up areas were identified on photos taken in late autumn or early spring due to the amount of trees around urban areas. The issue of rural habitation and water courses proved an impassable obstacle (Ciołkosz 1981).

From definitions of classes as well as from the analysis of the range of information presented on the map, we may conclude that derived classes consider both land use (e.g., meadows and pastures, industrial/storage areas) and land cover (e.g., coniferous forests, deciduous forests, mixed forests). Additionally, a combination of LULC in, for example, wasteland appears in the classification.

2.2.3 The CORINE Land Cover Base

The next charting of LULC in Poland was associated with the realization of the CLC program, created by the European Communities Commission in 1985. As part of the program a database of LULC was created for the member states of the European network EOINET, a partnership network of the European Environment Agency (EEA). The database included LULC data from 1990 and this was subsequently updated in 2000, 2006, and 2012. The land cover nomenclature is organized hierarchically into three levels: (1) five main land cover classes: artificial surface, agricultural areas, forests/semi-natural areas, wetland, and water; (2) 15 classes; and (3) 44 classes, used in developing CLC data sets at the national level (Table 2.2). The hierarchical structure of the CLC nomenclature allows original information to be combined in various ways to perform specific analyses in different thematic fields. As a result, it is possible to carry out spatiotemporal cross-border investigations and comparisons at European, regional, and national levels. CLC data sets are based on satellite images that determine not only the MMU (25 ha and linear features greater than 100 m), but also the scale of the cartographic products (CLC 1993; EEA and ETC/LC 1999). The consequence of this restriction is that even if the interpreter is able to distinguish land cover polygons smaller than 25 ha, the dominant class should be assigned. In the agriculture areas, the so-called mixed classes (all 24) represent typical LULC, with different shares of arable land, water, forest, and others.

These mixed land cover classes are the greatest drawback of these data, as their delineation is largely subjective. Analysis of the thematic reliability of CLC2000 has shown that the accuracy of delineation of these classes is 70% to 80% and is 10% to 15% lower than assumed (EEA 2006). There were many factors influencing that the most significant delimitation feature was the date (particularly month) of acquiring the image and the ambiguity of definitions of delineation, which leaves too much room for interpretation. It is worth noting that the CLC data were gathered as a result of visual interpretation. The definitions are therefore associated with a certain degree of subjectivity, depending on the understanding of class definitions by the interpreter and how they can identify each class on a satellite image (see Chapter 13). On the other hand, comparability, the ability to analyze temporal and cross-border changes, as well as the ability to generalize data at various levels, and well-documented thematic reliability, are the greatest advantages of CLC data.

According to CLC, land cover changes in Poland were relatively small, not exceeding 1% of the total surface area of the country (Ciołkosz and Bielecka 2005). Contradictory results came out from local and regional works where smaller MMU was used, including the two case studies described below. The very small size of some patches (especially of arable land) in the south of the country (Luc et al. 2009) leads to fragmentation in land cover type or land functionality. Classification and interpretation of such a mosaic is problematic, and only filtering processes or contextual classification is recommended. However, even these results are not accurate because of the blurring of boundaries between different land-cover units.

2.2.4 Topographic Database

The Polish Topographic Database (BDOT10k), maintained by the Head Office of Geodesy and Cartography, was initiated in 2003 and finally completed in 2013. BDOT10k is a spatially continuous vector database with the thematic scope and a level of detail corresponding to contemporary, civilian topographic maps at a scale of 1:10,000. It contains nine thematic categories, including LULC (compare Table 2.2). The nomenclature, which is hierarchical in nature, distinguishes 34 LULC classes at the second, more detailed, level. In accordance with the specifications of the data (Surveyor General of Poland 2003), each LULC class is stored as a separate object class and saved as a separate file. Furthermore each class is characterized by sets of attributes, for example, unique identifier, type (water course, water bodies, stands, built-up areas, bare rocks, etc.), date of acquisition, source material, and positional accuracy (Geoportal2 2013). Because of a 1 ha MMU, the LULC classes are mainly homogeneous, mapped on the basis of visual interpretation of orthophoto maps (panchromatic and color) with a spatial resolution of 1 m. On the basis of the BDOT10k methodology, it can be assumed that LULC classes were delineated according to criteria similar to those used for topographic maps at the scale 1:10,000 and level 3 in CLC.

2.2.5 A Common Ontology for National Land Use and Land Cover Maps

Polish maps and databases, despite titles that traditionally concern land use, contain information about both land use and land cover. It is common that meadows and pastures, industrial/commercial areas, and wasteland apply to land use, whereas forest and agricultural areas are associated with land cover. Described resources differ mostly in accuracy and base materials. As a consequence, MMU and number of classes vary. However, it is worth noting that definitions of land cover classes are invariable. They are based on criteria compatible to those used for topographic maps.

The use of satellite images as a basic data source for LULC mapping leads to a more effective updating and results in differentiation in precision and credibility in delimitation of some classes, as in the example of CORINE LC data. The issue of data harmonization in Polish LULC data involves the mapping of corresponding classes, and the conducted analysis indicates the possibility of mapping them to CLC data but on different hierarchical levels (Table 2.2). Since the 1990s, CLC has become a de facto standard, and for this reason a reference to its ontology in Polish cartographic sources enables the use of varied maps and data sets.

2.3 Dealing with Ontological Plasticity in Transborder Land Use and Land Cover Analysis

The topic of LULC has been discussed in the past few decades mostly in the context of the scale of changes, their causes and trends, and on a base of historical topographic maps and varied databases. In Central European countries, extensive transborder, scientific and application research are underway concerning changes in mountain ranges, border river beds, or large marshlands. Difficulties in comparison of data from various sources or land use/cover changes analysis lie, among others, in the variety of land characteristics, which results in ontological and semantic problems. Based on two quite different examples from Poland, we will illustrate the value of a pragmatic approach to manage varied source data, classification models, or land specificity.

2.3.1 Oder River Catchment

The Oder is one of the largest rivers in the watershed of the Baltic Sea (a total length of 854 km, river catchment of ca. 120k km^2) (Figure 2.1).

In 2000, the Institute of Geodesy and Cartography in Warsaw carried out an international project on modeling floods based on a retrospective LULC

FIGURE 2.1
Location of the Oder River drainage basin and the Polish Carpathians.

analysis. A catastrophic flood in 1997 in the Oder Valley has created interest in the study of changes that have occurred since the 1880s.

Floods have always been a threat to the inhabitants of the region, including in large urban areas in Poland; in the Czech Republic; and on the German side. Severe floods in 1501, 1515, and 1736 are reported by historical sources (Czerwinski et al. 1999) and after them systematic, large-scale hydrotechnical works were planned. The realization of these plans, which continued well into the twentieth century, completely changed the water regime of the Oder basin. The length of the river was shortened by 16% (IGIK 2000; Bielecka and Ciołkosz 2002). However, the regulation work has not prevented floods. The most severe and largest one, known as the "flood of the century," occurred in July 1997, when 1144 km² of land were submerged (Ciołkosz and Bielecka 1998); another large flood took place in May 2010. Analysis of their causes and effects revealed the LULC changes might have been responsible for the catastrophes.

Within three time frames—1900–1975, 1975–1990, and 1990–2006—changes in spatial distribution and surface area were analyzed on historical maps, as well as information derived from LULC databases (Land Cover Database 1975 [LCD75]; CORINE LC 90 and 2006). Because of the wide range scales of historical topographic maps (from 1:75,000 to 1:100,000), and the technology behind deriving information about the distribution and area of various land use classes (Ciołkosz and Bielecka 2005), it was assumed that MMU is 25 ha, which results in a geometrical precision equivalent to a map in 1:250,000 scale.

One of the most significant steps of the research was establishing the definitions of land use/cover types. The range of information included on historical German topographic maps and the thematic range of LULC databases determined the choice of main types of land use: built-up areas, arable lands, meadows and pastures, forests, marshes, and water bodies. In the case of LCD75, this required the regrouping of data from 11 to 6 classes; in the CLC database from 25 to 6 classes. Table 2.3 shows the rules of adaptation of data from varied sources.

The result of analysis of land use in the past more than 100 years suggests that both the means of using the land and the distribution of the main types of use (forests, arable lands, meadows, built-up areas, waters, and marshes) in the Oder River catchment (ORC) had not undergone major change (Table 2.4). This confirms the supposition that land use in this specific area is an exceptionally stable configuration. Therefore, the flooding was not caused by changes in land use but by errant hydrotechnical work.

2.3.2 The Polish Carpathian Mountains: An Example of Land Use and Land Cover Tessellation

The Carpathians are one of the most geographically, socially, and economically differentiated regions of Poland. They are also part of one of the longest mountain ranges in Europe (ca. 1300 km). This is why it is so important to study the changes this region has undergone in the past ca. 200 years. However, despite a few attempts, a complete LULC database covering this time period has not been created. A very detailed inventory of land use was carried out in 1988, when for each of the 1864 villages, 10 forms of land use were identified (Table 2.5). After 20 years, it was decided to reanalyze land use in this region and to compare the results with those of the previous inventory. This time images from the Indian environmental satellite were used (Table 2.1). Differing precision of the initial (1 ha) and the new inventory (5 ha) necessitated taking this into account in the comparison between the different time frames.

After the accession to the EU, Poland underwent significant social, political, and economic changes. As a consequence, fallow land became significant. This resulted in the creation of an additional LULC class defined as abandoned arable land, where local conditions determine the process of conversion into semi-natural woodland or grassland with weeds and shrubs (Ciołkosz et al. 2011).

TABLE 2.3

Principles of Data Adaptation from Historical Maps to LCD75 and CLC Databases

Accepted Land Use Nomenclature	Historical Topographic Maps	LCD 1975 Nomenclature	CORINE Land Cover Level 3
Built-up areas	Cities of >50k inhabitants, 2–50k, <2k and villages (4)	Built-up areas, industrial area (2)	Continuous urban fabric, discontinuous urban fabric, industrial or commercial units, road and rail networks and associated land, port areas, airports, mineral extraction sites, dump sites, construction sites, green urban areas, sport and leisure facilities, complex cultivation pattern with scattered houses (12)
Arable lands	Arable lands, fruit trees and berry plantations (2)	Arable lands, arable lands with dominance of large-scale land management (2)	Nonirrigated arable land, fruit trees and berry plantations, land principally occupied by agriculture with significant areas of natural vegetation (3)
Meadows and pastures	Meadows and pastures, meadows (1)	Meadows and pastures (1)	Pastures (1)
Forests	Forests, thickets and bushes, shrubs, swamps, sparse forests (5)	Deciduous forests, coniferous forest, mixed forests (3)	Broad-leafed forests, coniferous forests, mixed forests, transitional woodland shrubs, moors and heathlands (5)
Marshes	Swamps, dry swamps, bogs (3)	Wetlands (1)	Inland marshes (1)
Water bodies	Lakes, ponds, rivers, streams (4)	Water bodies, water courses (2)	Water bodies, water courses (2)

In brackets are the number of LULC classes in the source data.

Fallow land or wasteland, depending on the classification model, is an interesting example of disharmony and the lack of interoperability within one LULC class. CORINE has not distinguished it at all. Instead, there is an open space with little or no vegetation while what is called a "wasteland" on a satellite map is a mixture of vegetation of different kinds, wetlands, and rocks in BDOT (Table 2.2). Also during this time period, pastures became of marginal significance as livestock numbers declined, and it was impossible to differentiate them from meadows on satellite images of this resolution, so meadows and pastures were combined into one class termed "grassland" (Table 2.5).

TABLE 2.4

Land Use in the Oder River Basin in the Period of Nineteenth Century until 2006

	Nineteenth Century		1975		1990		2006	
	km²	%	km²	%	km²	%	km²	%
Built-up areas	2 741	4.4	5 532	8.8	5 774	9.2	5 875	9.4
Arable lands	35 752	57.0	31 023	49.4	31 000	49.4	30 990	49.4
Meadows	5 483	8.7	4 601	7.3	4 393	7.00	4 278	6.8
Forest	17 662	28.1	20 739	33.0	20 769	33.1	20 771	33.1
Marshes	282	0.4	132	0.2	143	0.2	143	0.2
Water bodies	865	1.4	742	1.2	689	1.1	711	1.1

TABLE 2.5

Comparison of LULC Classifications in the Carpathians Databases

Classes of 1988 database	Classes of 2006 database	Classes of NASA LCLUC database
Arable land	Arable land	Agriculture
Orchards	Orchards and plantations of fruit shrubs	Agriculture
Meadows	Grassland	Grassland and shrubs
Pastures	Grassland	Grassland and shrubs
Meadows	Grassland	Wetlands
Forests	Forests	Forest
Water	Water	Water
Mining fields	Mining fields	Undefined
Technological areas	Technological areas	Urban/built-up
Inhabited areas	Inhabited areas	Urban/built-up
Other areas	Other areas	Undefined
–	Fallow land	Bare land
–	–	No data

Generally, what this illustrates is that anyone who needs harmonized analyses is obliged to integrate classes of varied nomenclature oneself. It is an example of semantic plasticity (see Chapter 3), as every project has its own designation of land features, its own nomenclature.

Analysis of the changes revealed significant discrepancies between field observations and satellite images, mostly in the category of the smallest units of land, such as orchards, grasslands, and fallow lands. These small land units (a mosaic), which are very common in the Carpathians, led to the cessation of the use of the CLC databases and, despite high-resolution satellite images, produced dissatisfying results (Figure 2.2).

Currently, there are two parallel research projects being carried out at the Jagiellonian University in Krakow, both of which aim to show the changes

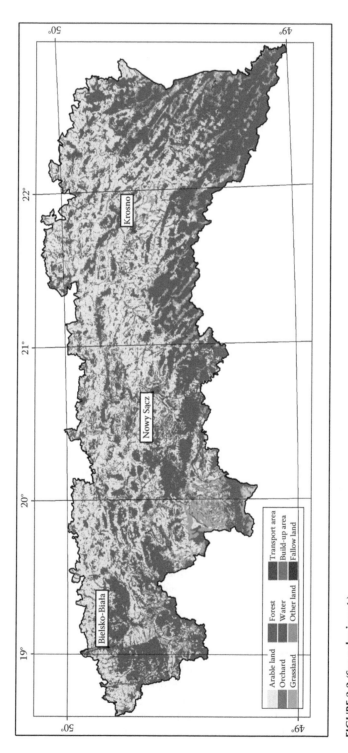

FIGURE 2.2 (See color insert.)
LULC tessellation in the Polish Carpathians in 2006.

in land use in the Carpathians within the past 250 years. The first, "Forest cover changes in mountainous regions—drivers, trajectories and implications (FORECOM)," concentrates only on woodland and presents changes in the Polish Carpathians (and Swiss Alps) in the context of climate changes. In the project, automatic extraction of forests from topographic maps and orthophoto maps is used. Historical maps undergo manual digitalization (FORECOM 2014).

The second project, "200 years of LULC changes and their driving forces in the Carpathian Basin," aims to create a database of land use focusing on two classes: agricultural areas and forests, from the peak of the Austro-Hungarian Empire time up to the Polish accession to the EU. The methodology assumes digitalization of basic LULC forms on old maps in a regular 2 × 2 km point grid, which matches the grid points used in the infrastructure for spatial information in Europe (INSPIRE) and LUCAS grids (NASA LCLUC 2014). Each of these points contains information about land cover in a very specific location, not for the entire neighboring area. Where possible, the highest (fourth) level of detail is used. It is presumed that, should a need arise, details can at any point be aggregated to a higher level. Subclasses or more detailed levels should be used only where information is certain. As a default, all points are certain. Uncertainty is recorded in a dedicated field of the sampling grid. Level 1 assumes seven fundamental classes plus one "undefined" and one "no data" (Table 2.5). For the analysis of changes, three classes were distinguished: cropland, grassland, and forests, although agriculture was identified "as a presence or absence of plowed signals interchanging with vegetation in the imagery" (Griffiths et al. 2013, p. 4).

The two cases of regional LULC database formation in Poland (ORC and Polish Carpathian Mountain [PCM]) demonstrate methodology and semantics unique to the applied aim and specificity of the area. In Poland, with its highly differentiated characteristic of land structures and forms, it is very challenging to create a single coherent classification model with fairly detailed categories. Each LULC project requires the preparation of a separate semantic guide, with the concept (definitions) and relations between classes. A reference to a commonly known database such as the CLC (Table 2.2) could be a helpful approach to enable a comparative analysis.

2.4 Harmonization and Interoperability of Land Use and Land Cover Data

In the presented case studies, as well as in all other cartographic and GIS projects conducted in Poland in the last century, LULC definitions were neither well established, nor were the relations between them distinguished. Usually the title of maps or projects suggests land use when single classes in fact show land cover (as in Uhorczak's map or PCM); in other cases land cover and use issues are mixed in one semantic entity (e.g., CLC, BDOT). Also changes in nomenclature have led to multiple difficulties and ambiguity in handling with existing spatial data (e.g., NASA LCLUC 2014). Working on data for numerous purposes and in international cooperation needs harmonization and interoperability. To establish ontological unity within the field of LULC and further on to simplify access to spatial information, the EU countries have been collaborating to create a single system. Finally, in 2007, the cooperation among countries (including Poland) led to the introduction of the aforementioned INSPIRE directive (INSPIRE 2007). The aim is to simplify the utilization of spatial data in the process of decision making at every administrative level (Tomas and Lutz 2013; Würriehausen et al. 2014).

Data are prepared by EU member countries according to an agreed application scheme, and the 31 themes of data within the system also include LULC. Data specifications for the two themes, land cover and land use, were created separately by different teams of experts. Land cover is understood as physical and biological use of the Earth's surface, including forests, agricultural areas, seminatural areas, wetlands, and water bodies. The ontology has the form of a classification model, is independent of mapping scales and data sources, and contains 18 pure land cover components. It does, however, allow the use of mixed classes under the condition of including a percentage of "pure land cover components." The specifications include a method of describing mixed classes of land use in the LCML (Land Cover Meta Language—ISO 19144-2:2012; LCML 2012). The proposed approach does not, however, solve the problem of varying legends and scales but only details the ontology of each class, thereby facilitating their use.

Land use data are subject to separate specifications, which define land use as territory characterized according to its current and future planned functional dimension or socioeconomic purpose (e.g., residential, industrial, commercial, agricultural, forestry, and recreational). It is divided into two groups: (1) existing land use, depicting the use and functions of a territory as it has been and effectively still is in real life; and (2) planned land use, which corresponds to plans defined by spatial planning authorities, depicting the possible future utilization of land. In Poland only the first group is applicable. Spatial planning on a single commune level takes place in plans consistent with the National Act of Spatial Planning, not with the INSPIRE directive. In practice, little attention is put on data in a context of land use.

Interoperability in an INSPIRE understanding is "a process of developing harmonised data product specifications and implementing the necessary arrangements to transform spatial data into interoperable spatial data" (INSPIRE 2007). Although there is a common awareness of this in Poland and the Ministry of Infrastructure and Development leads operations on terminological cohesion, there is still a lack of settled results. In the ORC project, the lack of interoperability and harmonization between data from varied sources was somehow replaced by the generalization process. In the PCM project, all classes were defined anew in a photointerpretation key. Presumptions of INSPIRE give hope for creation of one comprehensible and 100% applicable LULC ontology, but not before 2015.

According to Kubicek and Cimander (2009), there are four levels of interoperability: (1) technical—secure data transfer; (2) syntactic—processing of data; (3) semantic—processing and interpretation of data; and (4) organizational—linkage of processes among many systems. In Poland only the first two have already been developed, semantics still have implementation problems, and the last one is in a conceptualization clarity process.

2.5 Summary

Information about land cover and use has always depended on MMU, scale, data sources, and methods of classification. So far, in Poland no attempts have been made to create a classification system for LULC, which would be independent from the first two factors. Classification models very much rely on physical and biological characteristics of an analyzed region when land cover units are distinguished, and on functionality or socioeconomic factors for land use entities. However, as our presentation has shown, land use very often refers to the surface type or land utilization and sometimes a mixed land cover/use class is created (e.g., fallow land). This often causes confusion in analyses, especially of the land alterations, and results become incomparable with other regions.

Generally, definitions of the classes are vague, as much more attention is given to the rules of generalization. The twenty-first century brought about the realization by users of land use and cover data that the large volume of databases does not solve their problems with retrospective and transborder studies. Available materials are mostly incomparable and at best can be used to delineate a trend in land cover changes, but not to analyze it in detail.

It is very difficult in Poland to discuss the LULC ontology or semantics. Realization of the INSPIRE directive creates an optimistic perspective mostly within the existing land use, not so much in spatial planning aspects, and true harmonization and interoperability seem to remain a future prospect. However, building the relations in LULC classifications is much more realistic when class definitions are more adaptable to the land characteristics. A pragmatic approach toward nomenclature aspects was presented here as well as in a few European initiatives (e.g., ORC, FORECOM, and NASA LCLUC). However, because some of them focus on selected entities, they do not create a full LULC ontological view on Polish national land use and cover.

References

Bielecka, E. and A. Ciołkosz. 2002. Land-use changes during the 19th and 20th centuries. The case study of the Odra River Catchment area. *Geographia Polonica* 75: 67–83.

Burley, T. M. 1961. Land use or land utilization? *The Professional Geographer* 13: 18–20.

Ciołkosz, A. 1981. General map of land use in Poland elaborated on the basis of satellite images (in Polish). *Polski Przegląd Kartograficzny* 13: 2–7.

Ciołkosz, A. and E. Bielecka. 2005. Land cover in Poland. CORINE Land Cover data bases (in Polish). *Biblioteka Monitoringu Środowiska*. Warszawa, Poland: Inspekcja Ochrony Środowiska.

Ciołkosz, A. and E. Bielecka. 1998. Flood in Odra river valley interpreter on satellite images (in Polish). *Prace Instytutu Geodezji i Kartografii* 97: 81–95.

Ciołkosz, A., C. Guzik, M. Luc, and P. Trzepacz. 2011. *Land Use Changes in Polish Carpathians in a Period of 1988–2006* (in Polish). Kraków, Poland: IGiGP UJ.

CORINE Land Cover. 1993. *Technical Guide*. Brussels, Luxembourg: Office for Official Publications of European Communities.

Comber, A. J. 2008. The separation of land cover from land use using data primitives. *Journal of Land Use Science* 3: 215–229.

Czerwiński, J., B. Miszewska, and W. Pawlak. 1999. The history of Wroclaw and Odra river (in Polish). In *Wroclaw and the Odra River – the History*, edited by G. Roman, J. Waszkiewicz, and M. Miłkowski. Wroclaw, Poland: Wroclaw Town Board—the Office of Town Development.

EEA. 2006. The thematic accuracy of Corine land cover 2000 assessment using LUCAS (land use/cover area frame statistical survey). Technical Report 7/2006. ISSN 1725-2237. EEA, Copenhagen, Denmark.

EEA and ETC/Land Cover. 1999. Corine Land Cover Technical Guide. Technical Report No. 17/2007, accessed July 20, 2014, http://www.eea.europa.eu /publications/technical_report_2007_17.

EUROSTAT. 2001. *Manual of Concepts on Land Cover and Land Use Information Systems*. Luxembourg: Office for Official Publications of the European Communities.

FORECOM. 2014. FORECOM Forest Cover Changes in Mountainous Regions—Drivers, Trajectories and Implications, accessed July 20, 2014, http://www.wsl .ch/fe/landschaftsdynamik/projekte/FORECOM/index_EN.

Geoportal2. 2013. Baza danych obiektow topograficznych, accessed July 20, 2014, http://geoportal.gov.pl/dane/baza-danych-obiektow-topograficznych-bdot.

Goove, P. B., ed. 1993. *Webster's Third New International Dictionary of the English Language Unabridged*. Cologne, Germany: Könemann.

Griffiths, P., D. Müller, T. Kuemmerle, and P. Hostert. 2013. Agricultural land change in the Carpathian ecoregion after the breakdown of socialism and expansion of the European Union. *Environmental Research Letters* 8: 045024. doi:10.1088/1748-9326/8/4/045024.

IGiK. 2000. *Interpretation of Land Use and Land Cover Changes in the Odra River Catchment*. FINAL REPORT, Contract No. 15356-1999-10 F1ED ISP PL.

Jankowski, W. 1975. Land use mapping, development and methods. *Geographical Studies* No. 111. Institute of Geography and Spatial Organization. Warsaw, Poland: Polish Academy of Science.

Kubicek, H. and R. Cimander, 2009. Three dimensions of organizational interoperability. Insights from recent studies for improving interoperability frame works. *European Journal of ePractice*, accessed November 15, 2014, http://www.epractice.eu/files/6.1.pdf.

LCML. 2012. Geographic Information—Classification Systems—Part 2: Land Cover Meta Language (LCML). ISO 19144-2:2012, accessed December 3, 2014, http://www.iso.org/iso/home/store/catalogue_tc/catalogue_detail.htm?csnumber=44342.

Luc, M., J. B. Szmanda, and E. Bielecka. 2009. From land cover diversity to landscape variety in Poland. Raster data analysis. In *European IALE Conference 2009, European Landscape in Transformation Challenges for Landscape Ecology and Management: 70 Years of Landscape Ecology in Europe*, edited by J. Breuste, M. Kozová, and M. Finka. Salzburg (Austria), Bratislava (Slovakia). pp. 489–492.

Mark, D., B. Smith, M. Egenhofer, and S. Hirtle. 2004. Ontological foundations for geographic information science. In *A Research Agenda for Geographic Information Science*, edited by R. McMaster and E. Lynn Usery. Boca Raton, FL: Taylor & Francis.

Mücher, C. A., T. J. Stomph, and L. O. Fresco. 1993. *Proposal for a Global Land Use Classification*. Rome, Italy: FAO/ITU/WAU.

NASA LCLUC (Land Use and Land Cover Change). 2014. 200 Years of Land Use and Land Cover Changes and Their Driving Forces in the Carpathian Basin, accessed July 20, 2014, http://neespi.org/web-content/abstracts/Radeloff2_abstract.pdf and http://lcluc.umd.edu/project_details.php?projid = 215.

Tomas, R. and M. Lutz. 2013. Key pillars of data interoperability in Earth sciences—INSPIRE and beyond. *Geophysical Research Abstracts*, 15: EGU2013–13767.

Military Geographical Institute. 1925. Topographic Instruction. Part II—Technical (in Polish). Topographic Faculty II, accessed December 3, 2014, http://rcin.org.pl.

Uhorczak, F. 1969. Poland's General Land Utilization Map in Scale 1:1 000 000 (in Polish). *Prace Geograficzne IG PAN*, 17, A–Text volume, B–Cartographic volume.

Würriehausen, F., A. Karmacharya, and H. Müller. 2014. Using ontologies to support land use spatial data interoperability. In *Lecture Notes in Computer Science*. Vol. 8580, Springer International Publishing, Switzerland, pp. 453–468.

3

The Need for Awareness of Semantic Plasticity in International Harmonization of Geographical Information: Seen from a Nordic Forest Classification Perspective

Alexandra Björk and Helle Skånes

CONTENTS

ABSTRACT   The aim of this chapter is to address and clarify the important issues and challenges of semantic plasticity when it comes to forest classification and geographical information. Necessary improvements for international data harmonization and implementation are highlighted along with the need for increased awareness of the consequences for ecological modeling. We envisage a combination of thoroughly described metadata and controlled vocabularies as a means to ensure the future use of a wide range of regional and national classification systems in an ontological framework that enables crosswalks between classification systems and spatial comparisons between existing data sets. This would allow for a wide range of old, contemporary, and future data sets to be used together in landscape-related analyses.

KEY WORDS: *Land use, land cover, INSPIRE, landscape ecological modeling, landscape analysis, ontologies.*

3.1 Background

The ongoing changes to landscapes and terrestrial ecosystems worldwide are recognized as one of the most important processes leading to decline in habitat quality and biodiversity (Sala et al. 2000; Jenkins 2003; Butchart et al. 2010; European Commission 2010). The magnitude of changes is large-scale and the Fifth Assessment report of the Intergovernmental Panel on Climate Change (IPCC) illustrates visible global climate changes on all continents (IPCC 2013). Destructive landscape changes occur due to loss of habitats and isolation/fragmentation of many land cover categories such as forest. When, for example, forests decline in distribution or quality, many species depending on specific forest conditions are threatened. Changes might be gradual and slow, or abrupt and chaotic, and when catastrophes occur, such as earthquakes, landslides, tsunamis, and forest fires, the need for collaboration and good geographical information is crucial for crisis preparedness. These environmental issues know no administrative, national, or regional borders, which call for an increased cooperation between countries as well as cross-sectorial collaboration between authorities within society.

The Nordic region, except Iceland, has much in common regarding physical conditions as well as cultural historical development and biodiversity status, which in several aspects differ from the rest of Europe. Consequently, there is already a long Nordic tradition to collaborate on these issues (Påhlsson 1998), but we still have varying definitions, much to learn from each other, and possible synergy effects to gain from further collaboration and coordination of external contacts, such as harmonizing land cover data on a pan-European level, and participation in larger European Union (EU) projects where typically only one or a few Nordic countries participate (Skånes 2005; Jansen et al. 2008; Normander et al. 2012).

In support of international agreements on environmental assessment and policy making, decision makers need access to reliable, quality assessed, and up-to-date geographic information about the distribution, pattern, and key properties of land cover and land use (LC/LU), both at the regional and local levels (Brandt et al. 2002; Bunce et al. 2007; Feranec et al. 2007; Jansen et al. 2008). Knowledge is needed regarding present conditions as well as previous states and directions of ongoing successions (Käyhkö and Skånes 2006). To provide efficient methodologies and consistent data sets, it is crucial for a close and transdisciplinary collaboration between society and the research community. In this respect, it is important to emphasize that improved data alone have to be matched by knowledge about the drivers behind improved models and projections of LC/LU change (Lambin et al. 2001).

To meet the demands of uniform and homogeneous data covering larger areas than single countries, the Coordination of Information on the Environment (CORINE) program was initiated by the European Commission in 1985 (Heymann et al. 1994). Other attempts to create common classification systems or creating thesauri and translation paths between existing classification systems have been carried out on a European level. Examples of such attempts are EUNIS (Davies and Moss 2002; Moss and Davies 2002), LCCS (Di Gregorio 2005), Land Cover Meta Language (LCML) (Di Gregorio and O'Brian 2012), Natura 2000 (European Commission 2007), BioHab (Bunce et al. 2007), AGROVOC (Sini et al. 2010), EBONE (Bunce et al. 2013), LUCAS (Martino and Fritz 2008), EAGLE (Arnold et al. 2013; Arnold et al. 2015, Chapter 6), and the recently launched CadasterENV (Metria 2014).

In this context, we use and define *semantic plasticity* in parallel with how the term has been defined within linguistics and philosophy (Hawthorne 2006), particularly in line with Larsson (2007) who defines it as gradual and dynamic change in meaning of a linguistic construct (i.e., a word or a phrase) and adaptation of its usage patterns, both in terms of expansion, contraction, or shift. Larsson specifically points out the difference between semantic plasticity and formal semantics, where the latter gives a precise analysis of contextual meaning, but typically assumes that it is static and has paid little attention do the *dynamics* of meaning. Projected to our field, semantic plasticity is valid in terms of how definitions of concepts and classes have changed throughout history (years, decades, or even centuries), but also in the short-term time perspective, as in a single dialogue between practitioners from different sectors, or even in a given conversation between two persons in any constellation. In other words, classes in thematic maps or geodatabases might be the same or similar in naming, but the semantic definition of classes are not always obvious or equal (Ahlqvist 2004, 2005; Ahlqvist and Gahegan 2005).

One aggravating factor preventing a simple solution for handling semantic plasticity is that all landscape patterns are relative artifacts of our perception, selected ontologies, and measurement resolution (Käyhkö and Skånes 2006; Ahlqvist 2008a; Jepsen and Levin 2013). Jones' (1991) conclusion that all documentation is in fact interpretation, illustrates that all spatial landscape information is inherently a relative and subjective interpretation and simplification of reality. Acknowledging that landscape pattern is a relative concept, inevitably landscape change needs to be accepted as a relative concept as well (Käyhkö and Skånes 2006). This is the essence of semantic plasticity: accepting and coping with the intrinsic variability of definitions. It only comes naturally that many sectors need to handle and understand the landscape, starting from their own needs and conceptual frameworks, and thus landscape is perceived in many different ways (Skånes 1997). Consequently, there are a large number of different classification systems used in a variety of disciplines differing, among other things, in purpose, usage, class definitions, resolution, and scale. This works well when data are used for their intended purposes, but becomes problematic when used beyond the original context, merged, or compared with other data, unless awareness of semantic plasticity is included.

Another hampering factor is that the differences in data *meaning* needs to be assessed and conceptualized to properly realize integration benefits (Comber Fisher and Wadsworth 2005; Käyhkö and Skånes 2006; Ahlqvist 2008a; Eriksson and Skånes 2010). The wider and detailed meaning of information categories and the decisions made and by whom, and so forth, need to be communicated as metadata for the user to accurately determine the range and proper use for that particular data set. The growing number of user categories of geodata leads to an increased risk of errors in analyses and conclusions. This is greatly due to the lack of proper training in the particular field of study in which the data set was originally intended for, typically in combination with poor or inconsistent metadata.

The aim of this chapter is to address and clarify these important issues and challenges, and to highlight the need for awareness regarding semantic plasticity and necessary improvements for a successful international data harmonization and implementation. In this, we focus on the concept of forest that is so well known, and yet so elusive and complex when it comes to definitions and semantic plasticity. A second example is taken from the rapidly growing field of landscape ecological modeling, where map categories are directly translated into specific species habitat configurations and used to model species diversity, distribution probability, and patch dynamics over time.

3.2 The INSPIRE Directive, an Attempt to Harmonize Geographical Information

To address the different issues regarding the quality, organization, accessibility, and sharing of spatial information, the EU issued the Infrastructure for Spatial Information in the European Community (INSPIRE) directive in 2007. An EU directive sets up certain goals that must be followed by every Member State. They are used to harmonize national laws in matters that affect the operation of specific single market (Treaty of Amsterdam 1997). INSPIRE is, in its most basic form, a large European inventory of existing spatial data sets and other geographical information. All the Nordic countries—Sweden, Denmark, Finland, Norway, and Iceland—are implementing the INSPIRE directive (European Parliament 2007, EEA Joint Committee 2010).

The INSPIRE directive points to the importance of public authorities to have smooth access to relevant spatial data sets and services and for the EU's Member States to prevent practical obstacles for the sharing of data. Therefore, we focus on the need for harmonization and increased accessibility of geodata for users. It is an equal necessity to establish a measure of coordination between the users, providers, and producers of geographical information, and the need for combining information from different sectors and sources. The Directive pinpoints the most fundamental principles in regards to working with spatial information; that the data are stored, made accessible, and maintained at a proper level. The aim is also to avoid duplication of work that has already been done.

The Directive is not only meant to catalog geodata and make it more accessible but also to complement already existing initiatives, such as the Galileo Joint Undertaking and the Global Monitoring for the Environment and Security (GMES), and to enable interoperability (European Parliament 2007). Interoperability, simplified, is the ability to combine data from different sources and to share them between users and applications. Again, this is a situation where semantic plasticity needs to be assessed. Another key obstacle is the time wasted in search of existing geodata and trying to establish whether they may be used for a particular purpose (European Parliament 2007). Part of the Directive is, therefore, to provide metadata as descriptions of available spatial data sets and services (European Commission 2008).

So far the focus of INSPIRE has been primarily centered on the producer's perspective. In the implementation processes, and therefore in all 28 of the Member States, the term is currently used to delineate the descriptive properties of the data sets, mainly for search and identification purposes (European Commission 2008). Hence, the requirements on metadata regarding the semantic object level, which is necessary for accurate analysis and comparison, is still insufficient. This in turn leads to terminological confusion and hampers the intended increase in compatibility of data for pannational estimates and lowers the usability of data for spatial analysis and change detection. Another important issue to deal with is the lack of coordinated work toward saving data sets on a regular basis to enable landscape change detection. Currently, many data sets are continuously improved and updated but no temporal data are locked and saved for future retrospective studies.

In all Member States, major work until 2020 will aim toward implementing the INSPIRE directive. One of the challenges will be to organize data for sharing and creating geoportal interfaces. Another is to address the complexity of metadata and the need for more detailed and uniform descriptions for comparisons or extensions, quality assessment, and change analysis. Many of the Nordic and European countries have in light of this decided to progressively release their extensive data sets for public use. This option is unfortunately not available to everyone due to national budget constrictions. For example, the Swedish government has decided that the responsible authority Lantmäteriet has to finance its own area of business (Socialdepartementet 2013), therefore preventing such an elegant solution. However, Lantmäteriet announced in a press release on March 17, 2015, that they are releasing small scale map information as open data (Lantmäteriet 2015). This is an encouraging first step.

3.3 The Need for Semantic Awareness in Landscape Classification and Analysis

The validation of land cover changes has long overlooked the implications of semantic plasticity in the nomenclature of classes when using different classification systems for change detection. As previously described, there are a multitude of different classification systems that coexist and are in use around the world. One challenge is that the result of a survey or scientific query can vary widely, even when using the same input data, depending on which classification principles are used or how a classification system is translated or aggregated into broader categories. This, however, is sometimes an intentional effect when different sectors are using the same data set for different purposes, and by such, an inherent strength of a modern complex database or classification system. But when used beyond the original purpose, there is always a risk of violating the inherent limitations of data usability.

When using or comparing spatial data such as maps or databases, it is essential to be aware of the semantic structure and plasticity of a given land type such as forest (Ahlqvist 2008b). Bishr (1998) defines semantics as the relationship between the computer representation and the corresponding real-world feature within a certain context. The same phenomenon can be given different names by different user groups within academia and society (Morshed et al. 2010). The opposite applies when the same name can be given to a range of different phenomena (Skånes 1997). This is referred to as semantic heterogeneity or rather *plasticity* and occurs because of the variation of mental models in the different disciplines to describe phenomena in the real world and link between the producer and the users of data (Ahlqvist 2008a; Eriksson and Skånes 2010). It is also an effect of the inherent vagueness of many concepts such as forests (Bennet 2001).

In this chapter, we have selected two examples as an illustration of the need for semantic awareness in all classification and analysis in a landscape context. The first deals with the plasticity of the forest concept, and the second with the importance of semantic awareness in landscape ecological modeling.

3.3.1 The Semantic Plasticity of the Forest Concept

To illustrate the degree of complexity, we have selected to elaborate on a common concept that almost anyone can relate to, forest. When we say "forest" there is no unified perception of what that exactly means. When viewing the literature it becomes evident that it is far beyond a simple land cover (e.g., Comber et al. 2005). One can talk about forest, as opposed to open land, as a common divider of the landscape. To many it represents one of the most economically important land uses, where also a clear-cut without a single tree is still a part of the forest concept. Others think of forest as an important source of biodiversity that has to be untouched and continuously tree covered for centuries, or as a term that simply refers to solitude and recreational values. Hence, there are multiple ways to define forest and although we are using the same word, or variations of it, such as forest land, woodland, or wooded land, we mean a whole array of different things. The concept is therefore very dynamic, vague, and ambiguous (Bennet 2001). This is clearly illustrated in Lund (2014), where an extensive list of forest-related definitions from around the world is compiled.

In its most simple form, all Nordic countries, except Iceland, have now adopted the Food and Agricultural Organization (FAO) definition of forest requiring >0.5 ha with trees >5 m, and a crown cover (CC) of >10%, or trees able to reach these thresholds *in situ*. Land that is predominantly under agricultural or urban land use is excluded (FRA 2012). Furthermore, many countries have added a demand of productivity to the forest concept in a forestry perspective. For example, in Sweden, a forest has to produce >1 m^3 wood per hectare and year to be considered productive (Svensson 2006). This productivity demand influences, and partly conflicts with, the CC closure that will be expected, and it normally far exceeds 10% in a productive forest (cf. discussion in Glimskär and Skånes 2015, Chapter 8). In historical times, a multi-purpose utilization of forest land is known to have been generally extensive, including land uses such as grazing and haymaking (Aronsson 1980; Skånes 1997; Eriksson and Skånes 2010). This means that the forest cover in historical maps reflects different definitions of forest than do contemporary maps, where forest as land cover to a great extent refers to the single land-use forestry, or the more strictly morphological criteria of crown coverage and tree height. The historical example is relevant because of the fact that it relates to the long-term semantic plasticity that needs to be dealt with when using the longest timescale for spatial data in landscape change studies. Namely, the rich archive of historical maps available in some regions of the world from the mid-seventeenth century onward. These maps extend our spatial knowledge from approximately 80 years, including remote sensing based on satellite images and aerial photographs, to potentially 250–350 years (Skånes 1997; Skånes and Bunce 1997; Cousins 2001; Petit and Lambin 2002; Käyhkö and Skånes 2006).

The fact that modern mapping methods to an increasing degree are based on remote sensing also complicates things. These methods have apparent problems in dealing with the various soft aspects of the forest definition such as functions and processes. Existing data sets rarely describe how they have handled the criteria such as land use requirements, tree height, crown closure, and species composition. Mostly, the accuracy of applied thresholds of these criteria remains assumptions, and are typically not clearly assessed in their respective metadata. This becomes clear when studying the legends and other available metadata from existing maps and data sets. One complicating factor is that CC typically is gradually shifting in reality (Figure 3.1).

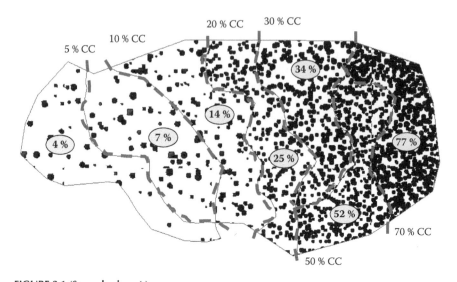

FIGURE 3.1 (See color insert.)
The effect of thresholds for forest mapping on gradual CC in a hypothetical landscape. (From Skånes, H.M., and Andersson, A., Flygbildstolkningsmanual för Uppföljningsprojektet Natura 2000 version 4.0. UF 19, *Naturvårdsverket*, p. 73 [in Swedish, unpublished], 2010, working document.)

This means that visual interpretation, both in the field and in remote sensing data, as well as automatic classification of the border between forest and nonforest in terms of openness, is still not easy to assess. Correspondingly, the ontological specifications in turn will rest on the perception of the concept of forest currently held by the producer of a forest map. Depending on thresholds set in the classification process, resolution, and other definitions and generalization principles, the border between forest and open land can be drawn in very different places on the ground.

It is difficult to establish the *true* and most accurate delineation because it all depends on the semantic definition that is used in combination with scale level, generalization, and minimum mapping unit. Today, with the increasing access to new data sets such as LiDAR point cloud data, CC can be estimated in a completely new and spatially detailed way (Lindberg et al. 2012; Nyström et al. 2012). This means that when classifying forest, for the first time we can divide it into more detailed classes containing more or less accurate estimations of variables such as CC. Statistical summaries (such as min, max, and variation) can also be calculated for individual forest patches, thus adding to the detail of information attached to each class definition and hopefully enabling more detailed analyses of class memberships as described by Ahlqvist (2008b).

Over the years, there have been Nordic workshops discussing the issues of vegetation mapping and classification semantics with the elusive forest concept as an important part of the discussions. In 2004, as part of the BioHab project (Skånes 2005), and in 2012 within Northscape, a Nordic network for land use and land cover monitoring, was the theme "approaching a common understanding, semantic and analytic differences" (Northscape 2014). Both workshops were in different contexts and financed by the Nordic Council of Ministers promoting collaboration within the Nordic region. This work has continued the tradition of Nordic collaboration on harmonizing vegetation mapping (Påhlsson 1998). Still, there is no real consensus to "forest," something which again reflects the inherent vagueness and ambiguity of the concept.

To gain some instant and wider insights into whether our experiences of the forest concept are unique or mainstream, we conducted an open Internet-based survey that was distributed to several hundred recipients, including NorthScape network participants, presumably engaged in forest-related work. Questions posed in the survey circled around the definition and complexity of the forest concept, both current and past uses, and what kind of forest-related data sets the people were working with. Also, questions were posed around the coverage and classification systems of forest-related data as well as metadata and resolution.

Most of the recipients were located in Sweden, but key contacts in Norway, Finland, Denmark, and Iceland were also selected to cover a wider region. The survey had 55 answers in a few weeks' time. A majority of answers, 76%, were from Sweden; the remaining were, in falling order, from Norway (13%), Finland (5%), Denmark (4%), and Iceland (2%). The uneven relationship between the countries is related to the fact that most of the initial distribution of the survey link went to Swedish stakeholders. The fact that only Denmark and Iceland have only a few responses can also be attributed to the fact that forest as a land cover only has minor, although increasing, distribution there. Note that this survey is not meant to enable statistically valid analysis of trends or in-depth knowledge but to provide some example experiences from practitioners.

More than half of the respondents (56%) work within different government agencies. Other respondents were affiliated with universities and higher education (31%), private sector (11%), and Nongovernmental organizations (2%). Almost half of the respondents (44%) work with forest-related tasks on a daily basis, 36% occasionally, whereas only 20% work indirectly with forest-related issues. As much as 80%–90% of the respondents work with either mapping/monitoring/statistics or with forest management/inventory. This means that the survey mainly reflects the views of users rather than producers.

When asked to depict in free text the preferred forest definitions in terms of criteria and purpose, the answers diverged remarkably, as exemplified below:

- No particular preference—depends on the purpose. Important not to mix definitions in discussions and that criteria are quantitative and measurable.
- Purpose should not influence the definition.

- Natural-generated forest under influence by natural disturbance regimes as, for example, wind, fire, grazing, water.
- A definition that clearly separates forest plantations from more naturally regenerating wooded areas with vertical layering and an ongoing natural succession.
- The term "forest" should not be used for areas where the main purpose is to produce timber, and so forth.
- There should also be other legal subdivisions of nonproductive forests, that is, forests set aside for nature conservation or recreation.
- The definition of forest needs to work together with definitions of other land use/land cover categories to avoid overlap.

The survey strengthened our views of the forest concept as complex and ambiguous, and that semantic plasticity enables flexible use but also adds confusion to the usage of forest-related data. All Nordic countries except Iceland have adopted the FAO 2000 definition of forest, but there are other previous and current definitions in use, as illustrated in Figure 3.2. Although the dashed red lines give an impression of temporal dynamics of the concept,

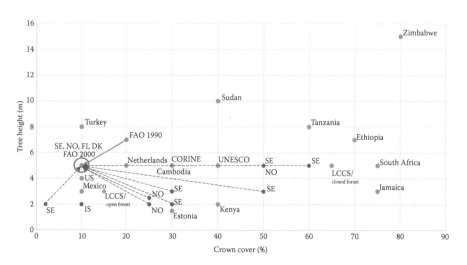

FIGURE 3.2 (See color insert.)
An illustration of the semantic plasticity of the forest concept, both in a global (blue dots) and in a Nordic (red dots) context. Forest classifications vary greatly on a global and national scale. The dashed lines indicate the general and conceptual shifts from previous Nordic definitions toward a harmonized use of the FAO 2000 definition. However, some of the other definitions are still in use in their respective fields of application. (Modified after Ahlqvist, O., *Environment and Planning B: Planning and Design*, 35, 169–186, 2008a, based on data from Lund, H.G., *Definitions of Forest, Deforestation, Afforestation, and Reforestation*, Forest Information Services, Gainesville, VA, 2014, and the Internet-based query for this article.) SE, Sweden; NO, Norway; FI, Finland; DK, Denmark; IS, Iceland.

it shows that several parallel forest definitions are still in use. The lines are instead indicative of the multiple definitions that exist in different sectors/ data sets, and their relation to the current official FAO definition. This also illustrates the potential problems with conducting change detection using existing maps and data sets where the results might actually show a shift in perception of the forest as concept rather than a change in actual forest cover.

Several countries, such as Finland, have not previously used the simplified classification definition, based on CC and tree height, but have still conformed with the FAO definitions. That explains their absence of red dots and dashed lines in Figure 3.2. This means that many older data sets, including forest classes, do not conform with the current definitions because these older maps and databases predate the implementation of the FAO definitions. Sometimes other contemporary data can help. For example, thanks to the collection of CC data in Swedish National Forest Inventory, direct calculation of forest according to the Forest Resources Assessment (FRA) categories and definitions can be backdated to 2000 (FRA 2010).

3.3.2 Implications of Semantic Plasticity on Landscape Ecological Modeling

Spatial modeling and landscape ecological assessment are rapidly growing fields in the intersection between ecology, geography, and computer science. The aim of these techniques is to analyze spatial configurations and content of the landscape to unravel patterns and processes of ecological significance to assist both nature conservation and urban planning (Mörtberg et al. 2007). GIS-based modeling tools are rapidly improving, which open up for a multitude of advanced spatial analyses, often in the form of a "black box" model, where many parameters are unknown. This sets both the users and the data sets under strong trial, and a combination of excellent computer skills and sound ecological knowledge are needed to ensure and improve the ecological significance in the output of any such modeling attempts.

When spatial modeling and analyses are performed, they rely on assumptions regarding spatial and functional relationships between objects in the landscape, described as the corridor concept (Forman and Godron 1981) or landscape connectivity (Merriam 1984). In this context, it is often addressed as the combination of core habitat areas in relation to the surrounding landscape (matrix). Usually a cost or resistance raster will be created to visualize the heterogeneity of suitability of the matrix to enable species dispersal, or posing barriers between core areas, and so forth (Taylor et al. 1993; With et al. 1997). These raster sets are often mechanistically calculated on the basic assumption that map categories can be readily combined and translated into specific species' habitat configurations.

This is, however, seldom the obvious case. A species habitat is built up by components that meet the requirements of a complete life cycle (UNEP 1995). Typically, a habitat comprises different biotopes that can be used in different seasons or in different life-history stages by a specific species or species group. This means that a habitat map always needs to be tailored to a specific species. As species operate on different scales in the landscape and because habitat requirements are often ambiguously described in the literature, the preparation of any resistance raster and other inputs from spatial analyses are inclined to contain major uncertainties and potential errors (Zeller, McGarigal, and Whiteley 2012), mainly due to a simplified view on ontological matters. Almost without exceptions, data sets used in such an operation need to be reclassified, aggregated, and evaluated before being used as variables in data, and this is often not clearly declared. To minimize these uncertainties and optimize analysis, we need well-defined, detailed and complex, flexible and consistent spatial data sets, closely accompanied by exhaustive and informative metadata on all aspects of the data, ranging from scope and methodological issues to semantic plasticity and classifiers.

We argue that it is crucial to acknowledge that semantic plasticity, here exemplified as CC requirements of forest definition, has serious spatial implications, which in turn lead to ecological implications (Figure 3.3). Running a landscape ecological model on these three maps for a species confined to forest conditions will give three completely different outcomes! All of them depend on how forest is defined in the spatial data and not how the forest species perceives its requirements on a suitable forest habitat. Consequently, there is no correct forest distribution map to use for any landscape ecological modeling. The best prerequisites for a successful modeling would be the existence of flexible data containing many additional attributes such as CC that would allow optimal aggregation and reclassification of data to suit the ecological question at hand. This would enable a flexible analysis design to fit the requirements for different species and increase its usability beyond its original context.

The field of spatial analysis holds an enormous complexity in itself and at the same pace as tools get increasingly accessible and deceptively easy to use, they are at the same time embedded in add-on or stand-alone software, where the mathematical algorithms are often hidden and the user has multiple settings to leave as default or start altering. This puts increasingly heavy demands on the users to cope with all aspects of the methodology and the transdisciplinary approach needed. This problem, however, goes far beyond the scope of this chapter, but we wish to emphasize the need for awareness of the semantic plasticity in geographical information science as discussed earlier and the complex implications this has on the outcome of any spatial analysis.

3.4 Discussion and Concluding Remarks

The IPCC describes a changing world, therefore the accurate and quality-assessed description of it becomes even more important for spatial analysis and change detection, so that policy makers can make accurate and informed

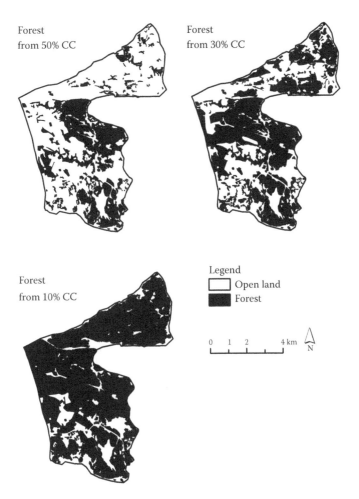

FIGURE 3.3
The figure shows a conceptual elaboration on the spatial implication of semantic plasticity in forest classification based on the variation of one single classifier—CC. The original map was a vector biotope map including CC information attached to each polygon. The three maps roughly represent the current and two known previous definitions of forest (see Figure 3.2). The FAO from 10%, previous NFI, CORINE, and vegetation mapping from 30%, and older maps where forest would be considered as areas dominated by tree cover from 50%.

decisions. The INSPIRE directive is both an excellent and necessary initiative for the harmonization of geographical data. At the same time, it is an example of terminological confusion because their definition of metadata is not on the same semantic level as we have been discussing in this chapter, defined as metadata/data set instead of metadata/object. The next challenge for INSPIRE is therefore to address the apparent lack of semantic awareness and the need for exhaustive metadata on class definitions and inherent qualities.

The substantial discrepancy regarding the subject of semantic plasticity between the research community and all parts of governments on the international arena is evident. Because it is the different levels of government that are responsible for the implementation of any new legislation, work to close this gap becomes even more crucial and calls for increased cross-sectorial cooperation between government officials and researchers. The challenge is equally true and necessary between different branches of the government and on the different levels of administrative authority. The wheel is continuously invented, which is at the very least an enormous waste of resources.

One of the fundamental challenges with categorization of the reality is that there are almost an infinite number of ways to define its content (Glimskär and Skånes 2015, Chapter 8). We illustrate this in our examples regarding the plasticity of the forest concept, where even the "simple" classification of *forest* remains ambiguous and elusive. The complexity of the problem becomes even more apparent when multiple data sources are used, and the difference in land cover nomenclature over time and space materializes (Käyhkö and Skånes 2006; Ahlqvist 2008a; Eriksson and Skånes 2010). Therefore, when creating ontologies for landscape classification, different people, depending on their background and current aim, will inevitably focus on different aspects and properties of each class. This leads to a lack of agreement in formalized meaning, but when taking awareness of semantic plasticity into account, this becomes manageable since it acknowledges a dynamic development of concepts and classifications over time (Larsson 2007).

One way of handling uncertainty between classification systems would be to set an International Organization for Standardization (ISO) standard that everyone needs to comply with. As pointed out by Ahlqvist (2008a), there are both pros and cons with an ambition to conform all classification into a fixed system. The main issue would be that it locks one single classification system and impedes further developments to keep up with new environmental issues in research and society. Also strict standardizations will impede national or regional need for specific details in areas that are not present in other places, or limit sectorial needs to use different classification systems in their respective applications.

Another way to minimize misunderstanding and misuse of data is again to work intensely toward exhaustive and detailed metadata. A more feasible approach, and probably most successfully implemented in combination with the previous, is to relate metadata to thesauri and controlled vocabularies such as AGROVOC (Sini et al. 2008; Morshed and Sini 2009; Sini et al. 2010) and ISO LCML (Di Gregorio and O'Brian 2012). To succeed with the purpose of turning thesauri and controlled vocabularies into full ontologies, we need to continue working on bridging these with the terms used by various end users (Morshed et al. 2010). Compiling lists of terms that are used (cf. Lund 2014) is not sufficient.

We conclude that semantic awareness and means to enable flexible use and integration of data with large semantic plasticity are the keys to successful monitoring of our changing world. Not until we accomplish that can we succeed in specific tasks such as landscape ecological modeling, geographical analysis, and urban planning. This would be one step closer to cut the Gordian knot of harmonizing complex and ambiguous data.

References

Ahlqvist, O. 2004. A parameterized representation of uncertain conceptual spaces. *Transactions in GIS* 8:493–514.

Ahlqvist, O. 2005. Using uncertain conceptual spaces to translate between land cover categories. *International Journal of Geographical Information Science* 19:821–857.

Ahlqvist, O. 2008a. In search of classification that supports the dynamics of science: The FAO land cover classification system and proposed modifications. *Environment and Planning B: Planning and Design* 35:169–186.

Ahlqvist, O. 2008b. Extending post-classification change detection using semantic similarity metrics to overcome class heterogeneity: A study of 1992 and 2001 U.S. National Land Cover Database changes. *Remote Sensing of Environment* 112:1226–1241.

Ahlqvist, O., and Gahegan, M. 2005. Probing the relationship between classification error and class similarity. *Photogrammetric Engineering & Remote Sensing* 71:1365–1373.

Arnold, S., Kosztra, B., Banko, G., Smith, G., Hazeu, G., Bock, M., and Valcarcel-Sanz, N. 2013. The EAGLE concept—A vision of a future European Land Monitoring Framework. In Lasaponara, R., Masini, N., and Biscione, M. (Eds.), *EARSeL Symposium Proceedings 2013*, "Towards Horizon 2020", Matera, Italy.

Aronsson, M. 1980. Markanvändning och kulturlandskapsutveckling i södra skogsbygden. In Resting, A. (Ed.), *Människan, kulturlandskapet och framtiden. KVHAA*, [In Swedish], konferens 4, februari 12–14, 1979, pp. 221–232. Stockholm, Sweden.

Bennet, B. 2001. What is a forest? On the vagueness of certain geographic concepts. *Topoi* 20:189–210.

Bishr, Y. 1998. Overcoming the semantic and other barriers to GIS interoperability. *International Journal of Geographical Information Science* 12:299–314.

Brandt, J. J. E., Bunce, R. G. H., Howard, D. C., and Petit, S. 2002. General principles of monitoring land cover changes based on two case studies in Britain and Denmark. *Landscape and Urban Planning* 62:37–51.

Bunce, R. G. H, Bogers, M. M. B., Evans, D., and Jongman, R. H. G. 2013. Field identification of habitats directive Annex I habitats as a major European biodiversity indicator. *Ecological Indicators* 22:105–110.

Bunce, R. G. H., Metzger, M. J., Jongman, R. H. G., Brandt, J., de Blust, G., Elena-Rossello, R., Groom, G. B., et al. 2007. A standardized procedure for surveillance and monitoring European habitats and provision of spatial data. *Landscape Ecology* 23:11–25.

Butchart, S. H. M., Walpole, M., Collen, B., van Strien, A., Scharlemann, J. P. W., Almond, R. E. A., Baillie, J. E. M., et al. 2010. Global biodiversity: Indicators of recent declines. *Science* 328:1164–1168.

Comber, A., Fisher, P., and Wadsworth, R. 2005. What is land cover? *Environment and Planning B: Planning and Design* 32:199–209.

Cousins, S. A. O. 2001. Analysis of land-cover transitions based on 17th and 18th century cadastral maps and aerial photographs. *Landscape Ecology* 16:41–54.

Davies, C. E., and Moss, D. 2002. *EUNIS Habitat Classification* February 2002. European Topic Centre of Nature Protection and Biodiversity, Paris, France.

Di Gregorio, A. 2005. *Land Cover Classification System (LCCS)*, version 2: Classification Concepts and User Manual. FAO Environment and Natural Resources Service Series, No. 8 - FAO, p. 208. Rome, Italy.

Di Gregorio, A., and O'Brien, D. 2012. Overview of land-cover classifications and their interoperability. In Giri, C. P. (Ed.), *Remote Sensing of Land Use and Land Cover Principles and Applications*, pp. 37–47.

EEA Joint Committee. 2010. Decision of the EEA Joint Committee No 55/2010 of 30 April 2010 amending Annex XX (Environment) to the EEA Agreement. *Official Journal of the European Union*, L 181, 15th July 2010, p. 23.

Eriksson, S., and Skånes, H. 2010. Addressing semantics and historical data heterogeneities in cross-temporal landscape analyses. *Agriculture, Ecosystems and Environment* 139:516–521.

European Commission. 2007. *Interpretation Manual of European Union Habitats*. EUR 27 http://ec.europa.eu/environment/nature/legislation/habitatsdirective/docs/2007_07_im.pdf.

European Parliament. 2007. Directive 2007/2/EC of the European Parliament and of the council of March 14, 2007, establishing an infrastructure for spatial information in the European Community (INSPIRE). *Official Journal of the European Union* L 108, April 25, 2007, p. 1–14.

European Commission. 2008. Commission Regulation (EC) No. 1205/2008 of December 3, 2008 implementing Directive 2007/2/EC of the European Parliament and of the Council as regards metadata (Text with EEA relevance). *Official Journal of the European Union* L 326, December 4, 2008, p. 12–30.

European Commission. 2010. Report from the Commission to the Council and the European Parliament. The 2010 assessment of implementing the EU Biodiversity Action Plan. Brussels, Belgium: The European Commission.

Feranec, J., Hazeu, G., Christensen, S., and Jaffrain, G. 2007. CORINE land cover change detection in Europe (case studies in the Netherlands and Slovakia). *Land Use Policy* 24:234–247.

Forman, R. T. T., and Godron, M. 1981. Patches and structural components for a landscape ecology. *Bioscience* 31:733–740.

FRA. 2010. *Global Forest Resources Assessment 2010*. Country Report Sweden. FRA2010/202 FAO UN, Rome, Italy. http://www.fao.org/docrep/013/al637e/al637e.pdf.

FRA. 2012. *FRA 2015—Terms and Definitions*. Food and Agriculture Organization of the United Nations. Forest Resources Assessment Working Paper 180 FAO UN, Rome, Italy. http://www.fao.org/docrep/017/ap862e/ap862e00.pdf.

Hawthorne, J. 2006. Epistemicism and semantic plasticity. In Hawthorne, J. (Ed.), *Metaphysical Essays*, pp. 185–210. Oxford, United Kingdom: Oxford University Press.

Heymann, Y., Steenmans, C., Croissille, G., and Bossard, M. 1994. *CORINE Land Cover Technical Guide*. Luxembourg. Office for Official Publications of the European Communities, EUR12585.

IPCC. 2013. Climate Change 2013: *The Physical Science Basis. IPCC Working Group I Contribution to the Fifth Assessment Report of the Intergovernmental Panel on Climate Change*, Stocker, T. F., Qin, D., Plattner, G. K., Tignor, M., Allen, S. K., Boschung, J., Nauels, A., Xia, Y., Bex, V., and Midgley, P M. (Eds.), pp. 1535. Cambridge, United Kingdom and New York, NY: Cambridge University Press.

Jansen, L., Groom, G., and Carrai, G. 2008. Land-cover harmonisation and semantic similarity: Some methodological issues. *Journal of Land Use Science* 3:131–160.

Jenkins, M. 2003. Prospects for biodiversity. *Science* 302:1175–1177.

Jepsen, M. R., and Levin, G. 2013. Semantically based reclassification of Danish land-use and land-cover information. *International Journal of Geographical Information Science* 27:2375–2390.

Jones, M. 1991. The elusive reality of landscape. Concepts and approaches in landscape research. *Norsk geografisk Tidsskrift* 45:229–244.

Käyhkö, N., and Skånes, H. 2006. Change trajectories and key biotopes—Assessing landscape dynamics and sustainability. *Landscape and Urban Planning* 75(3-4):300–321.

Lambin, E. F., Turner II, B. L., Geist, H., Agbola, S., Angelsen, A., Bruce, J. W., Coomes, O., et al. 2001. The causes of land-use and cover change: Moving beyond the myths. *Global Environmental Change* 11:5–13.

Lantmäteriet. 2015, accessed April 8, 2015, https://www.lantmateriet.se/sv/Nyheter-pa-Lantmateriet/pressmeddelande-lantmateriet-slapper-smaskalig-kartinformation-fri/.

Larsson, S. 2007. A general framework for semantic plasticity and negotiation. In Bunt, H. C., and Thijsse, E. C. G. (Eds.), *Proceedings of the 7th International Workshop on Computational Semantics (IWCS-7)* Tilburg ITK, Tilburg University, The Netherlands., p. 12.

Lindberg, E., Olofsson, K., Holmgren, J., and Olsson, H., 2012. Estimation of 3D vegetation structure from waveform and discrete return airborne laser scanning data. *Remote Sensing of Environment* 118:151–161.

Lund, H. G. 2014. *Definitions of Forest, Deforestation, Afforestation, and Reforestation.* Forest Information Services, Gainesville, VA, accessed April 8, 2015, http://home.comcast.net/~gyde/DEFpaper.htm.

Martino, L., and Fritz, M. 2008. Land use/cover area frame statistical survey: Methodology and tools. New insight into land cover and land use in Europe. *Eurostat Statistics in Focus* 33:1–8.

Merriam, G. 1984. *Connectivity: A fundamental Ecological Characteristic of Landscape Pattern.* Paper presented at the 1st International Seminar on Methodology in Landscape Ecological Research and Planning, Roskilde, Denmark.

Metria. 2014. CadasterENV Sweden. http://www.metria.se/CadasterENV/.

Morshed, A., and Sini, M. 2009. *Creating and Aligning Controlled Vocabularie,* pp. 50–53. Advance Technology for Digital Libraries, AT4DL, Trento, Italy. Bozen-Bolzano University Press, Bolzano, Italy, accessed September 11, 2014, ftp.fao.org/docrep/fao/012/ak563e/ak563e00 .pdf.

Morshed, A., Johannsen, G., Keizer, J., and Lei Zeng, M. 2010. Bridging end users' terms and AGROVOC concept server vocabularies. *Proceedings of International Conference on Dublin Core and Metadata Applications 2010,* pp. 186–189, accessed September 11, 2014, http://dcpapers.dublincore.org/pubs/article/view/1015.

Mörtberg, U. M., Balfors, B., and Knol, W. C. 2007. Landscape ecological assessment: A tool for integrating biodiversity issues in strategic environmental assessment and planning. *Journal Environmental Management* 82:457–470.

Moss, D., and Davies, C. E. 2002. *Cross-References between the EUNIS Habitat Classification, Lists of Habitats Included in Legislation, and Other European Habitat Classifications,* p. 176. NERC/Centre for Ecology and Hydrology, NERC Open Research Archive, accessed 8 April 8, 2015.

Normander, B., Levin, G., Auvinen, A. P., Bratli, H., Stabbetorp, O., Hedblom, M., Glimskär, A., and Gudmundsson, G. A. 2012. Indicator framework for measuring quantity and quality of biodiversity—Exemplified in the Nordic countries. *Ecological Indicators* 13:104–116.

Northscape. 2014. Northscape – Nordic Network for Land Use and Land Cover Monitoring, accessed April 8, 2015, http://ign.ku.dk/northscape.

Nyström, M., Holmgren, J., and Olsson, H. 2012. Change detection of mountain birch using multi-temporal ALS point clouds. *Remote Sensing Letters* 4:190–199.

Påhlsson, L. (Ed.) 1998. *Vegetationstyper i Norden.* TemaNord 1998:510. [In mixed Scandinavian languages].

Petit, C. C., and Lambin, E. F. 2002. Impact of data integration technique on historical land-use/land-cover change: Comparing historical maps with remote sensing data in the Belgian Ardennes. *Landscape Ecology* 17:117–132.

Sala, O. E., Chapin III, F. S., Armesto, J. J., Berlow, E., Bloomfield, J., Dirzo, R., Huber-Sanwald, E., Huenneke, L. F., et al. 2000. Global biodiversity scenarios for the year 2100. *Science* 287:1770–1774.

Sini, M., Lauser, B., Salokhe, G., Keizer, J., and Katz, S. 2008. The AGROVOC concept server: Rationale, goals and usage. *Emerald Group Publishing Limited, Journal: Library Review* 57:200–212.

Sini, M., Rajbhandari, S., Amirhosseini, M., Johannsen, G., Morshed, A., and Keizer, J. 2010. The AGROVOC concept server workbench system: empowering management of agricultural vocabularies with semantics. IAALD 2010 World Congress. Montpellier, France. *Proceedings from Session; Semantics,* p. 8. http://iaald2010 .agropolis.fr/proceedings/final-paper/SINI-2010-The_AGROVOC_Concept _Server_Workbench_System-IAALD-Congress-287_b.pdf.

Skånes, H. M. 1997. Towards an integrated ecological-geographical landscape perspective—a review of principal concepts and methods. *Norsk Geografisk Tidsskrift* 51:145–171.

Skånes, H. M. (Ed.) 2005. *Workshop on Harmonisation of Nordic Habitat Classifications in an EU Perspective.* Proceedings from the BioHab Stockholm workshop February 19–21, 2004, p. 61. Nordic Council of Ministers, ANP.

Skånes, H. M., and Andersson, A. 2010. Flygbildstolkningsmanual för Uppföljningsprojektet Natura 2000 version 4.0. UF 19. *Naturvårdsverket,* p. 73. [in Swedish, unpublished].

Skånes, H. M., and Bunce, R. G. H. 1997. Directions of landscape change (1741–1993) in Virestad, Sweden—characterised by multivariate analysis. *Landscape and Urban Planning* 38(1–2):61–75.

Socialdepartementet. 2013. Regleringsbrev för budgetåret 2014 avseende Lantmäteriet, S2013/8937/SAM. [In Swedish].

Svensson, S. A. 2006. Ägoslag i skogen. Förslag till indelning, begrepp och definitioner för skogsrelaterade ägoslag. *Skogsstyrelsen,* p. 48 [in Swedish] [Swedish Forest Agency], Rapport 20. Jönköping, Sweden.

Taylor, P. D., Fahrig, L., Henein, K., and Merriam, G. 1993. Connectivity is a vital element of landscape structure. *Oikos* 68:571–573.

Treaty of Amsterdam. 1997. Amending the treaty on European Union, the treaties establishing the European communities and certain related acts. *Official Journal of the European Union* C 340.

UNEP. 1995. *Global Biodiversity Assessment. Summary for Policy-Makers,* p. 46. Cambridge, MA: Cambridge University Press.

With, K. A., Gardner, R. H., and Turner, M. G. 1997. Landscape connectivity and population distributions in heterogeneous environments. *Oikos* 78:151–169.

Zeller, K. A., McGarigal, K., and Whiteley, A. R. 2012. Estimating landscape resistance to movement: a review. *Landscape Ecology* 27:777–797.

4

Parameterized Approaches to the Categorization of Land Use and Land Cover

Louisa J. M. Jansen

CONTENTS

ABSTRACT Categorization is part of human nature to facilitate communication of knowledge. It is actually the first step to model our environment. Categorizations arise out of social communication needs but they serve specific purposes. The user should be informed of what underlying principles are used in a categorization and what definitions are used for categories and classes that serve as the vehicles for the communication of meaning.

Two *a priori*, hierarchically organized, parameterized categorizations based on the standard set theory with crisp and mutually exclusive classes are discussed. A class is composed of measured or observed parameters with standard definitions. These categorizations are geared toward identification of the two main types of changes: conversions where large semantic differences between classes exist and modifications where small semantic differences exist.

The results clearly show that at aggregated data levels the local variability of spatially explicit land changes may be obscured, whereas patterns can be shown that at more detailed data levels may remain invisible, and vice versa. The use of a set of potential driving factors at various semantic levels shows that variation in semantic contents leads to different sets of spatial determinants.

To make scientific progress, the ambiguities of land use and land cover should be included in more innovative approaches to categorization using fuzzy set theory, thereby overcoming the traditional limitations on the exhaustiveness and mutual exclusivity of classes. The use of advanced mathematical theories in the development of parameterized categorizations will further improve our understanding of changes in land use and land cover.

KEY WORDS: *parameterized approach, categorization, land use, land cover, semantics, organizational hierarchy.*

4.1 Introduction

Why do we categorize or classify? "To classify is human" stated Bowker and Star (1999). Without categorization, phenomena would remain merely a bewildering multiplicity and the precise and unambiguous communication of ideas and concepts concerning these phenomena would be impossible. Categorization of relevant phenomena is essential if generalizations are to be made concerning these phenomena. The prime interest is in general truths, that is, truths related to classes or kinds rather than to their individual members. A truth discovered about such a member is always implicitly applied to the entire group to which the member in question belongs. Without categorization such generalizations would also be impossible. Then, finally, the evolution of a body of reliable knowledge concerning any set of phenomena through the process of accumulation would be extremely difficult without categorization (Shapiro 1959). However, few categorizations take formal shape or any formal algorithm; even fewer categorizations are standardized. Yet, we all use (in)formal categorizations on a daily basis, intentionally or inadvertently.

Categorization facilitates the communication of knowledge concerning specific phenomena, such as land use and land cover, between individuals. Ideally categorizations are able to travel across the borders of (scientific) communities, of which the individuals are part, and maintain some sort of constant identity. Categorizations can be tailored to meet the needs of any one community, though having, at the same time, common identities across settings. To represent multiple constituencies, categorizations leave terms open for multiple meanings across different worlds. Dante Alighieri wrote in 1320 that his work *Divina Commedia* is "polysemantic," that is, of many senses; the first sense in his *Divina Commedia* is that which comes from the letter, and the second is that which is signified by the letter. These multiple interpretations are neither by chance nor incidental. Thus as Ahlqvist (2008a) points out, categorizations are dynamic, ordered structures covered with ambiguity and vagueness. Operationally, though, categorization often makes a straightforward unproblematic leap from concept to class, eliminating any traces of concept ambiguity by stating mutually exclusive and crisp classes (Ahlqvist 2008b).

Sokal (1974) defined classification, or categorization, as "the ordering or arrangement of objects into groups or sets on the basis of relationships. These relationships can be based on observable or inferred properties." Another, and even earlier, definition by Shapiro (1959) that reads "the sorting of a set of phenomena composed of generally-alike units into classes or kinds, each class or kind consisting of members having definable characteristics in common" is also interesting but does not underline the importance of relationships. Categorization deals with the variation in semantic contents of data, expressed as differences in organizational hierarchy, and as such the definition "a spatio-temporal and organizational hierarchy based segmentation of the world" (Jansen 2010) would do justice to the importance of semantics; categorization is a manner to model the semantic aspects of our environment. Categorization is the basic cognitive process of arranging objects into classes or categories, as well as the act of distributing objects into classes or categories (i.e., a group of classes) of the same type. However, categorization is a simplification in the sense that it depicts a flawed representation of the reality (Di Gregorio and Jansen 2000), as models represent simplifications of the real world. Various categorizations can represent the same reality. The tangible results of categorization are classes and categories that serve as the vehicles for the communication of meaning (Ahlqvist 2008b).

Different perspectives, or so-called "scapes," to categorization can be taken that are all equally valid and valuable (Veldkamp 2009). One needs to recognize, therefore, that no categorization reflects accurately the social or the natural world (Bowker and Star 1999). Categorizations arise out of social communication needs but they serve specific purposes; not only do they reflect the ideas of a certain community or institution, but they can also be the end result of negotiating and reconciling individual, group, and institutional differences (Ahlqvist 2008b) (see also Chapter 9).

Few people realize how much impact a categorization of land use or land cover may have. At the level of policy, the type of category will have an impact on future economic decisions or on access to subsidies. There is a relation between categorization and policy and decision making that may be invisible but it is evidently powerful. A categorization is also a means for data standardization (for new data sets) and data harmonization (correspondence between existing data sets), as well as being an instrument in contributing to the harmonization of land-use and land-cover *change*, as we need to understand change processes to make informed decisions (McConnell and Moran 2001).

One should also realize that formal categorizations are made by certain groups that have created the design, description, and choice of categories and classes. Such categorizations embody choices that create people's identities. To know the underlying concepts and criteria is therefore important but such information is often difficult to obtain (Comber et al. 2005; Wadsworth et al. 2006).

Also, few people realize that categorizations change over time as knowledge advances, technology develops, and policy objectives change. Lack of knowledge on the differences in the naming of classes, changes in class definition, and adding or removing classes in data sets covering the same area in different periods will create difficulties in the interpretation of actual changes over time versus changes in category definition. Increasingly, data users become interested in understanding the wider meaning of data, that is, the concepts adopted and categorizations used. There is broad recognition that spatial data integration is an essential step in the modeling of changes in land use and land cover and initiatives (e.g., planning and decision making) that aim to respond to such changes (Comber et al. 2005).

The main objective of this chapter is to bring together specific international experiences gathered over time with land-use and land-cover categorizations using parameterized approaches. This means that the set of parameters, or criteria, used to define classes is explicitly mentioned. This chapter will not deal with the classical approach to categorization that consists of class names without explicit mention of criteria used to define classes. Two parameterized approaches to the categorization of land use and land cover (Section 4.2) have been developed. These will be briefly discussed, so as to demonstrate their use to understand change processes (Section 4.3). Furthermore, the use of a set of potential driving factors to explain land-cover change is used at various semantic levels (Section 4.4), to discuss the complementarity of the spatial, temporal, and semantic aspects (Section 4.4.4). The setup of the chapter is illustrated in Figure 4.1. This is followed by the "Way Forward" (Section 4.5).

4.2 International Efforts at Categorization

Land is at the center of sustainable development with its social (people), economic (profit,) and ecological (planet) dimensions (WCED 1987). The dynamics of changes in land use and land cover in the twentieth century were unprecedented. Consequently land use, what people do on the land, and land cover, what can be seen on the Earth's surface, gained increasing attention together with the categorization of land use and land cover.

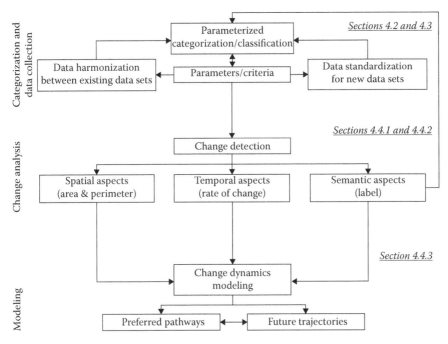

FIGURE 4.1
The setup of the chapter: parameterized categorization, the semantic aspects of change detection, and change dynamics modeling. (Modified from Jansen, L. J. M., "Analysis of Land Change with Parameterised Multi-Level Class Sets. Exploring the Semantic Dimension." PhD Thesis, Wageningen University, Wageningen, Netherlands, 2010.)

International efforts at improved categorization and data collection of land use and land cover emerged from the late 1980s through the 1990s. However, the number of people working on these conceptual approaches was limited. A first prototype for a global reference work for recording data on land use and land cover was developed as a basis for harmonization among systems. This resulted in the "Guidelines for Land-Use and Land-Cover Description and Classification" by the Institute of Terrestrial Ecology (ITE), Institute for Aerospace Survey and Earth Sciences (ITC), and World Conservation Monitoring Centre (WCMC) in a collaborative effort with the United Nations Environment Programme (UNEP) and the Food and Agriculture Organization of the United Nations (FAO) (Wyatt et al. 1998). These guidelines resulted from a number of activities undertaken in the period 1988–1996 (Remmerzwaal 1989; Stomph and Fresco 1991; Adamec 1992; Muecher et al. 1993; ITC/FAO/WAU 1996) and emphasized that a different approach was needed if harmonization and standardization were to be achieved within the land-use/land-cover community. The choice of United Kingdom–based institutions coincided with the United Kingdom government commissioning ITE to undertake a study on the development of systematic procedures for quantitative and qualitative intercomparison of categorizations used in the United Kingdom and Europe (Wyatt et al. 1994).

UNEP, through its Harmonization of Environmental Measurement Programme, and through more programs in collaboration with FAO, was working with other international agencies, including WCMC and the International Geosphere Biosphere Programme (IGBP) Global Change in Terrestrial Ecosystems Programme to promote greater consistency in approaches to the categorization of land and vegetation. They commissioned a survey of existing land-use and land-cover categorizations (Young 1994). The UNEP/FAO Expert Consultation in Geneva in the same year resulted in the understanding that the aim of a single globally applicable categorization was unrealistic (UNEP/FAO 1994). At the same time, the needs of the global climate change research community led to a number of activities coordinated through the IGBP and the International Human Dimensions of Global Environmental Change Programme (IHDP), such as the IGBP Data and Information Systems that compiled a global land-cover data set at 1 km resolution from satellite remote sensing (Loveland et al. 2000) and the IGBP-IHDP Land Use Land Cover Change (LUCC) project (Turner et al. 1995) followed up in 2006 by the Global Land Project (GLP 2005).

At FAO, the Africover project aimed at establishing a digital georeferenced land-cover database and strengthening of the national capacities related to mapping and monitoring (Kalensky 1998). The Africover project established the International Working Group on Classification and Legend (FAO 1997). Contacts between the U.S. Federal Geographic Data Committee's (FGDC) Earth Cover Working Group and Vegetation Subcommittee with the UNEP/FAO initiatives were established in 1996. Contacts with the European Commission were established in the late 1990s through the LANES Concerted Action (LANES 1998). Contacts among all these groups aimed to facilitate joint concept development on the systematic recording of land use and land cover and for the intercomparison of existing systems.

From 2000 onward, the majority of the research community focused on modeling, thereby forgetting that categorization is actually the first step to model our environment.

4.3 Developed and Tested Parameterized Approaches

4.3.1 A Parameterized Approach to Categorize Land Cover

An *overarching concept* for a universally applicable system for land-cover categorization and spatially (geographically) explicit data collection based on a *structural-physiognomic approach* was developed by FAO after analysis of existing systems (Danserau 1961; Fosberg 1961; Eiten 1968; UNESCO 1973; Mueller-Dombois and Ellenberg 1974; Kuechler and Zonneveld 1988; UNEP/

FAO 1994; CEC 1999). The FAO/UNEP Land-Cover Classification System (LCCS) is an *a priori*, hierarchically organized, parameterized categorization where a class is composed of measured or observed parameters with standard definitions (Figure 4.2; Di Gregorio and Jansen 2000). Explicit definition of the overarching concept assists users in understanding the concepts of the categorization system and the meaning of classes. All land covers can be described, though the level of detail for vegetated and cultivated areas is more elaborated than that for bare areas and water bodies. An example of class formation is provided in Table 4.1.

The LCCS categorization methodology (versions 1 and 2) has been tested, modified, and validated in several international projects to evaluate its applicability in different environmental settings, its use at different data collection scales and with different means of data collection, and its usefulness for data harmonization and in land-cover change analysis. LUCC endorsed the methodology (McConnell and Moran 2001). LCCS has been applied by the European Commission's Global Land Cover 2000 (GLC2000) project (Mayaux et al. 2004, 2006), the FAO Africover project (10 countries) (Kalensky 1998) and FAO projects (e.g., Azerbaijan, Bulgaria [Travaglia et al. 2001], Romania and Moldova [Jansen et al. 2014]), in projects financed by the Italian Ministry of Foreign Affairs (e.g., Niger [Jansen et al. 2003c], The Gambia, Mozambique [Jansen et al. 2008a] and Senegal [Jansen and Ndiaye 2006]), by the Nordic Council of Ministers' Nordic Landscape Monitoring project (e.g., Estonia, Denmark, Norway, and Sweden) (Groom 2004), and by a World Bank-financed project in Albania (Jansen et al. 2006a). In March 2014, FAO released the "Global Land Cover-SHARE" land-cover database at the global

TABLE 4.1

Formation of LCCS Class in the "Cultivated & Managed Terrestrial Areas" by Use of a Set of Parameter Options with Increasing Level of Detail of the Class

Parameters Used	Boolean Formula[a]	Standard Class Name
A. Life Form	A4	Graminoid crop(s)
B. Spatial Aspects:		
Field Size	A4B1	Large- to medium-sized field(s) of graminoid crop(s)
Field Distribution	A4B1B5	Continuous large- to medium-sized field(s) of graminoid crop(s)
C. Crop Combination	A4B1B5C1	Monoculture of large- to medium-sized field(s) of graminoid crop(s)
D. Cover-Related Cultural Practices:		
Water Supply	A4B1B5C1D1	Rain-fed graminoid crop(s)
Cultivation Time Factor	A4B1B5C1D1D8	Rain-fed graminoid crop(s) with fallow system

[a] String of parameter codes selected; each code comprises a letter referring to the parameter and a figure referring to the parameter option selected.

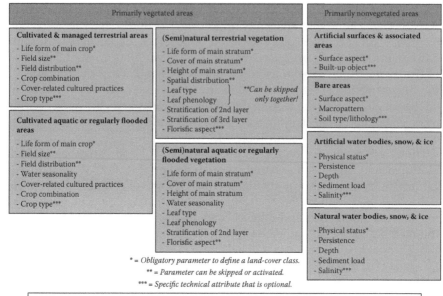

FIGURE 4.2

The major land-cover categories of LCCS (version 2.0) grouped under the primarily vegetated and primarily nonvegetated area distinction. (From Jansen, L. J. M., "Analysis of Land Change with Parameterised Multi-Level Class Sets: Exploring the Semantic Dimension." PhD Thesis, Wageningen University, Wageningen, Netherlands. 2010.)

level using LCCS (see http://www.fao.org/news/story/en/item/216144/icode/). It provides a set of eleven major thematic land-cover layers, resulting in a combination of "best available" high-resolution national, regional, and/or subnational land-cover databases.

4.3.2 A Parameterized Approach to Categorize Land Use

A conceptual approach for land-use categorization and spatially explicit data collection was developed at FAO based on analysis of existing class sets and systems in various sectors (Guttenberg 1959 and 1965; Urban Renewal Administration 1965; Anderson et al. 1976; IGU 1976; Kostrowicki 1977, 1983a, 1983b, 1992a, 1992b; UN-ECE 1989; UN 1989; CEC 1993; UN 1998; Duhamel 1998; Wyatt et al. 1998; APA 1999). The overarching concept combines two key criteria: *function*, grouping all land used for a similar purpose, with *activity*, grouping all land undergoing a certain process resulting in a homogeneous type of product that may serve different functions (UN 1989). The "function" approach relates both to the intended (or primary) and unintended (or secondary) land-uses.

These concepts were tested by FAO in Kenya (Jansen and Di Gregorio 2003) and, in a more advanced version, by an EU PHARE project in Albania in the developed "Land-Use Information System for Albania" (LUISA), (Jansen 2003, 2006) (Figure 4.3).

4.4 Use of Parameterized Categorizations for Change Detection

The LCCS and LUISA categorizations are geared toward identification of the two main types of changes, indicating the type of process taking place and enabling their detailed description and in-depth analysis:

- *Conversions*, where evident changes occur that cannot be (easily) reversed. In terms of semantics, a conversion means large semantic differences between classes. For example, the change from "pasture" into "residential area" in which each class is defined with a different set of parameters.

- *Modifications*, where changes can be reversed. In terms of semantics, modification means small semantic differences. Thus, the set of parameters is almost identical but one or more options within the parameters may be different or one class is defined with more parameters, thus having more detail than the other. An example is the change from "low-density residential area" into a "high-density residential area" where the parameter density has changed.

Contrary to conversion, modification is not as well studied, and, at the global scale, often ignored. The ecological consequences, however, are as important in the case of conversion as in the case of modification (Jansen and Di Gregorio 2002). These two processes are driven by the interaction in space and time between biophysical and human dimensions (Turner et al. 1995). To know the type of changes is important for policy and decision makers, as the type of change has implications for the choice of intervention.

The advantages of the parameterized approach are that change detection becomes possible at the level of the used parameters. Land-cover modifications occur within a category, and land-cover conversions between categories. LCCS will register modifications within the land-cover type, that is, from one domain to another (e.g., from "Forest" to "Woodland," from "Shrubland" to "Sparse vegetation" or from "Tree crops" to "Herbaceous crops") or within the domain (e.g., from "Multi-layered forest" to "Single-layered forest," from "Small-sized fields of graminoid crops" to "Large-sized fields of graminoid crops") (Jansen and Di Gregorio 2002).

In LUISA, land-use modifications occur within a land-use category and land-use conversion occurs between land-use categories. The exceptions are

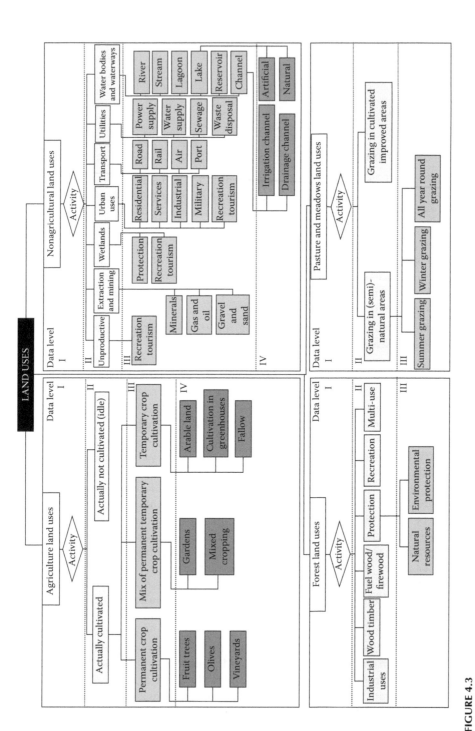

FIGURE 4.3

The four main categories of the "Land-Use Information System for Albania" class set with the data levels indicating the different semantic levels. (From Jansen, L. J. M., *A Standard Methodology for Land-Use Description and Harmonisation of Data Sets*, p.81. Technical report 2. EU PHARE Land-Use Policy II project. AGROTEC S.p.A. Consortium, Rome, Italy, 2003.)

the Nonagricultural Land-use classes, where modifications occur within one group (e.g., within "Urban Uses," within "Transport," and within "Utilities") and conversions between groups (e.g., from "Unproductive" to "Urban Uses," or from "Water Bodies and Waterways" to "Extraction and Mining"). In the Agricultural, Forests and Pasture, and Meadows Land-use categories, conversions occur between categories, whereas modifications occur within a single category within and between groups (within the Agricultural Land-uses, modifications exist within "Permanent Crop Cultivation" or between "Temporary Crop Cultivation" and "Permanent Crop Cultivation," etc.) (Jansen 2003).

4.4.1 Use of the Land-Cover Classification System for Land-Cover Change

The Albanian National Forest Inventory (ANFI) project provided an analysis of spatially explicit land-cover change dynamics in the period 1991–2001, that is, before and after the land reform, at national and district levels. The first year, 1991, shows land cover under a centrally planned economic system in which the state owns the land; 2001 shows the land cover under a market-oriented economic system in which agricultural land is privately owned. The data set was compiled using the FAO/UNEP Land-Cover Classification System (LCCS) (Di Gregorio and Jansen 2000) for codification of classes. This hierarchical data set comprises four levels (described in full detail in Jansen et al. 2003a, 2003b, 2006):

1. At the most aggregated level, the eight categories shown in Figure 4.2
2. That can be disaggregated into 12 land-cover domains
3. That can be further disaggregated into 16 land-cover groups
4. At the most disaggregated level 35 land-cover classes

To gain a better insight in what happened with the semantics, the data are compiled at land-cover group level (the two groups of Water bodies were merged). The land-cover changes in the country are compared with those in Tirane District (Figures 4.4a and b). The figures show the changes from 1991 to 2000/2001 in percentages; areas without changes are not shown in the figures. Each column is linked to what the land-cover group was in 1991 and what it has become in 2000/2001. The higher the column, the more change occurred. Figure 4.4a indicates that the land-cover changes concern mainly two types: (1) from "Broadleaved forests" (FOB) into "Broadleaved woodlands" (WLB) (15.3%); and (2) from "Broadleaved forests" into "Herbaceous crops" (HC) (17.3%). These changes are followed by changes of a more limited extent, such as "Broadleaved woodlands" into "Grasslands" (GL) (6.3%), "Broadleaved forests" into "Grasslands" (4.4%), and "Herbaceous crops"

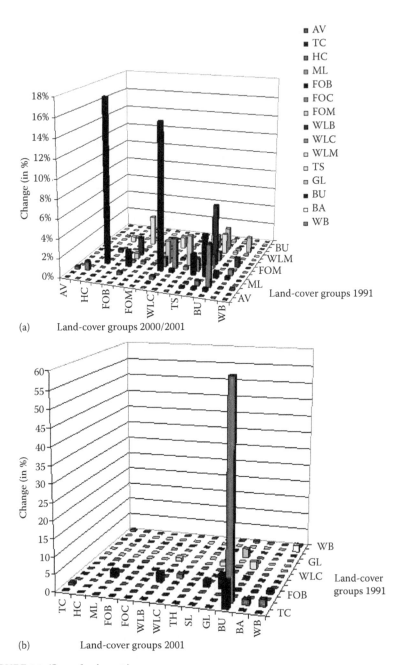

(a)   Land-cover groups 2000/2001

(b)   Land-cover groups 2001

FIGURE 4.4 (See color insert.)
Land-cover changes 1991–2001. (a) At National level. (b) In Tirane District. AV, Aquatic vegetation; TC, Tree & shrub crops; HC, Herbaceous crops; ML, Managed lands; FOB, Broadleaved forests; FOC, Coniferous forests; FOM, Mixed forests; WLB, Broadleaved woodlands; WLC, Coniferous woodlands, WLM, Mixed woodlands; TS, Thickets and shrublands; TH, Thickets; SL, Shrublands; GL, Grasslands; BU, Built-up areas; BA, Bare areas; WB, Water bodies.

into "Built-up areas" (BU) (4.3%). "Broadleaved forests" are the land-cover group with the largest spatio-temporal aspect of change dynamics (Jansen et al. 2006). Figure 4.4b shows that many land-cover changes occurred in Tirane District in the period 1991–2001, but that most changes comprised relatively small areas (less than 5% of total area) with the exception of three types of land-cover changes: (1) the conversion from "Herbaceous crops" (HC) to "Built-up areas" (BU) with 60.9%; (2) the conversion from "Tree and shrub crops" (TC) to "Built-up areas" (BU) with 8.0%; and (3) 13.0% of the "Broadleaved forests" (FOB) changed into another land-cover group. The latter change is very pronounced at the national level. This comparison shows clearly that at the national level the most marked changes are around 15%, whereas at district level a change can be as manifest as more than 60% using the same level of semantics but changing the spatial dimension from national to district.

Analysis of the semantic aspect reveals that in all cases the parameter "canopy cover" of the "life form" trees has changed either from closed to open, closed to sparse, or from open to sparse. This tree loss in the land-cover type could be logically explained by the fact that these natural resources were depleted as a result of deforestation (e.g., illegal cutting) (Jansen et al. 2006). Tirane District is located near the urban and economic center of the country; this is also a market for fuel wood. The increase in built-up area may be explained by the many persons that moved to the (peri-)urban centers, but it may also be related to the change in economic system and land ownership, as constructing a house may be the ultimate sign of land ownership.

Moving from one observational scale to another, that is, from the national to district, alters the behavior of the categories chosen for studying the phenomenon. The differences in behavior may arise for a variety of reasons: variations in intensity, new or unexpected elements, actors, or factors that appear at some observational scales but not at others (Turner et al. 1989). The analysis at the district level shows that some of the changes occur in the whole country, whereas others are more location specific. These data seem to justify a decentralized approach to decision making as, at the national level, policies are normally formulated, and districts are the level at which resource management takes place and laws should be enforced.

4.4.2 Use of the "Land-Use Information System for Albania" to Explain Land-Use Decisions

The Land-Use Policy II (LUP II) project developed the "Land-Use Information System for Albania" (LUISA) in three pilot areas that are linked to the object-oriented "Land-Use Change Analyses" (LUCA) methodology that groups land-use changes in the period 1991–2003 at commune (i.e., municipality) level (Jansen 2003; Jansen et al. 2007). The function and activity approaches were combined in LUISA. The hierarchical data set comprises four levels with increasing levels of detail: four main land-use categories that represent the key categories of the Albanian law on the land (e.g., Agricultural, Forest, Pastures and Meadows, and Nonagricultural Land Uses), 17 classes at level II, 33 classes at level III, and 48 classes at level IV (Figure 4.3).

Not only land-use changes were identified with LUCA, as either conversions or modifications, but also the categorization hierarchy was used to distinguish, for each type of modification, three levels of intensity (i.e. low, medium, and high). The degree of modification, depends on the level of the class (e.g., at Level IV modification is small, at Level III medium, and at Level II high). It is surprising that the hierarchy of class sets, often carefully constructed, is hardly ever used in the analysis of change (Jansen 2006).

As physical and social characteristics of communities vary in space and time, so do land-use choices (Cihlar and Jansen 2001). To gain an insight in the processes, Figure 4.5 shows the land-use conversions and modifications at category level. This allows easy identification of areas in Preza Commune that were more prone to changes than were others. The colors used in Figure 4.5 show changes from temporary to permanent vegetation types (e.g., Agriculture-to-Forests, Agriculture-to-Pasture) as environmentally favorable and conversions to Nonagricultural Land Uses as unfavorable because it means loss of the limited amount of agricultural suitable land in the country. Each of the three land-use data sets available represents a critical moment in time: before the land reform under a centrally-planned economy, the moment of distribution of agricultural lands to the rural households (privatization), and after the land reform in the market-oriented economy. Figure 4.5 shows that

- The change dynamics in the period 1991–1996 were much less compared to 1996–2003, with the exception of cadastral zone 3049, where mainly favorable changes occurred (shown in green).
- The type of land-use changes evolved over time, as undesired conversions (shown in red and yellow) occurred mostly in the period 1996–2003.
- The area affected by land-use changes varies over time because in the period 1996–2003 almost the whole territory is involved.
- If the period 1991–2003 is analyzed then the number of changes is even higher compared to the periods 1991–1996 and 1996–2003 and covers the whole territory of the commune. The different developments in the two periods have disappeared. Favorable changes occur in cadastral zone 3049 (shown in green colors) and unfavorable changes occur mainly in cadastral zone 2862 and less in cadastral zone 1669.

The changes in Preza seem to be divided clearly over the territory: most conversions are found in the western part that consists mainly of hills, whereas most modifications occur in the eastern part that consists of foothills and a plain (indicated by the channel system). Figure 4.5 shows that

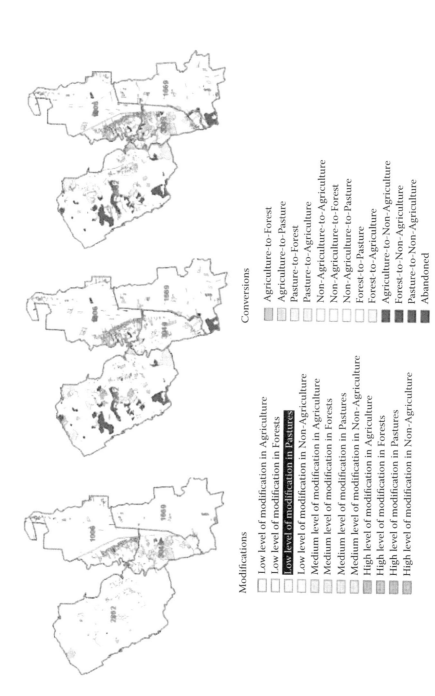

Modifications

☐ Low level of modification in Agriculture
☐ Low level of modification in Forests
☐ Low level of modification in Pastures
☐ Low level of modification in Non-Agriculture
☐ Medium level of modification in Agriculture
☐ Medium level of modification in Forests
☐ Medium level of modification in Pastures
☐ Medium level of modification in Non-Agriculture
☐ High level of modification in Agriculture
☐ High level of modification in Forests
☐ High level of modification in Pastures
☐ High level of modification in Non-Agriculture

Conversions

☐ Agriculture-to-Forest
☐ Agriculture-to-Pasture
☐ Pasture-to-Forest
☐ Pasture-to-Agriculture
☐ Non-Agriculture-to-Agriculture
☐ Non-Agriculture-to-Forest
☐ Non-Agriculture-to-Pasture
☐ Forest-to-Pasture
☐ Forest-to-Agriculture
☐ Agriculture-to-Non-Agriculture
☐ Forest-to-Non-Agriculture
☐ Pasture-to-Non-Agriculture
☐ Abandoned

FIGURE 4.5 (See color insert.)
Land-use change in Preza Commune in Albania: from left to right in the period 1991–1996, 1996–2003, and 1991–2003 (numbers refer to the four cadastral zones).

the history of land-use change is composed of periods for which the within-period change rate is quite stationary but the cross-period change rate is considerably different (Liu and Anderson 2004; Bakker and Van Doorn 2009). The land reform had a definite impact on the number and type of land-use changes. Different parts of the territory are affected by specific modifications or conversions over time. The temporal aspect is important, as the changes before privatization are different in number and nature than they are subsequently. Before 1996, the observed changes seemed still influenced by a central planning policy. With the collapse of central government and the absence of any planning authority, land uses were mainly preserved where environmental conditions were more favorable, and degradation occurred where environmental conditions were less favorable. However, the areas most suitable to agriculture have in general maintained their production characteristics.

4.4.3 Use of Potential Driving Factors to Explain Change at Different Semantic Levels

Categorization produces (hierarchical) data sets comprising classes that have different semantic contents (e.g., class labels). The difference in categorization, and thus in semantic contents of data, is another often ignored issue related to model parameterization (Feng and Flewelling 2004). Semantic contents could be considered another aspect of change dynamics besides space and time (Jansen et al. 2006a; Wu and Li 2006). The semantic contents of data also represent a model, or simplification, and hence a flawed representation of reality (Foody 2001). So the classes present in data sets and used in analysis of land change can also affect the type of explanation given to observed phenomena. Differences in categorization may influence the analysis of land change, and subsequently this might strongly affect the analysis of preferred pathways and future trajectories (see Figure 4.1).

The country-wide parameterized land-cover data set for Romania compiled with LCCS was used in combination with the potential driver data from the EURURALIS project (described in detail by Verburg et al. 2006, 2008). Three levels of variation in semantic contents in the LCCS data set were used: from the most aggregated level comprising 8 land-cover categories, to 13 domains and to the less aggregated level with 21 groups. These three levels were used to analyze what drivers would best explain a certain land-cover type. The set of potential driving factors from the EURURALIS project (e.g., various accessibility, biogeographical, demographic, geomorphology, and soil variables) was identical for the three semantic levels. No assumptions as to what driver would be most important at what semantic level (category, domain, or group) were made *a priori*. Statistically, the relationships between driving factors and land cover were determined, as was the sign of this relationship (negative or positive) (see for details Jansen and Veldkamp 2012).

Table 4.2 shows the results for the (semi)natural vegetation land-cover types at the three semantic levels. The driving factors at category level do not coincide with any domain or group. Only "Grasslands" and "Closed trees" at domain level are explained by the same set of driving factors with the same sign. The analysis results show that there is no overall preferred semantic level that furnishes a statistical model with the highest explanatory power for all land-cover classes at that level. It depends very much on the type of land-cover class distinguished. At each of the three distinguished semantic levels, some land-cover classes are better explained than at the other two levels. This may be surprising because, at the level of spatial scale, the coefficients of determination at the national level explained substantially more than at the regional level, and they performed better at coarse resolution compared to fine resolution (Kok and Veldkamp 2001). Variation in semantic contents thus leads to different sets of spatial determinants.

Thus for categorization, the ecological fallacy applies: relationships at more aggregated semantic levels are not necessarily found among less aggregated levels (e.g., the relationship with impermeable layer at category level is not found at domain level). Future policy and decision making depend to a great extent on which semantic contents are present in the data set used to formulate a policy or to make an informed decision.

TABLE 4.2

The Three Most Important Variables in Terms of Standardized Betas and Their Sign in Multiple Regression Equations at Three Semantic Levels

Semantic Level	Category	Domain		Group			
	(Semi-) Natural Vegetation	Forests and Woodlands	Grasslands	Closed Trees	Tree-Dominated Cover Types	Closed Grass Cover	Grassland-Dominated Cover Types
Salinity	1–	1–	2–	2–			3–
Environmental region[a] "Alpine South"					1–	1–	
Environmental region "Continental"					3–		
Environmental region "Pannonian"	2+	3+					
Impermeable layer	3+					2+	
Geomorphology "flat" (average height difference of 0–20 m)		2–	3–	3–	3–		1–
Geomorphology "very mountainous" (average height difference of >400 m)			1–	1–	2+		2–

[a]*Source:* Metzger, M. J., et al., *Journal of Global Ecology and Biogeography*, 14, 549–563, 2005, for explanation of the environmental regions.

4.4.4 Discussion of the Different Examples

There is an obvious need to make complementary analyses comprising the spatial, temporal, and semantic aspects of changes in land use and land cover. The described examples also make clear that the temporal aspect, that is, the period for which data is available, plays a role. Change is not a continuous process but comprises periods for which the within-period change rate is quite stationary but the cross-period change rate is considerably different. Also the spatial aspect is important, as analysis at district and national levels provides different results. The results clearly show that at aggregated data levels the local variability of spatially explicit land changes may be obscured, whereas patterns can be shown that at more detailed data levels may remain invisible, and vice versa (Veldkamp et al. 2001). The same holds true for the different semantic levels: variation in semantic contents leads to different sets of spatial determinants. The parameterized approach facilitates the identification of conversions and modifications. The organizational hierarchy can be used to distinguish different levels of intensity of modifications.

The described examples make clear that the inventory of land-change types—their location, extent, and distribution—and an understanding of the dynamics in a certain period at different organizational levels provide crucial information to policy and decision makers. Complementarity exists not only at the different aspects: the interpretation of land-cover change is also strengthened by land-use change, and vice versa. Land cover is an aid in understanding patterns, whereas land use helps understanding processes.

4.5 Way Forward

The most commonly used categorization systems are hierarchically structured (e.g., plant taxonomy). To many ecologists, it has been long apparent that ecological systems are structured as such (Egler 1942; Schultz 1967). Early on it was also acknowledged that "it is not to be assumed that some one classification will one day be found, and all others will then be abandoned. Each classification serves a certain purpose, and will continue to exist by its own right" (Egler 1942). Thus, there is not one categorization that best characterizes land use or land cover. In addition, it seems not fruitful to go in search of the one hierarchy, because there is no single *a priori* parameter for developing such a hierarchy. Instead, a number of different hierarchies may be used to address different problems.

With standardization, one runs the risk of adopting a categorization with a determined hierarchy that fits a predetermined purpose. Adopting such a categorization for another purpose involves working with a system with a bias that might force our thinking into the framework (e.g., overarching concept) it was designed for, and is probably more appropriate for, rather than another problem area. Currently, there is no *a priori* designation of hierarchy imposed by the social and biophysical sciences in such a way that no other manner of looking at either land use or land cover is feasible or useful. The hierarchy theory also includes that principles developed at one hierarchical level cannot be transposed to higher and lower levels. Clear distinction of type and category within the hierarchy will not lead to more scientific progress.

Definitions are the main, and usually the only, descriptions of categories and classes, since other elements that could contribute to the semantic definition (e.g., the parameters or criteria used) are often absent. Rich narratives are needed that further specify and clarify what is included in a parameter, class, or category, because anyone using a parameter, class, or category will have to interpret their semantic definition and may therefore introduce bias. A parameter, class, or category needs to be understood in a similar manner by the data producer (the generator of data sets), the distributor (the subsequent distribution of data sets), and user (in the end the user of data sets).

Definitions expressed in natural language associated by hierarchical relationships are called terminological ontologies (Sowa 2000). Almost all land-use and land-cover categorizations to date are terminological ontologies. Ontology is an explicit specification of a conceptualization to represent shared knowledge (Gruber 1993; Ahlqvist 2008b). Semantic information can be determined from the definitions of the ontology, and the representation of categories can be enriched with semantic properties (e.g., purpose, time, and location) and relations (e.g., "is-a," "is-a-part-of," and "associated-with") to reveal similarities and heterogeneities (Kavouras et al. 2005). Semantics often form a problem due to the limited description of how exactly class labels should be understood (Comber et al. 2004a) and expert opinions by definition differ (Comber et al. 2005). Moreover, data sets from the same area but from different times often need to be integrated in a geo-database while at the same time each is based on a (slightly) different categorization (Comber et al. 2004b). Similarity in terms does not necessarily imply equivalent category terms. Recognition of semantic heterogeneity is the basis for creating sound data linkages between multiple data sets that are needed for analysis of changes in land use and land cover, for monitoring and modeling, and for planning, policy, and informed decision making.

It is the inherent and awkward ambiguities of land use and land cover that should be included in more innovative approaches. Semantic uncertainty is an inseparable companion of almost any information. In particular, when assessing change, one could estimate the semantic uncertainty similar to thematic accuracy assessments (Ahlqvist et al. 2000). If one looks at change analysis from the semantic perspective, then one can observe that in both LCCS and LUISA it is performed in a rather straightforward manner based on the *standard set theory*: the crisp class A has either changed in another crisp class or crisp class A remains unchanged. Changes of crisp class A into crisp class B or into crisp class C are treated in an identical manner, though one change type may relate to a conversion and the other to a modification. A more sophisticated approach is to consider the notion of vagueness in the categorization system using *fuzzy set theory*. The notion of category semantics and category similarity metrics (e.g., overlap and distance) is concerned with the vagueness inherent in category definitions and semantic relations between categories (Ahlqvist 2008b), thereby overcoming the traditional limitations on the exhaustiveness and mutual exclusivity of classes (Rocchini and Ricotta 2007). The use of advanced mathematical theories in categorizations has become nearly compulsory to make scientific progress in the understanding of *change* in land use and land cover.

Furthermore, in the hierarchical categorizations relations had to be decided once for all the time of original creation. Today, one can incorporate object-oriented views whereby different parameters can be selected and combined on the fly for different purposes (Bowker and Star 1999). Such upcoming categorizations, such as the EIONET Action Group on Land Monitoring in Europe (EAGLE) concept discussed in Chapter 6, allow an unprecedented flexibility and capability in the design and use of very complex information systems. The improved understanding of land use and land cover requires such a system.

Acknowledgments

Because this chapter is based on a compilation of institutional and project work, the author acknowledges

- FAO for having been able to develop LCCS and the concepts for land-use categorization.
- The Directorate General of Forests and Pastures of the Ministry of Agriculture and Food in Albania, and AGROTEC S.p.A. (Italy), which executed the "Albanian National Forest Inventory" (Loan/credit 2846 ALB) project in association with the Department of Forest Science and Environment of the University La Tuscia (Italy).
- The Ministry of Agriculture and Food and the EU PHARE Programme Management Unit in Albania, and AGROTEC S.p.A. (Italy), which executed the "Land-Use Policy II" (EU PHARE AL98-0502) project in consortium with Kadaster (The Netherlands).
- The Ministry of Agriculture, Food and Forestry in Romania, the Romanian Space Agency and the Romanian Centre for Remote Sensing Applications in Agriculture for making available the data from the FAO "Land-Cover/Land-Use Inventory by Remote Sensing for the Agricultural Reform" (TCP/ROM/2801) project.

- The Land-Dynamics Group of Wageningen University in the Netherlands for making available the elaborated data for Romania used in the EURURALIS project.

References

Adamec, J. 1992. *Land-Use Classification Study*. First draft. Internal working paper. FAO, Rome, Italy.

Ahlqvist, O., Keukelaar, J., Oukbir, K. 2000. Rough classification and accuracy assessment. *International Journal of Geographic Information Science* 14(5): 475–496.

Ahlqvist, O. 2008a. Extending post-classification change detection using semantic similarity metrics to overcome class heterogeneity: A study of 1992 and 2001 US National Land-Cover database changes. *Remote Sensing of Environment* 112(3): 1226–1241.

Ahlqvist, O. 2008b. In search for classification that supports the dynamics of science: The FAO Land-Cover Classification System and proposed modifications. *Environment and Planning B* 35: 169–186.

Anderson, J. R., Hardy, E. E., Roach, J. T., Witmer, R.E. 1976. *A Land-Use and Land-Cover Classification System for Use with Remote Ssensor Data*. U.S. Geological Survey Professional Paper No. 964. USGS, Washington, DC.

APA [American Planning Association]. 1999. *Land-Based Classification Standard*. Second draft. Research Department, Chicago, IL.

Bakker, M. M., Van Doorn, A. M. 2009. Farmer-specific relationships between land-use change and landscape factors: Introducing agents in empirical land-use modelling. *Land-Use Policy* 26: 809–817.

Bowker, G. C., Star, S. L. 1999. *Sorting Things Out: Classification and Its Consequences*. MIT Press, Cambridge, MA.

CEC. 1993. *Nomenclature des activités de la communauté européenne (NACE)*. Première révision. Règlement 761/93 du Conseil, Brussel, Belgium.

CEC. 1999. *CORINE Land Cover*. Brussels, Belgium.

Cihlar, J., Jansen, L. J. M. 2001. From land cover to land use: A methodology for efficient land-use mapping over large areas. *The Professional Geographer* 53(2): 275–289.

Comber, A. J., Fisher, P. F., Wadsworth, R. A. 2004a. Assessment of a semantic statistical approach to detecting land-cover change using inconsistent data sets. *Photogrammetric Engineering & Remote Sensing* 70(8): 931–938.

Comber, A. J., Fisher, P. F., Wadsworth, R. A. 2004b. Integrating land-cover data with different ontologies: Identifying change from inconsistency. *International Journal of Geographical Information Science* 18(7): 691–708.

Comber, A., Fisher, P., Wadsworth, R. 2005. Comparing statistical and semantic approaches for identifying change from land-cover data sets. *Journal of Environmental Management* 77: 47–55.

Danserau, P. 1961. Essai de représentation cartographique des éléments structuraux de la végétation. In *Méthodes de la cartographie de la végétation*, edited by Gaussen, H., 233–255. Centre National de la Recherche Scientifique. 97th International Colloquium, Toulouse, Midi-Pyrénées, France, 1960.

Di Gregorio, A., Jansen, L. J. M. 2000. *Land-Cover Classification System (LCCS): Classification Concepts and User Manual*, p.177. FAO/UNEP/Cooperazione italiana, Rome, Italy. CD-ROM.

Duhamel, C. 1998. *First Approximation for a Land-Use Classification*. CESD-Communautaire, Luxembourg. FAO commissioned study (contains an analysis of previously FAO commissioned studies).

Egler, F. E. 1942. Vegetation as an object of study. *Philosophy of Science* 9: 245–260.

Eiten, G. 1968. Vegetation forms. A classification of stands of vegetation based on structure, growth form of the components, and vegetative periodicity. *Boletim do Instituto de Botânica*. Vol. 4, San Paulo, Brazil.

FAO. 1997. *Africover Land-Cover Classification*. FAO, Rome, Italy.

Feng, C. C., Flewelling, D. M. 2004. Assessment of semantic similarity between land use/land cover classification systems. *Computers, Environment and Urban Systems* 27: 229–246.

Foody, G. M. 2001. GIS: The accuracy of spatial data revisited. *Progress in Physical Geography* 25(3): 389–398.

Fosberg, F. R. 1961. A classification of vegetation for general purposes. *Tropical Ecology* 2: 1–28.

Global Land Project [GLP]. 2005. *Science Plan and Implementation Strategy*. Edited by Ojima, D., Moran, E., McConnell, W., Stafford Smith, M., Laumann, G., Morais, J., Young, B., p. 64. IGBP Report No. 53/IHDP Report No. 19. IGBP Secretariat, Stockholm, Sweden.

Groom, G. (Ed.). 2004. *Developments in Strategic Landscape Monitoring for the Nordic Countries*, p.167. ANP2004:705. Nordic Council of Ministers, Copenhagen, Denmark.

Gruber, T. R. 1993. Towards principles for the design of ontologies used for knowledge sharing. *International Journal Human-Computer Studies* 43: 907–928.

Guttenberg, A. Z. 1959. A multiple land-use classification system. *Journal of the American Institute of Planners* 25: 143–50.

Guttenberg, A. Z. 1965. *New Directions in Land-Use Classification*. American Society of Planning Officials, Chicago, IL.

IGU [International Geographic Union]. 1976. World land-use survey. Report of the Commission to the General Assembly of the International Geographic Union. *Geographica Helvetica* 1: 1–28.

ITC/FAO/WAU. 1996. *The Land-Use Database*. A knowledge-based software program for structured storage and retrieval of user-defined land-use data sets. Draft version, January, Enschede.

Jansen, L. J. M. 2003. *A Standard Methodology for Land-Use Description and Harmonisation of Data Sets*, p.81. Technical report 2. EU PHARE Land-Use Policy II project. AGROTEC S.p.A. Consortium, Rome, Italy.

Jansen, L. J. M. 2006. Harmonisation of land-use class sets to facilitate compatibility and comparability of data across space and time. *Journal of Land-Use Science* 1(2-4): 127–156.

Jansen, L. J. M. 2010. *Analysis of Land Change with Parameterised Multi-Level Class Sets: Exploring the Semantic Dimension*. PhD Thesis, Wageningen University, Wageningen, the Netherlands.

Jansen, L. J. M., Badea, A., Milenov, P., Moise, C., Vassilev, V., Milenova, L., Devos, W. 2014. The use of the Land-Cover Classification System in Eastern European countries: Experiences, lessons learnt and the way forward. In *Land-Use and*

Land-Cover Mapping in Europe: Practices & Trends, edited by Manakos, I., Braun, M., 297–325. Vol. 18 of *Remote Sensing and Digital Image Processing*. Springer Publishers, Berlin Heidelberg, Germany.

Jansen, L. J. M., Bagnoli, M., Focacci, M. 2008a. Analysis of land-cover/use change dynamics in Manica Province in Mozambique in a period of transition (1990-2004). *Forest Ecology & Management* 254: 308–326.

Jansen, L. J. M., Carrai, G., Morandini, L., Cerutti, P. O., Spisni, A. 2003a. *Analysis of the Spatio-Temporal and Semantic Aspects of Land-Cover/Use Dynamics*, p. 40. Technical report 2. World Bank Albanian National Forest Inventory (ANFI) project. AGROTEC S. p.A., Rome, Italy.

Jansen, L. J. M., Carrai, G., Morandini, L., Cerutti, P. O., Spisni, A. 2006a. Analysis of the spatio-temporal and semantic aspects of land-cover/use change dynamics 1991–2001 in Albania at national and district levels. *Environmental Monitoring & Assessment* 119: 107–136.

Jansen, L. J. M., Carrai, G., Petri, M. 2007. Land-use change dynamics at cadastral parcel level in Albania: An object-oriented geo-database approach to analyse spatial developments in a period of transition (1991–2003). In *Modelling Land-Use Change—Progress and Applications*, edited by Koomen, E., Bakema, A., Stillwell, J., Scholten, H., 24–44. Vol. 90 of *GeoJournal Library*. Springer Publishers, Berlin Heidelberg, Germany.

Jansen, L. J. M., Cerutti, P. O., Spisni, A., Carrai, G., Morandini, L. 2003b. *Interpretation of Remote Sensing Imagery for Land-Cover/Use Mapping*, p.62. Technical report 1. World Bank Albanian National Forest Inventory (ANFI) project. AGROTEC S.p.A, Rome, Italy.

Jansen, L. J. M., Di Gregorio, A. 2002. Parametric land-cover and land-use classifications as tools for environmental change detection. *Agriculture, Ecosystems & Environment* 91(1-3): 89–100.

Jansen, L. J. M., Di Gregorio, A. 2003. Land-use data collection using the 'Land-Cover Classification System' (LCCS): Results from a case study in Kenya. *Land-Use Policy* 20(2): 131–148.

Jansen, L. J. M., Mahamadou, H., Sarfatti, P. 2003c. Land-cover change analyses using LCCS. *Journal of Agriculture & Environment for International Development* 97(1–2): 47–68.

Jansen, L. J. M., Ndiaye, D. S. 2006. Land-cover change dynamics 1978–1999 of (peri) urban agriculture in the Dakar region. *Journal of Agriculture & Environment for International Development* 100(1–2): 29–52.

Jansen, L. J. M., Veldkamp, A. 2012. Evaluation of the variation in semantic contents of class sets on modelling dynamics of land-use changes. *International Journal of Geographical Information Science* 26(4): 717–746.

Kalensky, Z. D. 1998. Africover land-cover database and map of Africa. *Canadian Journal of Remote Sensing* 24(3): 292–297.

Kavouras, M., Kokla, M., Tomai, E., 2005. Comparing categories among geographic ontologies. *Computers & Geosciences* 31: 145–154.

Kok, K., Veldkamp, A. 2001. Evaluating impact of spatial scales on land use pattern analysis in Central America. *Agriculture Ecosystems and Environment* 85: 205–222.

Kostrowicki, J. 1977. Agricultural typology concept and method. *Agricultural Systems* 2: 33–45.

Kostrowicki, J. 1983a. Land-use survey, agricultural typology and land-use systems: Introductory remarks. *Rural Systems* 1(1): 1–23.

Kostrowicki, J. 1983b. Land-use systems and their impact on environment: An attempt at a classification. *Advances in Space Research* 2(8): 209–215.

Kostrowicki, J. 1992a. A hierarchy of world types of agriculture. In *New Dimensions in Agricultural Geography*, edited by Mohammed, N., 163–203. Vol. I of *Historical Dimensions of Agriculture*. Vedams Books International, New Delhi, India.

Kostrowicki, J. 1992b. Types of agriculture map of Europe: Concept, method, techniques. In *New Dimensions in Agricultural Geography*, edited by Mohammed, N., 227–249. Vol. I of *Historical Dimensions of Agriculture*. Vedams Books International, New Delhi, India.

Kuechler A.W., Zonneveld, I.S., (Eds.). 1988. Vegetation mapping. In *Handbook of Vegetation Science*. Vol. 10. Kluwer Academic Publishers, Dordrecht, the Netherlands.

LANES. 1998. *Development of a Harmonised Framework for Multi-Purpose Land-Cover/Land-Use Information Systems Derived from Earth Observation Data*. Final report. Concerted Action of the European Commission—research programme on the environment and climate. CESD-Communautaire, Luxembourg.

Liu, X., Anderson, C. 2004. Assessing the impact of temporal dynamics on land-use change modelling. *Computers, Environment & Urban Systems* 28: 107–124.

Loveland, T. R., Reed, B. C., Brown, J. F., Ohlen, D. O., Zhu, Z., Yang, L., Merchant, J. W. 2000. Development of a global land-cover characteristics database and IGBP DISCover from 1 km AVHRR data. *International Journal of Remote Sensing* 21(6–7): 1303–1330.

Mayaux, P., Bartholomé, E., Fritz, S., Belward, A. 2004. A new land-cover map of Africa for the year 2000. *Journal of Biogeography* 31: 861–877.

Mayaux, P., Eva, H., Gallego, J., Strahler, A. H., Herold, M., Agrawal, S., Naumov, S., et al. 2006. Validation of the global land-cover 2000 map. IEEE *Transactions on Geoscience & Remote Sensing* 44(7): 1728–1739.

Metzger, M. J., Bunce, R. G. H., Jongman, R. H. G., Mücher, C. A., Watkins, J. W. 2005 A statistical stratification of the environment of Europe. *Journal of Global Ecology and Biogeography* 14: 549–563.

McConnell, W. J., Moran, E. F., (Eds.). 2001. Meeting in the middle: The challenge of meso-level integration. In *An International Workshop on the Harmonisation of Land-Use and Land-Cover Classification*. October 17–20, 2000, Ispra, Italy. LUCC Report Series No. 5. Anthropological Center for Training and Research on Global Environmental Change—Indiana University and LUCC International Project Office, Louvain-la-Neuve, Belgium.

Muecher, C. A., Stomph, T. J., Fresco, L. O. 1993. *Proposal for a Global LandUuse Classification*. Internal final report, February. FAO/ITC/WAU, Wageningen, the Netherlands.

Mueller-Dombois, D., Ellenberg, H. 1974. *Aims and Methods of Vegetation Ecology*, p. 547. John Wiley & Sons, New York and London, United Kingdom.

Rocchini, D., Ricotta, C. 2007. Are landscapes as crisp as we may think? *Ecological Modelling* 204(3–4): 535–539.

Schultz, A. M., 1967. The ecosystem as a conceptual tool in the management of natural resources. In *Natural Resources: Quality and Quantity*, edited by Cieriacy-Wantrup, S.V., 139–161. University of California Press, Berkeley, CA.

Shapiro, I. D. 1959. Urban land-use classification. *Land Economics* 35(2): 149–155.

Sokal, R. 1974. Classification: Purposes, principles, progress, prospects. *Science* 185(4157): 1115–1123.

Sowa, J. F. 2000. *Knowledge Representation: Logical, Philosophical and Computational Foundations*, p. 594. Brooks Cole Publishing Co., Pacific Grove, CA.

Stomph, T. J., Fresco, L. O. 1991. *Procedures and Database for the Description and Analysis of Agricultural Land Use.* A draft. FAO Rome/ITC, Enschede/University of Wageningen, the Netherlands.

Travaglia, C., Milenova, L., Nedkov, R., Vassilev, V., Milenov, P., Radkov, R., Pironkova, Z., 2001. *Preparation of Land-Cover Database of Bulgaria through Remote Sensing and GIS.* FAO Environment and Natural Resources Working Paper No. 6. FAO, Rome, Italy.

Turner, M. G., O'Neill, R., Gardner, R., Milne, B. 1989. Effects of changing spatial scale on the analysis of landscape pattern. *Landscape Ecology* 3: 153–162.

Turner, B. L., Skole, D., Sanderson, S., Fischer, G., Fresco, L. O., Leemans, R. 1995. *Land-Use and Land-Cover Change Science Research Plan.* IGBP Global Change Report No.35/ IHDP Report No. 7. IGBP/IHDP, Stockholm/Geneva, Sweden/Switzerland.

UN [United Nations]. 1989. *International Standard Classification of all Economic Activities (ISIC).* Third revision. United Nations Statistics Division, Statistical Classifications Section, New York, NY.

UN. 1998. *Central Product Classification (CPC).* Version 1.0. United Nations Statistics Division, Statistical Classifications Section, New York, NY.

UN-ECE [United Nations Economic Commission for Europe]. 1989. *Proposed ECE Standard International Classification of Land Use.* Geneva, Switzerland.

UNEP/FAO. 1994. Report on the UNEP/FAO expert meeting on harmonizing land-cover and land-use classifications. Geneva, November 23–24, 1993. GEMS Report Series No. 25. Nairobi, Kenya.

UNESCO [United Nations Educational, Scientific and Cultural Organization]. 1973. *International Classification and Mapping of Vegetation.* Paris, France.

Urban Renewal Administration. 1965. *Standard Land-Use Coding Manual: A Standard System for Identifying and Coding Land-Use Activities (SLUCM).* Urban Renewal Administration, Housing and Home Financing Agency, and Bureau of Public Roads, Department of Commerce. Washington, DC.

Veldkamp, A., 2009. Investigating land dynamics: Future research perspectives. *Journal of Land-Use Science* 4(1): 5–14.

Verburg, P. H., Eickhout, B., Van Meijl, H. 2008. A multi-scale, multi-model approach for analysing the future dynamics of European land use. *The Annals of Regional Science* 42(1): 57–77.

Verburg, P. H., Schulp, C. J. E., Witte, N., Veldkamp, A. 2006. Downscaling of land-use change scenarios to assess the dynamics of European landscapes. *Agriculture, Ecosystems & Environment* 114: 39–56.

Wadsworth, R. A., Comber, A. J., Fisher, P. F. 2006. Expert knowledge and embedded knowledge: Or why long rambling class descriptions are useful. In *Progress in Spatial Data Handling: 12th International Symposium on Spatial Data Handling,* edited by Riedl, A., Kainz, W., Elmes, G., 197–213. Springer Publishers, Berlin Heidelberg, Germany.

WCED [World Commission on Environment and Development]. 1987. *Our Common Future,* p. 247. United Nations, New York, NY.

Wyatt, B. K., Greatorex-Davies, J. N., Hill, M. O., Parr, T. W., Bunce, R. H. G., Fuller, R. M. 1994. *Countryside Survey 1990: Comparison of Land Cover Definitions.* Department of the Environment, London, United Kingdom.

Wyatt, B. K., Billington, C., De Bie, K., De Leeuw, J., Greatorex-Davies, N., Luxmoore, R. 1998. *Guidelines for Land-Use and Land-Cover Description and Classification.* ITE/WCMC/ITC/UNEP/FAO Internal unpublished working document. ITE, Monkshood, United Kingdom.

Wu, J., Li, H. 2006. Concepts of scale and scaling. In *Scaling and Uncertainty Analysis in Ecology: Methods and Applications*, edited by Wu, J., Jones, K. B., Li, H., Loucks, O. L., 1–13. Springer Publishers, the Netherlands.
Young, A. 1994. *Towards International Classification Systems for Land Use and Land Cover*. Report of the UNEP/FAO expert meeting on harmonising land cover and land use classifications, 23–25 November 1993. Earthwatch—Global environment Monitoring System, Geneva, Switzerland.

5

Eliciting and Formalizing the Intricate Semantics of Land Use and Land Cover Class Definitions

Margarita Kokla, Alkyoni Baglatzi, and Marinos Kavouras

CONTENTS

ABSTRACT Land Use and Land Cover (LU/LC) are important aspects for the interpretation and monitoring of the geospatial world. This has led to the development of several LU/LC classifications by different organizations, for various purposes, and at different levels of abstraction. Semantic heterogeneities among LU/LC classes impede information exchange, reuse, and integration. To resolve semantic heterogeneities, top-down approaches have been developed for the formalization and comparison of existing LU/

LC classifications based on attributes, qualities, or features of LU/LC classes. In this chapter, a bottom-up approach is undertaken: the elements that determine LU/LC class semantics are not defined *a priori* but based on the semantic analysis of existing LU/LC nomenclatures. The proposed approach does not aim at substituting top-down formalization or comparison approaches but at supporting such approaches in eliciting and formalizing the variant and intricate semantics of LU/LC class definitions.

KEY WORDS: *Land use, land cover, semantic elements, semantic relations, semantic information extraction, definitions.*

5.1 Introduction

Land Use and Land Cover (LU/LC) are important aspects for the interpretation and monitoring of the geospatial world. This has led to the development of several LU/LC classifications by different organizations, for various purposes, and at varying levels of abstraction, resulting in different LU/LC classes and definitions. Semantic heterogeneities among LU/LC classes impede information exchange, reuse, and integration, and call for elaborate approaches for (a) modeling LU/LC semantics, (b) computing semantic similarity among LC classes (Ahlqvist 2005a; Ahlqvist 2008; Feng and Flewelling 2004; Jansen et al. 2008), and (c) integrating information from different classifications (Comber et al. 2010; Chapter 12). Semantic issues in LU/LC are also considered critical for the identification and explanation of change processes (Chapter 4).

LU/LC classifications are commonly developed using a top-down approach, that is, by specifying the attributes or classifiers subsequently used to form classes. Similar top-down approaches have been developed for the formalization of existing LU/LC classifications to enable their comparison in a meaningful way and thus information exchange across different mapping projects. The Land Cover Meta-Language (LCML) specified by ISO 19144-2:2012 provides a framework for describing different LC classification systems for enabling both the comparison of information from existing LC classifications and the development of consistent future classifications. LCML defines a metalanguage to describe land cover features based on physiognomic aspects that can be part of different land cover nomenclatures. Land cover classes are defined by a combination of a set of land cover elements (basic meta-elements, properties, and qualities) (Di Gregorio et al. 2011). For example, basic meta-elements for biotic features are trees, scrubs, herbaceous vegetation, and so forth.

Another approach with a similar motivation with LCML is the EAGLE concept (Chapter 6). The EAGLE group has developed the EAGLE matrix and the EAGLE data model to provide the conceptual basis for supporting a European Land Monitoring System and facilitating the semantic translation among different European land monitoring initiatives such as CORINE Land Cover (CLC), Copernicus High Resolution Layers, and the INSPIRE data specifications. Three collections of landscape descriptors are defined: (a) land cover components, (b) land use attributes, and (c) landscape characteristics (e.g., land management type, status, spatial pattern, biophysical characteristics, parameters, ecosystems types).

Besides formalization approaches, various approaches have been developed to bridge the semantic differences among different nomenclatures based on features, attributes properties, or quality dimensions used to define LU/LC classes. Ahlqvist (2005b) introduced uncertain conceptual spaces with a use case from the LC domain using attributes, for example, cover, leaf type, leaf phenology, water salinity, edaphic conditions, and so forth. Deng (2008) used properties and relations such as hypernym, material/cover, use/ purpose, and time, whereas Feng and Flewelling (2004) used more specific features such as land cover dominance, life form dominance, inclusion, persistence coverage, and so forth, to compute semantic similarity among geospatial concepts. Baglatzi and Kuhn (2013) used quality dimensions such as cover, life form, tree canopy cover, height, water quality, and so forth, to form conceptual spaces for LC classification systems.

Most of these approaches follow a top-down approach by identifying a priori the features, attributes, properties, and so forth, based on which the description and comparison of LU/LC semantics are accomplished. In this chapter, a bottom-up approach is undertaken; the elements that determine LU/LC class semantics are not defined a priori, but based on the semantic analysis of existing LU/LC nomenclatures. The proposed approach does not aim at substituting top-down formalization or comparison approaches but at supporting such approaches in eliciting and formalizing the variant and intricate semantics of LU/LC class definitions.

The remainder of the chapter is organized as follows: Section 5.2 describes the process of semantic information extraction, Section 5.3 analyzes the semantic elements used for the definition of LU/LC classes and presents examples from different LU/LC classifications, Section 5.4 concludes this chapter and discusses possible future research directions.

5.2 Semantic Information Extraction from Definitions

Semantic information extraction is a field of Natural Language Processing (NLP) that consists of the identification of semantic relations in text, that is, relations between words such as cause, goal, location, part, time, manner, and so forth. Semantic relations may be automatically extracted on the basis of statistical methods applied in large corpora. Pattern matching is another approach, which focuses on the identification of linguistic patterns systematically used to express specific semantic relations. For example, phrases containing the preposition "for" (e.g., for [the] purpose[s] of, used for, intended for) followed by a noun phrase, present participle, or infinitival clause are patterns for expressing the purpose semantic relation. The phrase [X] is used for [Y], which signifies the purpose semantic relation between the terms X and Y.

Auger and Barrière (2010) identify the steps involved in pattern-based semantic relation extraction:

1. Definition of the semantic relation of interest
2. Discovery of the actual patterns that explicitly express the semantic relation in text and the syntactic conditions that realize the meaning of the specific relation
3. Search of instances of the relation using the patterns
4. Structuring of new instances as part of a new or existing ontology (or terminological database)

This work deals with the first and partly the second step of semantic relation extraction. To match with the terminology commonly used in the geospatial domain and to be able to distinguish between different types of semantic information, such as properties, relations, functions, and so forth, the wider term "semantic elements" is used thereinafter instead of the term "semantic relations."

The focus of this work is not on free text in general, but on definitions that are considered as structured resources, since they have special syntax and language. Although there is a debate about their importance and relevance, definitions are important sources of general and scientific knowledge (Jensen and Binot 1987; Klavans et al. 1993; Swartz 1997) and are widely used to describe and communicate the meaning of concepts. However, in literature, there is no complete list of semantic information that can be extracted from definitions (Barrière 1997), because this endeavor is dictionary and domain specific.

The aim of this chapter is to analyze LU/LC definitions to identify the specific semantic elements that characterize LU/LC class semantics. An advantage of extracting semantic information for a limited domain such as LU/LC is that a more detailed analysis may be achieved focusing on specific semantic information relevant to the domain. Also, word senses are rather restricted in a specific domain in contrast to general lexicons and free text, where an important issue is word sense disambiguation. For example, the word "nursery" has a specific meaning within an LC classification, that of a place for the propagation and care of young plants, but in a general context, it may also refer to a baby's room.

Previous research (Kokla and Kavouras 2005) has exploited the power of definitions as a rich source of concept semantics to perform semantic integration among geographic categories from different ontologies and categorizations. The methodology adopted for analyzing definitions and extracting immanent semantic information was introduced by Jensen and Binot (1987) and further pursued by Vanderwende (1995) and Barrière (1997). This approach is based on the following:

1. Parsing (syntactic analysis) of definitions to identify the form, function, and syntactic relations among each part of speech. Syntactic analysis is usually performed with a specialized tool called a parser.

2. Application of rules that locate certain syntactic and lexical patterns (or defining formulas) in definitions.

For example, the rule for extracting the semantic element "PURPOSE" from definitions is as follows (Vanderwende 1995):

If the verb used (intended, etc.) is post-modified by a prepositional phrase with the preposition "for," then there is a PURPOSE semantic relation with the head (s) of that prepositional phrase as the value.

GeoNLP (Kavouras and Kokla 2008; Kokla 2008; Mourafetis 2005) is a tool for the extraction and formalization of semantic information from definitions of geospatial concepts. The tool is used to analyze the definition of each concept and based on the identification of specific lexicosyntactic patterns, extract semantic elements, and their corresponding values that describe the concept's semantics. GeoNLP includes rules for extracting the following semantic elements and their values: agent, cover, direction, flow, is-a, has part, is part of, nature, point in time, proximity, purpose, separation, adjacency, connectivity, intersection, containment, exclusion, surrounding, extension, relative position, shape, size, start, and destination. For example, Figure 5.1 shows the semantic elements, and their corresponding values for the definition *Canal: artificial waterway made for boats or for irrigation*.

The output may be used for several tasks, such as concept formalization, comparison, and integration. GeoNLP has been developed for the extraction of the above semantic elements from rather simple definitions. LU/LC definitions on the other hand are rather extensive and intricate and they include various specific semantic elements.

<Definition>

<DefinitionString = Canal: artificial waterway made for boats or for irrigation/> <ISA = waterway/>

<PURPOSE = for boats or irrigation/>

<NATURE = artificial/>

</Definition>

FIGURE 5.1
Extraction of semantic elements and values for the definition *Canal: artificial waterway made for boats or for irrigation.*

Thus, the first step of the proposed approach is the semantic analysis of LU/LC class definitions from various classifications to identify frequently occurring semantic information. The classifications that were used for such an analysis are the Anderson LU/LC Classification System (LCCS) (Anderson LULC) (Anderson et al. 1976), CLC (EEA 2000), Land-Based Classification Standards (LBCS) (American Planning Association 2001), FAO LCCS (Di Gregorio and Jansen 2005), LU/LC Classification of United Kingdom's National Land Use Database (NLUD) (ODPM 2006), and Land Use and Cover Area frame Survey (LUCAS) (Eurostat 2013). The classifications were selected indicatively. At this point, the idea is not to compare the selected classifications and their class descriptions per se but to acquire the knowledge about the types and facets of semantic information inherent in LU/LC definitions. The second step is the identification of the patterns that explicitly express these semantic relations in text and the syntactic conditions that instantiate these specific relations. This knowledge is essential for subsequent identification of the patterns and the formulation of the rules for automatically extracting this information from definitions.

5.3 Semantic Information in Land Use/ Land Cover Class Definitions

This work deals with the first and partly the second step of semantic relation extraction outlined by Auger and Barrière (2010): (a) the definition of the semantic relations of interest and (b) the discovery of the actual patterns that explicitly express the semantic relation in LU/LC definitions. The subsequent sections examine the semantic elements that commonly occur in the selected LU/LC classifications. LU/LC class definitions include a wealth of semantic information such as (a) spatial and nonspatial properties (e.g., nature, shape, size, and arrangement); (b) subclass–superclass, part–whole, distance, orientation, and topological relations; (c) function; and (d) time. This list is by no means exhaustive but may be used complementarily to LU/LC attributes and classifiers. The semantic elements are analyzed based on the existing literature and examples of their instantiation in definitions are given.

5.3.1 Properties

Properties (or attributes, qualities, features, characteristics) are those entities that can be attributed to things (Swoyer and Orilia 2011) and are extensively used to define the semantics of concepts. Properties are an interdisciplinary matter of discourse (philosophy, semantics, linguistics, information science) and a subject of controversy over their existence, their nature, their types, and the functions they fulfill.

On the basis of different criteria, various kinds of properties have been proposed in literature (Swoyer and Orilia 2011): characterizing, particularizing, and mass properties, intrinsic and extrinsic properties, primary and secondary properties, essential and accidental properties, natural and artificial kinds. Relevant to this work is the categorization of properties into (a) determinables, that is, general properties such as color and shape, and (b) determinates which are more specific, such as yellow and triangular (Funkhouser 2006). In this work, properties are considered as determinables although their values are seen as the determinates. For example, in the definition "ford: a shallow part of a river, stream, etc., that may be crossed," the value (determinate) "shallow" is the value of the determinable (property) "depth." LU/LC class definitions include two major types of properties: (a) spatial properties and (b) nonspatial properties.

5.3.1.1 Spatial Properties

According to psychologists, humans perceive geographic entities primarily through their external characteristics. According to Johansson (2005), qualities such as mass and length are referred to as monadic physical qualities. We use the term "spatial properties" to refer physical qualities such as shape, size (including dimensions width, height, length, and depth), slope, and arrangement.

These properties constitute an important subset of the characteristics of LU/LC classes and for this reason they are extensively used in their definitions. The following examples illustrate the use of the spatial properties shape, size, height, slope, and arrangement.

Shape

> *Schoolhouse churches look like one-room schoolhouses, typically frame-built in a **rectangular shape** with a double row of pews to define the cruciform aisle, and the pulpit centered at the head of the main aisle (LBCS).*

> *Malls, shopping centers, or collection of shop: … Typically the layout of stores are in a **straight line, "U", or "L" shaped**…(LBCS).*

Size

> *Refrigerated warehouse or cold storage: **Large** industrialized warehouse structures with specialized cold storage and climate control facilities (LBCS).*

> *Merchandise marts also serve the same purpose as trade centers but also have permanent exhibit space (**30,000 to 50,000 square feet**) with lower ceilings than exhibition halls (LBCS).*

Height

> *Shrubs are woody perennial plants with persistent woody stems and without any defined main stem (Ford-Robertson, 1971), being less than **5 m tall** (LCCS).*

> *Traditional churches refer to the standard rectangular plan with steep roof pitches, masonry built, and sometimes having **tall** bell towers or steeples (LBCS).*

> *Buildings with more than three floors: Roofed constructions with more than three floors or **more than 10 meters of height** in total (LUCAS).*

Slope

> *Beaches are the smooth sloping accumulations of sand and gravel along shorelines. The surface is stable inland, but the shoreward part is subject to erosion by wind and water and to deposition in protected areas (Anderson LULC).*

Arrangement

According to WordNet, arrangement is the spatial property of "the way in which something is placed" (Princeton University 2010). Geospatial entities may be arranged in lines, grids, or in any other pattern. Linear arrangement describes the organization of geospatial entities in lines such as

> *Dunes: ...onshore wind-carried sand deposits **arranged in cordons of ridges** parallel to the coast (NLUD).*

5.3.1.2 Nonspatial Properties

Artificiality/Naturality

A straightforward distinction in the geospatial domain is between geographic entities created by nature (e.g., forest, river) and those which are created by human intervention (e.g., bridge, road). This distinction refers to the artificiality/naturality of geographic entities seen, for example, in

> *Semi-natural and Natural areas not in use: This class includes areas which are in **natural/semi-natural** state and no signs of any use are visible (LUCAS).*

> *Multiple surface: Any composite surface comprising a mixture of **artificial and natural** elements, for example, a garden or landscaped area adjacent to a building (NLUD).*

> *Cultivated and Managed Terrestrial Area: This class refers to areas where the natural vegetation has been removed or modified and replaced by other types of vegetative **cover of anthropogenic origin** (LCCS).*

Color

The semantic property color is not very common in LU/LC definitions, as it is usually not an identity criterion of a geographic entity. Nevertheless, some examples are found, such as

> *Burnt areas: Areas affected by recent fires, still mainly **black** (CORINE LC).*

> *Evergreen Forest Land includes all forested areas in which the trees are predominantly those that remain **green** throughout the year (Anderson LULC).*

5.3.2 Relations

Relations are crucial for knowledge organization, representation, and reasoning. They are considered to be the glue connecting concepts and creating structured knowledge (Khoo and Na 2006). Because of their central role in human thought, they are a matter of discourse for disciplines such as philosophy, cognitive science, linguistics, and information science. From a linguistic perspective, concepts are defined only relatively to other concepts; this implies the existence of relations (De Saussure and Bally 1986). Relations may be considered as a kind of property: a main distinction is that common properties are monadic (one-place, nonrelational), whereas relations are polyadic (multi-place) (Swoyer and Orilia 2011).

In LU/LC definitions, four types of relations are identified: (a) subclass–superclass relations, (b) part–whole relations, (c) causal relations, and (d) spatial relations.

5.3.2.1 Superclass–Subclass Relations (Hypernymy–Hyponymy)

Hypernymy–hyponymy (superclass–subclass, supertype–subtype, kind-of) defines the relation between a concept and the broader concept it belongs to, for example, river is a kind of watercourse. The specific concept (hyponym) inherits all the characteristics of the broader concept (hypernym) with the addition of at least one more that distinguishes it from the generic concept and its siblings. Examples of hypernymy in LU/LC definitions are as follows:

> Naturally regenerated forest: **Forest** predominantly composed of trees established through natural regeneration (LCCS).

> Land temporarily fallow: **Arable land** that is not seeded for one or more growing seasons (LCCS).

"Arable land" is the more general concept (hypernym) to which "land temporarily fallow" refers to.

The inverse relation of hypernymy is hyponymy or inclusion. Inclusion is commonly used in LU/LC definitions to define subclasses of the defined class, as illustrated in the following definitions:

> Buildings with one to three floors: Roofed constructions with one to three floors or less than 10 m of height in total. This class **includes: single-family houses, mobile homes, summer cottages**…(LUCAS)

> Farming, tilling, plowing, harvesting, or related activities: Agricultural activities, such as farming, plowing, tilling, cropping, seeding, cultivating, and harvesting for the production of food and fiber products. Also **includes sod production, nurseries, orchards, and Christmas tree plantations**… (LBCS)

5.3.2.2 Part–Whole Relations (Meronymy)

Meronymy is the part–whole relation (e.g., walls and roof are parts of building). Parts pertain to every entity in the world, physical or abstract, spatial or temporal, and are intuitively used for the description of objects. Mereology, the formal theory of part and whole, has been deeply studied by philosophy (Simons 2000; Varzi 2014) but is also fundamental to cognitive science, linguistics, and knowledge representation.

Different classifications of the part–whole relation exist in literature (Gerstl and Pribbenow 1995; Iris et al. 1988; Simons 2000; Varzi 2014; Winston et al. 1987). Winston et al. (1987) distinguish six types of meronymic relations from a linguistic point of view: (a) component–integral object (e.g., engine–vehicle), (b) member–collection (e.g., tree–forest), (c) portion–mass (scoop–ice cream bucket), (d) stuff–object (e.g., glass–vase), (e) feature–activity (e.g., take off–air travel), and (f) place–area (e.g., Acropolis–Athens, living room–house).

Meronymy is a crucial relation in the geospatial domain, since geospatial entities are widely defined via their constituting parts. In the context of this research, many facets of the meronymic relation are observed in LU/LC definitions. By adopting the categorization of Winston et al. (1987), three types of meronymic relations are identified in definitions: (a) component–integral object, (b) member–collection, (c) stuff–object, and (d) feature–activity.

5.3.2.2.1 Component–integral object

In the component-integral object meronymic relation the component is not a self-existent entity as shown in the following definition of a building (integral object) that consists of a roof and walls (components):

> Building: a substantial and permanent construction **with a roof and walls** for giving shelter …(NLUD)

The components of an integral object are arranged according to a particular structure and hold specific functional relations to each other and to the integral object. The roof and walls of a building can be placed only on specific positions to properly perform their functions; the roof can be placed only on the top of a building to act as a protective covering.

5.3.2.2.2 Member-collection

Land cover may also be described by the meronymic relation member–collection. Members of a collection do not perform a particular function nor are arranged according to a particular structure to each other and to their whole. This relation refers to a set of distinct entities (the members) forming a compound entity (the collection). In the case of geospatial entities, membership in a collection is determined by spatial proximity. A prominent example of the member–collection relation is that between trees and forest. A tree is a self-existent entity which is also a member of the forest; trees that are parts of a forest are spatially close to each other:

> Planted forest: forest predominately **composed of trees** established through planting and/or deliberate seeding (LCCS).

5.3.2.2.3 Stuff–object

Another aspect of land cover may be described using the stuff–object meronymic relation. This relation specifies not the parts of a whole but the stuff this whole is made of, for example, the road is made of asphalt. In contrast to components of an integral object and members of a collection, staff cannot be separated from the whole. Examples of the stuff–object relation are as follows:

> *Natural Waterbodies, Snow and Ice: This class refers to areas that are naturally* **covered by water, such as lakes, rivers, snow, or ice** *(LCCS).*

5.3.2.2.4 Feature–activity

Although not particularly often, sometimes meronymic relations in LU/LC definitions do not refer to parts of a physical entity but rather indicate phases of an activity or process and thus are best described by the feature–activity meronymic relation as in the following definition:

> *Aircraft takeoff, landing, taxiing, and parking: These* **activities encompass all aspects of air travel and transportation** *that occur at ground facilities, such as airports, hangars, and similar facilities (LBCS).*

5.3.2.3 Causal Relations

The causal (or cause–effect) relation usually establishes the interlinkage between an event (happening) and its results. It is immanent in human nature to "perform causal inferencing... to make sense of events in the world" (Khoo et al. 2002).

  In the geographic domain, the cause is often a physical power or process resulting in the emergence of a phenomenon or geospatial entity. For example, in the following definition the attraction of the moon acts as the physical power causing the regular rise and fall of the sea level:

> *Tidal: A regular rise and fall in the level of the sea,* **caused by the attraction of the moon**, *leads to various combinations of water cover and substrate exposure (LCCS).*

> *Soil Surface—This class includes the naturally occurring unconsolidated material on the earth's surface, which may* **result from weathering of parent material, climate (including the effects of moisture and temperature) and macro- and microorganisms** *(LCCS).*

5.3.2.4 Spatial Relations

Spatial relations are essential for defining geographic entities. Sometimes they are considered to be equally important as the entities themselves (Casati et al. 1998; Klien and Lutz 2005). They constitute an interdisciplinary field of research studied by cognitive science, linguistics, and artificial intelligence. Language and space in particular are well bonded, since "space has a privileged position as a foundational ontological category in language, a position which most other domains do not share" (Regier 1995).

Various categorizations of spatial relations exist in literature according to different criteria and for different purposes (see Pullar and Egenhofer 1988; Pellens 2003; Hudelot et al. 2008). Shariff et al. (1998) outline the major categories of spatial relations distinguished in geographic information system (GIS) literature: (a) topological relations, which remain invariable under consistent topological transformations such as rotation, translation, and scaling; (b) cardinal direction relations, which change under rotation of the frame of reference; and (c) distance relations, which change under scaling since they incorporate a notion of metric. This chapter adopts this categorization with the difference that the category of cardinal direction relations is broadened to include various orientation relations, depending on the adoption of a spatial frame of reference.

5.3.2.5 Topological Relations

Topological relations constitute an important tool for the definition of geographic concepts, since many geographic entities may be specified on the basis of their topological relations to other entities. For example, to define the concept "island" it is necessary to use the topological relation "surrounded by" the sea, since this is a fundamental characteristic that distinguishes islands from other geographic entities.

According to the nine-intersection model (Egenhofer and Franzosa 1991) in the two-dimensional space, 8 topological relations between spatial regions, 19 between lines and surfaces, and 33 between lines can be defined. The notion of topological relations is also applied to the three-dimensional space (Pigot 1991; Zlatanova et al. 2002; Egenhofer 2009), resulting in the following eight topological relations between spatial objects: (1) disjoint, (2) contain, (3) inside, (4) equal, (5) meet, (6) cover, (7) covered by, and (8) overlap.

Geographic definitions use various linguistic expressions to express a wealth of topological relations, such as connect, separate, surround, intersect, and parallel, as shown in the following definitions:

Connect

> *National parkway: A parkway is a roadway in combination with adjacent parkland paralleling the roadway that often **connects cultural or historic sites**. The primary activity here is scenic motoring along a protected corridor (LBCS).*

> *Coastal lagoons: Unvegetated stretches of salt or brackish waters separated from the sea by a tongue of land or other similar topography. These water bodies can be **connected with the sea at limited points**, either permanently or for parts of the year only (CORINE LC).*

Separate

> *Lagoons: **cut off from the sea** by coastal banks or other forms of relief with, however, certain possible openings (LUCAS).*

Coastal lagoons: Unvegetated stretches of salt or brackish waters separated from the sea by a tongue of land or other similar topography. These water bodies can be connected with the sea at limited points, either permanently or for parts of the year only (CORINE LC).

Surround

Sport and leisure facilities: Camping grounds, sports grounds, leisure parks, golf courses, racecourses, and so forth. Includes formal parks not surrounded by urban areas (CORINE LC).

Intersect

Road and rail networks and associated land: composed mainly of large road intersections with associated infrastructure and planted areas, and large marshalling yards (CORINE LC).

Parallel

Longitudinal dunes: Long, narrow, symmetrical dunes running parallel with the prevailing wind direction (LCCS).

Dunes: …onshore wind-carried sand deposits arranged in cordons of ridges parallel to the coast (NLUD).

5.3.2.6 *Orientation Relations*

Orientation relations in geographic definitions refer both to the vertical and the horizontal direction. Orientation relations rely on the adoption of a spatial frame of reference, which functions as a coordinate system for their identification. Several distinctions of spatial frames of reference exist in various disciplines (philosophy, psychology, linguistics, visual perception, etc.), such as relative versus absolute, egocentric versus allocentric, deictic versus intrinsic, and viewer-centered versus object-centered versus environment-centered (Levinson 1996). This chapter adopts the categorization of frames of reference or coordinate systems introduced by Levinson (2003), according to which three major frames of reference in language and cognition may be distinguished: (a) intrinsic, (b) relative, and (c) absolute. In the intrinsic frame of reference, an object is located with regard to a reference object, for example, the expression "in front of the hotel." In the relative frame of reference, the location of an object is determined on the basis of a reference object as seen from the location of a perceiver, for example, "to the left of the hotel." In the absolute frame of reference, the location of an object is determined according to a fixed coordinate system. The three frames of reference refer to both the horizontal and the vertical directions. However, the three frames of reference tend to coincide for the vertical dimension due to its unproblematic nature (Levinson 2003).

Geographic definitions include orientation relations based on only the intrinsic and absolute frames of reference; since there is no perceiver involved when defining LU/LC classes, orientation relations based on relative frames of reference are not included. Therefore, according to the frame of reference used, orientation relations both for the horizontal and vertical dimensions are further distinguished into absolute and intrinsic orientation relations.

5.3.2.7 Horizontal Relations

Absolute orientation relations or cardinal direction relations such as north, south, east, west, northeast, southwest, and so forth, are based on an absolute frame of reference. These relations are common in large-scale spaces (Frank 1996) and are widely used in different human languages (Levinson 2003). An absolute frame of reference is independent of the observer; as a result, cardinal direction relations remain invariant under changes of the observer. For example,

> *Forest dominated by thermophilous deciduous oaks, under local microclimatic or edaphic conditions, are found also far **north** in the Atlantic region, Pannonic and Continental regions (LUCAS).*

As it concerns the intrinsic orientation relations, a concept plays the role of the reference object based on which the relation is defined. The following definitions illustrate the use of intrinsic orientation relations:

> *Fir woods (Abies) are distributed **along the rim** of the southern Mediterranean basin and western Anatolia (LUCAS).*

> *Dunes are defined as low ridges or hillocks of drifted sand mainly moved by wind. They occur in deserts or **along coasts** (LCCS).*

> *Malls are enclosed and built in various shapes and sizes. Strip centers are a row of stores or service outlets managed as one retail entity that does not have enclosed walkways. Most have on-site parking **in front of** stores…(LBCS)*

For example, in the earlier definition, the concept "stores" plays the role of the reference object according to which the concept "on-site parking" is defined.

5.3.2.8 Vertical Relations

As already mentioned, the intrinsic and absolute frames of reference tend to coincide with regard to the vertical direction, since the intrinsic top of an object is aligned with the gravitational field (Levinson 2003). The use of the earth's surface to express vertical orientation relations is quite intuitive in large-scale spaces. Gibson (1979) states that "[t]he ground is the basis of behaviour of land animals and it is also the basis of the visual perception." The ground and the water level are commonly used as reference to describe concepts lying under, on, or over them.

> *Water storage: Not related to utilities, but may be related to an industrial or commercial enterprise. This may include tanks, tank farms, open storage, and so forth, **above or below ground** (LBCS).*

> *Sand—This class includes… gravel or sand banks **above water level** (LUCAS).*

5.3.2.9 Distance Relations

Distance relations in LU/LC definitions are specified using mainly qualitative expressions such as near to, close, far away, etc. Similar to orientation relations, distance relations also require a reference concept on the basis of which the relation is defined, for example, concept A is near to/far from concept B. In the definition of floodplain forests, European river channels function as the reference objects to define the distance relation "close to."

> *Floodplain forests: Alluvial and riparian woodlands and galleries **close to** main European river channels (LUCAS).*

> *Confined Feeding Operations have a built-up appearance, chiefly composed of buildings, much fencing, access paths, and waste-disposal areas. Some are located **near an urban area** to take advantage of transportation facilities and proximity to processing plants (Anderson LULC).*

5.3.3 Function

Function is "a special activity or purpose of a person or thing" (Hornby 2000). Function is central to LU/LC definitions, since it is closely related to use and purpose. The notion of function may be further analyzed in two categories: (a) primary functions and (b) secondary functions. Primary functions refer to the intention behind the emergence of an entity. From a philosophical viewpoint they are called "proper functions" (Millikan 1989) and are synonymous to "purpose," whereas from an environmental viewpoint they imply human intention (Kitamura et al. 2007). Primary functions are used mostly in geographic definitions to describe artificial entities and, together with scale, are substantial for the understanding of the geographical world (Couclelis 1992). For instance, the primary function of a building is to give shelter:

> *Building: a substantial and permanent construction with a roof and walls **for giving shelter,** for example, house, office, shop, warehouse, factory, church, barn (NLUD).*

> *Libraries, museums and galleries: buildings, places, or institutions **devoted to the acquisition, conservation, study, exhibition, and educational interpretation of objects having scientific, historical, or artistic value**...(NLUD).*

Secondary functions refer to the notion of use, which may be considered as incidental or accidental function (Kitamura et al. 2006) because it does not always coincide with the primary function.

> *Developed site—no buildings and no structures: Site is not in natural state, but is **used for a variety of purposes, such as outdoor storage, parking, and a whole host of other functions and activities** (LBCS).*

Use is also considered for natural or seminatural geographic entities in which some kind of activity takes place. The following definition illustrates the use of the concept "agriculture":

> *Agriculture: Enclosed unimproved or little-improved grasslands with little or no management **used for grazing** (NLUD).*

5.3.4 Time

Top-level ontologies deal with time either independently, such as Mizoguchi's Content Ontologies (Gómez-Pérez et al. 2004), or as subsumed, in processes as in Basic Formal Ontology (BFO) (Grenon and Smith 2004). The World Wide Web Consortium (W3C) has proposed Time Ontology in OWL,[*] an ontology for the representation of time concepts. In LU/LC definitions, two aspects of time may be distinguished: (a) the point in time or time period when something occurs and (b) the duration of the occurrence or existence of something. The following definition illustrates the first aspect of time:

> *Low-lying land usually flooded **in winter**, and more or less saturated by water **all year round** (CORINE LC).*

Duration in LU/LC definitions is defined either quantitatively or qualitatively, as shown in the following definitions:

> *Permanent meadows and pastures: land used **permanently (for five years or more)** to grow herbaceous forage crops through cultivation or naturally (LCCS).*

> *Land under temporary crops: land used for crops with **a less than one-year** growing cycle, which must be newly sown or planted for further production after the harvest (LCCS).*

The above semantic elements are organized in a taxonomy, as shown in Figure 5.2. These semantic elements together with specific LU/LC classifiers and attributes may be used to analyze, formalize, and compare the semantics of LU/LC classes across different classifications. For example, Figure 5.3 shows the formalization in a graphical form of the class "building" as defined by NLUD.

5.4 Conclusion

The aim of this chapter was neither to develop an ontology to describe LU/LC classes per se, nor to develop an ad hoc set of semantic elements; rather it was to provide a bottom-up semantic analysis of LU/LC definitions to identify the specific semantic elements immanent in them. In general, definitions are a rich source of semantic information that, however, cannot be used as such for the formalization and comparison of class semantics. In particular, LU/LC class definitions are rather extensive and intricate, including a wealth of semantic elements. The semantic analysis of definitions showed that there are a number of semantic elements frequently and

---

[*]http://www.w3.org/TR/owl-time/

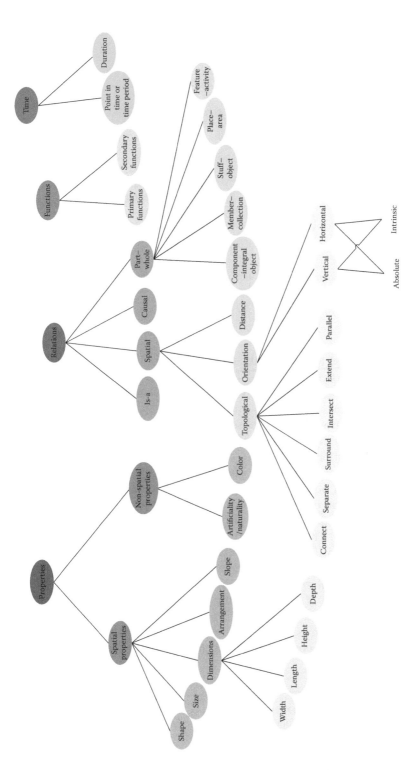

FIGURE 5.2. (See color insert.)
The taxonomy of semantic elements.

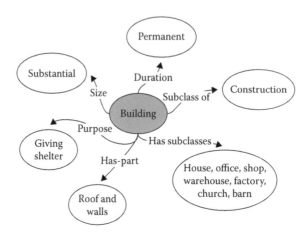

FIGURE 5.3.
Graphical representation of the formalization of the class building: A substantial and permanent construction with a roof and walls for giving shelter, for example, house, office, shop, warehouse, factory, church, barn (NLUD).

systematically used to describe the meaning of LU/LC classes: spatial and nonspatial properties, functions, time, subclass–superclass, part–whole, and causal relations, as well as a wealth of spatial semantic elements that express topological, orientation, and distance relations among concepts. The bottom-up extraction of these semantic elements from definitions may support the formalization and comparison of LU/LC class semantics across different classification systems.

This work deals with the first and partly the second step of semantic relation extraction outlined by Auger and Barrière (2010): the focus was on the identification of semantic elements that frequently occur in LU/LC definitions and on the discovery of linguistic patterns that express these semantic elements. The future step of the proposed approach consists in completing the identification of linguistic patterns and syntactic conditions that instantiate the semantic elements in LU/LC definitions. The completion of this step will enable the formulation of rules for supporting the automatic extraction of these semantic elements and their corresponding values from definitions. Another future direction is to explore whether an existing ontology such as DOLCE,[†] OpenCyc,[‡] and BFO[§] can provide a solid framework for the organization and formalization of the set of semantic elements.

[†]http://www.loa.istc.cnr.it/old/DOLCE.htm
[‡]http://www.cyc.com/platform/opencyc
[§]http://www.ifomis.org/bfo/

References

Ahlqvist, O. 2005a. Using semantic similarity metrics to uncover category and land cover change. *GeoSpatial Semantics*, Lecture Notes in Computer Science 3799: 107–119.

Ahlqvist, O. 2005b. Using uncertain conceptual spaces to translate between land cover categories. *International Journal of Geographical Information Science*, 19(7): 831–857.

Ahlqvist, O. 2008. Extending post classification change detection using semantic similarity metrics to overcome class heterogeneity: A study of 1992 and 2001 U.S. National land Cover Database changes. Remote Sensing of Environment, 112(3): 1226–1241.

American Planning Association. 2001. Land-Based Classification Standards, accessed July 31, 2014, https://www.planning.org/lbcs/standards/pdf/InOneFile.pdf.

Anderson, J. R., E. E. Hardy, J. T. Roach, and R. E. Witmer. 1976. *A Land Use and Land Cover Classification System for Use with Remote Sensor Data.* U.S. Geological Survey Professional 964. United States Government Printing Office, Washington, DC, accessed April 9, 2015, http://landcover.usgs.gov/pdf/anderson.pdf.

Arnold, S., G. Banko, M. Bock, G. Hazeu, B. Kosztra, C. Perger, G. Smith, and T. Soukup, 2015. The EAGLE concept—a paradigm shift in land monitoring. In *Land Use and Land Cover Semantics: Principles, Best Practices and Prospects*, edited by O. Ahlqvist, K. Janowicz, D. Varanka, and S. Fritz. Boca Raton, FL: CRC Press.

Auger, A., and C. Barrière. 2010. Pattern-based approaches to semantic relation extraction: A state-of-the-art. In *Probing Semantic Relations: Exploration and Identification in Specialized Texts*, edited by A. Auger and C. Barrière, pp. 1–18. Amsterdam, The Netherlands: John Benjamins Publishing Co.

Baglatzi, A., and W. Kuhn. 2013. On the Formulation of Conceptual Spaces for Land Cover Classification Systems. In *Proceedings of the 16th AGILE Conference on Geographic Information Science*, May 14–17, Leuven, Belgium.

Barrière, C. 1997. "From a Children's First Dictionary to a Lexical Knowledge Base of Conceptual Graphs." PhD Thesis, School of Computing Science, Simon Fraser University, British Columbia, Canada.

Casati, R., B. Smith, and A. Varzi. 1998. Ontological tools for geographic representation. In *Formal Ontology in Information Systems: Proceedings of the First International Conference (FOIS '98)*, p. 77, June 6–8, Trento, Italy.

Comber, A., A. Lear, and R. Wadsworth. 2010. A comparison of different semantic methods for integrating thematic geographical information: The example of land cover. In *Proceedings of the 13th AGILE International Conference on Geographic Information Science 2010*, Guimarães, Portugal.

Couclelis, H. 1992. People manipulate objects (but cultivate fields): Beyond the Raster-Vector Debate in GIS. In *Proceedings of the International Conference GIS—From Space to Territory: Theories and Methods of Spatio-Temporal Reasoning on Theories and Methods of Spatio-Temporal Reasoning in Geographic Space*, pp. 65–77. London, United Kingdom: Springer-Verlag.

De Saussure, F. 1986. *Course in General Linguistics*, C. Bally and A. Sechehaye , eds. (translated by R. Harris). La Salle, Illinois: Open Court Publishing.

Deng, D. 2008. Measurement of semantic similarity for land use and land cover classification systems. International Conference on Earth Observation Data Processing and Analysis (ICEODPA), ed. D. Li, J. Gong, H. Wu, Proc. of SPIE Vol. 7285, 72850J. Wuhan, China, December 28.

Di Gregorio, A., and L. J. M. Jansen. 2005. Land Cover Classification System: Classification concepts and user manual, Food and Agriculture Organization of the United Nations, accessed July 31, 2014, http://www.glcn.org/downs /pub/docs/manuals/lccs/LCCS2-manual_en.pdf.

Di Gregorio, A., G. Jaffrain, and J. L. Weber. 2011. ISSUE 3: Land cover mapping, land cover classifications, and accounting units—Land cover classification for ecosystem accounting. *Expert Meeting on Ecosystem Accounts*, December 5–7, 2011, London, United Kingdom, accessed November 10, 2014, http://unstats .un.org/unsd/envaccounting/seeaLES/egm/Issue3_EEA_FAO.pdf.

Egenhofer, M., and R. Franzosa. 1991. Point-set topological spatial relations. *International Journal of Geographical Information Science*, 5(2): 161–174.

Egenhofer, M. 2009. A reference system for topological relations between compound spatial objects. *Advances in Conceptual Modeling—Challenging Perspectives*, 5833: 307–316.

European Environment Agency (EEA). 2000. *CORINE Land Cover Technical Guide: Addendum*. Technical report No 40, accessed April 9, 2015. http://www.eea. europa.eu/publications/tech40add.

Eurostat. 2013. LUCAS 2012 Land Use/Cover Area frame statistical survey accessed July 31, 2014, http://epp.eurostat.ec.europa.eu/portal/page/portal/lucas /documents/LUCAS2012_C3-Classification_20131004_0.pdf.

Feng, C., and D. Flewelling. 2004. Assessment of semantic similarity between land use/land cover classification systems. *Computers, Environment and Urban Systems*, 28(3): 229–246.

Frank, A. 1996. Qualitative spatial reasoning: Cardinal directions as an example. *International Journal of Geographical Information Science*, 10(3): 269–290.

Funkhouser, E. 2006. The determinable-determinate relation. *Nous*, 40(3): 548–569.

Gerstl, P., and Pribbenow, S. 1995. Midwinters, end games, and body parts: A classification of part-whole relations. *International Journal of Human Computer Studies*, 43: 865–890.

Gibson, J.J. 1979. *The Ecological Approach to Visual Perception*. Boston, MA: Houghton Mifflin.

Gómez-Pérez, A., M. Fernández-López, and O. Corcho. 2004. *Ontological Engineering: With Examples from the Areas of Knowledge Management, e-Commerce and the Semantic Web*. London, United Kingdom: Springer Verlag, Germany.

Grenon, P., and B. Smith. 2004. SNAP and SPAN: Towards dynamic spatial ontology. *Event-Oriented Approaches in Geographic Information Science: A Special Issue of Spatial Cognition and Computation*, 4(1): 69–104.

Hornby, A. S. 2000. *Oxford Advanced Learner's Dictionary of Current English*. Oxford: Oxford University Press.

Hudelot, C., J. Atif, and I. Bloch. 2008. A spatial relation ontology using mathematical morphology and description logics for spatial reasoning. In *ECAI 2008 Workshop on Spatial and Temporal Reasoning*, pp. 21–25. Patras, Greece.

Iris, M., B. Litowitz, and M. Evens. 1988. *The Part-Whole Relation in the Lexicon: An Investigation of Semantic Primitives*. Cambridge, United Kingdom: Cambridge University Press.

ISO 19144-2:2012. Geographic information: Classification systems— Part 2: Land Cover Meta Language (LCML), accessed April 9, 2015, https://www.iso.org/ obp/ui/#iso:std: 44342:en.

Jansen, L. J. M. 2015. *Land Use and Land Cover Semantics: Principles, Best Practices and Prospects*, edited by O. Ahlqvist, K. Janowicz, D. Varanka, S. Fritz. Boca Raton, FL: CRC Press.

Jansen, L. J. M., G. Groom, and G. Carrai. 2008. Land-cover harmonisation and semantic similarity: Some methodological issues. *Journal of Land Use Science*, 3(2–3): 131–160.

Johansson, I. 2005. *Qualities, Quantities, and the Endurant-Perdurant Distinction in Top-Level Ontologies, 3rd Conference Professional Knowledge Management WM 2005*, pp. 543–550. Springer Verlag, Germany.

Jensen, K., Binot, J. L. 1987. Disambiguating prepositional phrase attachments by using on-line dictionary definitions. *Computational Linguistics* 13 (3–4): 251–260.

Kavouras, M., and M. Kokla. 2008. *Theories of Geographic Concepts: Ontological Approaches to Semantic Integration*. Boca Raton, FL: CRC Press.

Khoo, C., S. Chan, and Y. Niu. 2002. The many facets of the cause-effect relation. In A. Bean and S.H. Myaeng (eds.) *The Semantics of Relationships: An Interdisciplinary Perspective*, 51–71, Dordrecht: Kluwer.

Khoo, C., and J. Na. 2006. Semantic relations in information science. *Annual Review of Information Science and Technology*, 40, 157.

Kitamura, Y., S. Takafuji, and R. Mizoguchi. 2007. Towards a reference ontology for functional knowledge interoperability. In *Proceedings of the International Design Engineering Technical Conferences & Computers and Information in Engineering Conference IDETC/CIE 2007, Vol.6, p. 111–120, September 4–7, 2007, Las Vegas, Nevada, USA*.

Kitamura, Y., N. Washio, M. Ookubo, Y. Koji, M. Sasajima, S. Takafuji, and R. Mizoguchi. 2006. Towards a reference ontology of functionality for interoperable annotation for engineering documents. In *Proceedings of Posters and Demos of ESWC 2006*, 75–76.

Klavans, J., M. Chodorow, and N. Wacholder. 1993. Building a knowledge base from parsed definitions. In *Natural Language Processing: The PLNLP Approach*, edited by K. Jensen, G. Heidorn, and S. Richardson, pp. 119–133, Vol. 193. Boston: The Kluwer International Series in Engineering and Computer Science.

Klien, E., and M. Lutz. 2005. The role of spatial relations in automating the semantic annotation of geodata. *Spatial Information Theory*, Lecture Notes in Computer Science, ed. A.G. Cohn and D.M. Mark, LNCS 3693, Springer-Verlag Berlin Heidelberg 133–148.

Kokla, M. 2008. GEONLP: A Tool for the Extraction of Semantic Information from Definitions. In *Proceedings of the ISPRS 2008 Congress*, Vol XXXVII, Part B2, Commission II. Beijing, China.

Kokla, M., and M. Kavouras. 2005. Semantic information in geo-ontologies: Extraction, comparison, and reconciliation. *Journal on Data Semantics III*, 125–142.

Levinson, S. 1996. Frames of reference and Molyneuxs question: Crosslinguistic evidence. In *Language and Space*, eds. P. Bloom, M. Peterson, L. Nadel, and M. Garrett. pp. 109–169. Cambridge, MA: The MIT Press.

Levinson, S. 2003. Space in language and cognition: Explorations in cognitive diversity. Cambridge University Press, Cambridge, United Kingdom.

Millikan, R. 1989. In defense of proper functions. *Philosophy of Science*, 56(2): 288–302.

Mourafetis, G. 2005. *Automated Extraction and Comparison of Geographic Information from Definitions*. MSc Thesis, Geoinformatics Postgraduate Course. National Technical University of Athens, Greece (in Greek).

ODPM. 2006. National Land Use Database: Land Use and Land Cover Classification, Version 4.4. London, United Kingdom. Office of the Deputy Prime Minister, accessed November 10, 2014, https://www.gov.uk/government/uploads/system/uploads/attachment_data/file/11493/144275.pdf.

Pellens, B. 2003. Specifying Spatial and Part-Whole Relations for Virtual Reality: A Language Perspective. Master Thesis, Department of Computer Science, Faculty of Science, Vrije Universiteit Brussel, Belgium, accessed July 31, 2014, https://wise.vub.ac.be/sites/default/files/theses/PellensB-thesis.pdf.

Pigot, S. 1991. Topological models for 3d spatial information systems. In *Autocarto Conference*, 6: 368–368.

Princeton University. 2010. About WordNet. WordNet. Princeton University, accessed July 31 2014, http://wordnet.princeton.edu.

Pullar, D., and M. Egenhofer. 1988. Towards formal definitions of topological relations among spatial objects. In *Proceedings of the 3rd International Symposium on Spatial Data Handling*, pp. 225–242. Sydney, Australia.

Regier, T. 1995. A model of the human capacity for categorizing spatial relations. *Cognitive Linguistics*, 6(1): 63–88.

Shariff, A., M. Egenhofer, and D. Mark. 1998. Natural-language spatial relations between linear and areal objects: The topology and metric of English-language terms. *International Journal of Geographical Information Science,* 12(3): 215–246.

Simons, P. 2000. *Parts: A Study in Ontology*. Oxford: Oxford University Press.

Swartz, N. 1997. Definitions, Dictionaries, and Meanings, accessed November 10, 2014, http://www.sfu.ca/~swartz/definitions.htm.

Swoyer, C., and F. Orilia. 2011. Properties. Stanford Encyclopedia of Philosophy, accessed November 10, 2014, http://plato.stanford.edu/entries/properties/.

Vanderwende, L. 1995. *The Analysis of Noun Sequences Using Semantic Information Extracted from On-Line Dictionaries*. Ph.D. Thesis, Faculty of the Graduate School of Arts and Sciences, Georgetown University, Washington, DC.

Varzi, A. 2014. *Mereology, The Stanford Encyclopedia of Philosophy*, edited by E. N. Zalta, accessed April 10, 2015, http://plato.stanford.edu/archives/fall2014/entries/mereology/.

Wiegand, N., G. Berg-Cross, and N. Zhou. 2015. Resolving Semantic Heterogeneities in Land Use and Land Cover. In *Land Use and Land Cover Semantics: Principles, Best Practices and Prospects*, edited by O. Ahlqvist, K. Janowicz, D. Varanka, S. Fritz. Boca Raton, FL: CRC Press.

Winston, M., R. Chaffin, and D. Herrmann. 1987. A taxonomy of part-whole relations. *Cognitive Science: A Multidisciplinary Journal*, 11(4): 417–444.

Zlatanova, S., A. Rahman, and W. Shi. 2002. Topology for 3D spatial objects. In *International Symposium and Exhibition on Geoinformation*, October 22–24, Kuala Lumpur, Malaysia.5.3.2.5 **Topological Relations**

6

The EAGLE Concept: A Paradigm Shift in Land Monitoring

Stephan Arnold, Geoffrey Smith, Gerard Hazeu, Barbara Kosztra, Christoph Perger, Gebhard Banko, Tomas Soukup, Geir-Harald Strand, Nuria Valcarcel Sanz, and Michael Bock

CONTENTS

ABSTRACT   The multitude of mapping surveys and applications using land cover (LC) and land use (LU) data has led to a broad variety of land classification systems. Each of them may emphasize different aspects of LC and LU, related to its specific requirements, drivers, and heritage, and may contain a mixture of both land cover and land use information. Integration of multiple data sources and data mining has become key factor in a globalized world of information. However, variations bound to different criteria or thresholds in the descriptions of classes as well as semantic gaps and overlaps in the class definitions hamper the exchange and integration of data between different classification systems. To overcome the difficulties of transferring LC/LU between systems, the EAGLE group has created a feature-oriented data model including an explanatory documentation and a machine-readable version of the model in UML. The targeted approach of the EAGLE concept is to apply a descriptive view on landscape or class definitions respectively by decomposing land units or class definitions into components, attributes, and characteristics instead of classifying them. Like this, the EAGLE concept can be used (a) as a semantic translation tool between classification systems, (b) for semantic content analysis of class definitions, (c) as a guideline for the design of classification systems, or (d) as a mapping guide. As for future steps, the EAGLE group is working on a physical database, its semantic and geometric testing, as well as general aggregation rules and a data model population tool. In the long run, the aim is to provide the basis for a future European land monitoring framework.

KEY WORDS: *Land cover/land use, land monitoring, feature-oriented data modelling, EAGLE.*

6.1 Introduction

The current environmental challenges require the interconnection of ecological, economic, and social factors at local to global scales. Therefore, it is a fundamental need to monitor these factors, their impact on land, their spatial distribution, and changes over time in the form of land cover (LC) and land use (LU) observations. To work effectively across temporal and spatial scales, land monitoring observations need to be modeled in a consistent and machine-readable way. Integration of data originating from different land classification systems into one application is currently difficult.

Numerous classification systems and nomenclatures have been developed to feed the multitude of applications using LU and LC data (Di Gregorio & O'Brien 2012). Class definitions from different systems that are addressing the same topic are determined by differing classifiers or thresholds. In addition, different data collection methods, different scales, narrow and tailored-to-purpose definitions, and the lack of completeness for either LC or LU information impede the direct transfer of an entire data set from one application to another.

Another challenge is that most existing classification systems contain a pragmatic mixture of LC and LU information. A particular application may even emphasize LC aspects in some classes (e.g., in extensively used areas such as forest or natural vegetation) and LU aspects in other classes (e.g., in intensively used areas such as settlements or cropland) due to specific requirements, drivers, and heritage. The mix of LC and LU aspect within one classification system makes it difficult to extract comparable information from available data sources (Comber, 2008). To solve this problem, a basic criterion must be the consequent distinction between LC and LU information, mutually exclusive classes and thematically exhaustive taxonomies. As an example, the classification used in the LUCAS field survey of Eurostat (2014)* represents a step in the right direction, but even this system has potential for enhancement.

To overcome these obstacles to data integration between different systems, there have been global initiatives such as FAO's Land Cover Classification System (LCCS) (Di Gregorio & Jansen 2000) that finally led to the ISO standard 19144-2—Land Cover Meta Language (LCML) (Comber et al. 2010; Herold & Di Gregorio 2012). LCML provides a standardized way to describe an LC classification system, and can therefore also be used to compare and also possibly translate between systems.

Transformation between classification systems will usually be based on identifying similarities between classes, accepting a certain degree of uncertainty. Examples are the translation of the European nomenclature of CORINE Land Cover (CLC) into LCCS (Herold et al. 2009) and the comparative analysis of GLC2000 and CORINE (Neumann et al. 2007). The CLC nomenclature has also in a similar fashion been compared to a North American LC classification system (Ahlqvist 2005). The globalization of information on land requires a certain level of interoperability, which so far was achieved by spatial and thematic generalization, resulting in coarser aggregated data. With their generalizing effect on geometry and thematic content, classifications are not well suited to store descriptive parameters and more differentiated and detailed description of landscape at the object level (Herold et al. 2006b; Comber et al. 2007). This chapter will present the EAGLE concept that provides a framework for integration and comparison of data from different classifications by decomposing the LU/LC classes. Section 6.2 provides background information on the EAGLE initiative and the closely related project Harmonised European Land Monitoring (HELM), both in the context of current European policy requirements. In Section 6.3, the criteria for a data model will be presented, whereas Section 6.4 will deal with the EAGLE model itself, its structure and content. The application of

---

* URL: http://ec.europa.eu/eurostat/statistics-explained/index.php/LUCAS_-_Land_use _and_land_cover_survey.

the EAGLE concept in enhancing the CLC nomenclature is described in Section 6.5. Section 6.6 addresses the relation between the EAGLE concept and other existing standards. The database merging and grid approach are introduced in Section 6.7 as techniques to physically integrate datasets. Section 6.8 will present an outlook for the future in three steps: (1) an explanatory documentation of the EAGLE concept and Unified Modeling Language (UML) application schema, (2) the implementation of the EAGLE Model Population and Comparison Tool (EMPACT), and (3) the development of a Common Integrated Generalization and Aggregation Rules Set (CIGARS). This chapter finalizes with a summary and some conclusions.

6.2 Background

6.2.1 European Policy Requirements

Information on LC, as a result of interaction between natural environment and human activities, is a key input to strategic analysis and planning, general assessment of the state-of-play, and to policy making. Although to date in the European Union's (EU) legislation, there is no legal obligation to derive LC information in general, there are several European policies and many directives that require information on the status and changes of LC and LU (Blanes Guardía et al. 2014). The European Commission's Communication on "Land as a Resource," which is prepared by the Directorate General for Environment and is foreseen to be published in 2015, may lead the land monitoring activities to a more orchestrated and legal-based phase. It is also related to the Communication on the "Roadmap to a Resource Efficient Europe" (EC COM/2011/0571). The European Environment Agency (EEA) is one of the major European users of LC information in fulfilling its mandate to inform European citizens and policy makers about the state of environment in form of its regular reports* (EEA 2010). Several other reporting obligations—among them international ones—to be accomplished by member states (MS) require timely access to LC information, such as the Kyoto Protocol (UNFCCC 1997) and the Convention on Biological Diversity (UNEP 1992). Directives in need of LC information are, among others, the Birds Directive,[†]

---

* The geographical scope of EEA's acitivities is broader than the territory of the European (political) Union. EEA has 33 MS (28 EU MS together with Iceland, Liechtenstein, Norway, Switzerland, and Turkey) and 6 cooperating countries (Albania, Bosnia, and Herzegovina, the former Yugoslav Republic of Macedonia, Montenegro, and Serbia as well as Kosovo under the UN Security Council Resolution 1244/99), covering much of geographical Europe.
† Directive 2009/147/EC of the European Parliament and of the Council of 30 November 2009 on the conservation of wild birds.

Habitats Directive,* Water Framework Directive,† Renewable Energy Directive,‡ and Floods Directive.§ Other EU policies and strategies linked to land monitoring are the Common Agriculture Policy (CAP), Environmental Action Plan (EAP), Biodiversity Action Plan (BAP), Forest Action Plan (FAP), and the Biodiversity Strategy (BD). It lays within the nature of topics that all these different drivers with their specific fields of work or technical and political constraints may require LC information with a particular emphasis on certain aspects of LC, which results in variations of scale, level of detail, or degree of abstraction among the mapping initiatives.

6.2.2 The EAGLE Group

The EAGLE group Environmental Information and Observation Network (EIONET Action Group on Land Monitoring in Europe)¶ was established as an open and self-initiated assembly of national land monitoring and data modeling experts from several European countries. Mainly—but not exclusively—these experts are representatives of National Reference Centers (NRCs) on LC from the European EIONET under the umbrella of the EEA. The objectives of the EAGLE group are (1) to elaborate a conceptual data modeling solution that would support a European information capacity for land monitoring built on existing or future national data sources, (2) to describe landscape in an object-oriented way by using a characterizing decomposition approach instead of classifying (Villa et al. 2008), and (3) facilitate semantic translation between different land monitoring initiatives. As an outcome of the group's work, a first version of an object-oriented data model was drafted (Arnold et al. 2013).

Within this surrounding of different stakeholders and organizational settings, the EAGLE group is active and contributing with its expertise. Further details about the EAGLE group are attached to this chapter as an annex in Section 6.10.

6.2.3 Harmonized European Land Monitoring: The HELM Project

Land monitoring in Europe is to some extent inefficiently organized owing to lack of coordination on LC and LU data production and on exchange between the national, subnational, and European levels. Mostly due to historically evolved responsibilities or data access restrictions, data collection

---

* Council Directive 92/43/EEC of 21 May 1992 on the conservation of natural habitats and of wild fauna and flora, ("FFH-Directive" Natura2000).
† Directive 2000/60/EC of the European Parliament and of the Council of 23 October 2000 establishing a framework for community action in the field of water policy.
‡ Directive 2009/28/EC of the European Parliament and of the Council of 23 April 2009 on the promotion of the use of energy from renewable sources.
§ Directive 2007/60/EC of the European Parliament and of the Council of 23 October 2007 on the assessment and management of flood risks.
¶ See Annex 1 "Who is EAGLE?": http://sia.eionet.europa.eu/EAGLE.

efforts are partly duplicated, and opportunities for mutual sharing of resources are not used exhaustively. To address this situation, the HELM project* was launched within the frame of the seventh EU's Research and Innovation Funding Programme (FP7, EC 2014). The project was placed under the FP7 call for the theme Space, and the project had a 3-year working period from 2011 to 2013. HELM was established as a network of representatives of national authorities concerned with land monitoring from 27 European countries. The project has initiated a move that aims at making European land monitoring more productive by increasing the alignment of national and subnational land monitoring endeavors and by enabling the integration of the national data into a coherent European land monitoring system characterized by high-quality LC/LU data and resource-efficient production.

The objectives addressed by HELM were Existing LULC activities within the MS, Best practices in LC/LU mapping, Common strategic issues and requirements, Operational commonalities and differences among MS, European actors, Societal implications of land monitoring, Financial and legal constraints, Already achieved coordination between different MS, Already achieved bottom-up approaches within MS, Gaps in national systems, Relation to INSPIRE, Data/service requests toward the European Land observation program "Copernicus," Databases merging/Synchronization and Aggregation methods, and Nomenclature/Data model development.

The long-term goal of HELM—which lies beyond the project's time frame—is to increase the maturity of European land monitoring along five sequential steps: (1) mutual interest in achieving reciprocal knowledge, (2) shared visions and planning for the future, (3) joint activities by taking on tasks collectively, (4) alignment of national systems involving the mutual adaptation of data interpretation methods and of the timing of data gathering, and (5) lasting integration and combining data across all administrative levels.

To reach this goal, recommendations were made in form of task reports by the HELM consortium; one of these addresses the collection of criteria for a data model capable of being the conceptual basis for an integrated land monitoring system. These criteria were then collated against the already existing first draft of the EAGLE data model.

In parallel to the various European or national land surveying or monitoring initiatives, also the data production itself is heterogeneously organized, with two main approaches being identifiable. One is the top-down approach, where a central institution such as the EEA is the main driver in steering the requirements and production conditions independent from national data sets. The other one is a bottom-up approach, where the MS derive contributions to a pan-European data set based on existing national initiatives.

---

* HELM website URL: http://www.fp7helm.eu/.

Ideally, a bottom-up solution also would follow a common agreement about methodology among MS, but in practice European data sets are derived from nonharmonized sources, which all have their own particular properties and taxonomy. Normally, those national data sets have a higher spatial resolution and precision compared to the centrally organized European wall-to-wall productions. Three aspects are here of great importance: technical access, interoperability, and comparability of information content. With the INSPIRE directive* (Infrastructure for Spatial Information in the European Community) a legal framework has been installed to foster the interoperability and technical harmonization of spatial data; its implementation is planned to be fully finalized in 2020.

The future vision of an integrated European land monitoring system has the EAGLE concept as a centerpiece that functions as a vehicle for semantic translation and data integration between data sets and nomenclatures (see Figure 6.1). Once the concept is implemented, LC/LU information can be expressed by using the descriptive decomposition of the EAGLE data model for the source data, which then can be recomposed according to the taxonomy of a target nomenclature. It shall be applicable horizontally between national approaches, but also vertically between the national and the European level. As both the INSPIRE data specification (DS) on LC and on LU have been taken into account, the concept is INSPIRE compliant.

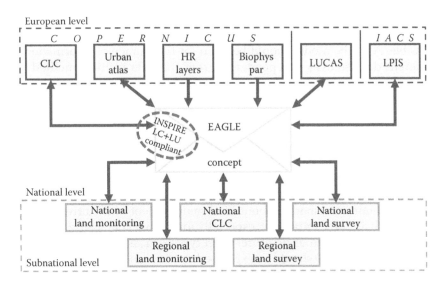

FIGURE 6.1 (See color insert.)
The integration scheme of the European Land Monitoring Framework as envisioned by the EAGLE concept.

---

* DIRECTIVE 2007/2/EC of the EU parliament and of the council of 14 March 2007 establishing an Infrastructure for Spatial Information in the European Community (INSPIRE).

6.3 Criteria of Future Data Model

A harmonized land monitoring process in Europe should accommodate the integration of data from different sources. To facilitate this process, an integrative data model is needed. A list of criteria for such a data model was elaborated and consolidated (Arnold & Kosztra 2013) as part of the HELM project. A collection of the most important criteria is listed below, categorized by conceptual aspects on one side and by thematic content on the other side.

From the *conceptual point of view,* a data model should enable the separation between LC and LU (real and intended). The LC part should be mutually exclusive and have a good balance between level of detail and being generic. The model should be object oriented in terms of landscape description. The landscape should be characterized by spatial units that are populated with attribute information. It should be possible to store quantitative attributes as parameterized information (e.g., soil sealing degree or crown cover density). Also temporal aspects of the landscape should be captured in the model (e.g., seasonal alterations, frequency or duration of a certain state). Furthermore, the model should be scale independent (data storage at any level of abstraction) and applicable on national as well as European level in such a way that it supports bottom-up and top-down approaches. Another criterion of the model is that it should be flexible in its structure or hierarchy; the integration of new model elements or modules should be possible. Also the model should be open for the storage of information for different data sources independently from the data capture method (not only remote sensing but also in situ data and measures). And of course, the data model should be INSPIRE compliant in terms of thematic and technical specifications.

According to the requirements that arise from the European policy requirements and user needs, a data model should contain *thematic information* on soil sealing degree (Copernicus High Resolution Layer [HRL] Imperviousness) and tree cover density (Copernicus HRL Tree cover density) (EEA 2013) as two important parametric attributes in the pan-European context. The model should contain information about land management and cultivation practices, which are very important, especially for further characterization of the LU types forestry and agriculture. Crop types (e.g., according to Eurostat's LUCAS classification) should be expressed by the model; also building types or mining product types would be of benefit. To be able to put LC types into a spatial and temporal relation to each other, spatial patterns and transitional phenomena (seasonal variation and frequencies) are needed as model elements, as well as the details on the status of LC (like "Burnt area," "Under construction," or "Abandoned"). The model should also give room for habitat and ecosystem types (handled separately from LC). Regarding habitat types or ecosystem types, it is a common practice to also use them as LC classes in the taxonomy. This leads constantly to an overlap between those habitat types and pure land cover components (LCCs). Therefore it should be strictly separated between habitats or ecosystem types (e.g., wetlands) and LC (e.g., reeds, surface water).

The summarized outcome of this criteria assessment was that the EAGLE data model can already tackle the majority of the criteria listed in the HELM task 4.3. As the data model is still under development and a living document, as well as surrounding requirements, the list of criteria might be modified at a later stage.

6.4 Terminology and Content of the EAGLE Concept

6.4.1 Terminology

Before going into details of the concept, a few technical terms shall be clarified.

6.4.1.1 Land Cover

According to the INSPIRE Directive, land cover is seen as the "physical and biological cover of the earth's surface including artificial surfaces, agricultural areas, forests, (semi-)natural areas, wetlands, water bodies." The thematic working group for the theme land cover (TWG LC) added that it is an *abstraction* of reality as the earth's surface is actually populated with landscape elements that may or may not be of relevance from each application's point of view. The landscape elements are physical objects such as buildings, roads, trees, plants, water bodies, and so forth. Inside a land unit, the combination of these landscape elements together with their (bio-)physical characteristics form the LC type of that unit. Mapping and describing LC within a certain classification system, however, usually is different from the mapping of the individual landscape elements and concerned with the portrayal of a continuous surface and not with the individual elements that comprise this surface. In this sense, classified LC types are to be understood already as an abstraction of the reality.

In terms of the EAGLE concept, the abstracted representations of the real-world landscape elements that are relevant for LC modeling are called "land cover components." These LCCs are mostly arranged in a typical spatial constellation with a specific spatial distribution. In conventional classification systems, a name is given to these constellations of LCCs by capturing them as LC classes. To express it with an analogy: the LCCs are the ingredients of the "dish" Land Cover Unit (LCU), or the parts of a puzzle that—once put together—give the entire picture on the unit.

6.4.1.2 Land Use

According to the INSPIRE Directive, land use is defined as the "territory characterized according to its current and future planned functional dimension or socio–economic purpose (e.g., residential, industrial, commercial, agricultural, forestry, recreational)." LU (INSPIRE Directive Annex III, INSPIRE 2013c) is different from LC (INSPIRE Directive Annex II, INSPIRE 2013a), which is dedicated to the description of the surface of the earth by its (bio)physical characteristics. LU indicates human influence on the land. Human behavior that has a spatial impact on land can be categorized into seven fundamental principles of existence, namely living, working, education, recreation, supply services, participation in mobility (including communication), and participation in community life (Partzsch 1970). Any kind of LU is present to serve the fulfillment of these principles. One aspect that has gained existential importance can be added here, namely nature conservation in a broader sense, which is the precondition for equilibrium of all other principles of existence.

6.4.1.3 The Landscape Characteristics

"Landscape characteristics" are used here as a third important term. They express further information about the properties of a particular land unit, which cannot be listed neither among "land cover" nor among "land use" in the narrower sense but still are connected to LC or LU because they can specify in more detail activities on a piece of land (fertilization, irrigation, drainage), quantitative parameters (soil sealing degree) or the status (under construction, damaged). According to Comber (2008), this kind of atomic land characters are addressed in a similar way, as so-called "data primitives for land cover and land use." The above clarified meaning of the terms as used in this context is important to explain the concept in the following.

6.4.2 Content of the EAGLE Concept

The EAGLE concept embodies a methodological approach that contains two corresponding representations: the matrix and the UML application schema (UML model).

EAGLE matrix: A tool for semantic comparison between the class definitions of different classification systems by using the EAGLE conceptual model to decompose them to LCCs, land use attributes (LUA), and further landscape characteristics, in the form of a spreadsheet.

EAGLE UML model: A UML model representation of the conceptual data model, visualized in the form of a graphical UML chart. It follows the ISO standard 19109 (Geographic information—Rules for application schema) similar to that applied for INSPIRE.

The two representations of the data model are designed to store the same information and are based on the same considerations; each data model element has its corresponding element in the matrix. The matrix is more a specific form of the data model as a tool for semantic analysis and comparison; the UML data model suits better for being implemented as a system for mapping or monitoring purposes. According to the application purpose, the users can decide to either choose to work with the matrix or with the data model. Both the EAGLE matrix and the EAGLE UML data model can be downloaded from EAGLE website.*

6.4.2.1 The EAGLE Matrix

The EAGLE matrix is subdivided into three blocks as follows:

1. LAND COVER components (LCCs)
2. LAND USE attributes (LUA)
3. Landscape CHARACTERISTICS (CH) (e.g., land management type, status, spatial pattern, biophysical characteristics, parameters, ecosystems types)

In the matrix, these three blocks stand in parallel beside each other, where each column in the spreadsheet table is occupied by one single matrix element (a LCC, a LUA, or a CH). According to the adopted hierarchical structure of the matrix, the matrix elements are grouped together on higher levels.

In the EAGLE model, the LCCs (see Table 6.1), that make up a certain land surface unit or the respective abstracted LC class are the basis for the description of landscape. For better readability, the matrix is here displayed horizontally instead of in vertical columns. The LCC block is structured hierarchically until the fifth level, starting with Abiotic Surfaces, Biotic Surfaces, and Water Surfaces, which are then further subdivided. "Abiotic" here is not used as a synonym for "Artificial," but as the super category for both Artificial and (Nonvegetated) Natural Material Surfaces. Through the hierarchy, the LCC block aims at being mutually exclusive and exhaustive, which means it should be possible to address any kind of LC with a LCC from the matrix. The LCC are not meant to be part of another list of classes, but rather a set of atomic features. They can be used as stand-alone components for homogenous surfaces (e.g., needle-leaved trees, grassland) or must be used in combination with each other for more complex landscape situations (e.g., mixed forest, wetlands, urban built-up areas). In doing so, it is up to the user which hierarchical level he wants to choose for describing the land, also depending on the available LC information. For example, from the remote sensing perspective in case of uncertainty whether the surface is consolidated (rock) or unconsolidated (sand), the LCC "Natural Material Surface" can be used; for semantic translation from one class to another the difference between trees and bushes may not be relevant, so the LCC "Woody Vegetation" can be used.

---

* http://sia.eionet.europa.eu/EAGLE

TABLE 6.1

Land Cover Components of the EAGLE Matrix, Hierarchically Structured from Left to Right

Abiotic/ nonvegetated	Artificial surfaces and constructions	Sealed	Buildings	Conventional buildings
				Specific buildings
			Other constructions	Specific structures and facilities
				Open sealed surface
		Nonsealed	Waste materials	
			Other artificial surfaces	
	Natural material surface	Consolidated surface	Bare rock	
			Hard pan	
			Mineral fragments	
			Bare soil	
		Unconsolidated surface	Natural deposits	Inorganic deposits
				Organic deposits
Biotic/ vegetation	Woody vegetation	Trees	Broadleaved trees	
			Needle-leaved trees	
			Palm trees	
		Bushes, shrubs	Regular shrubs	
			Dwarf shrubs	
	Herbaceous plants (grasses and forbs)	Graminaceous (grass-like)	Regular graminaceous	
			Reeds (high growth)	
		Nongraminaceous (forbs, ferns)		
	Succulents and others			
	Lichens and mosses	Lichens		
		Mosses		
Water	Liquid	Inland water	Water courses	
			Water bodies	
		Coastal water	Estuaries	
			Lagoons	
		Open sea		
	Solid	Permanent snow		
		Ice and glaciers		

With the LCC as the basis, a land unit or a LC/LU class can then be further specified by attaching a LU type (LUA) as an attribute. Other than the LCC block, the LUA block and also the CH block are aiming at mutual exclusiveness; being completely exhaustive is ideal but is not compulsory, as long as there is an "unkown/other" position. The LUA block (Table 6.2) is very much in line with the INSPIRE Hierarchical Land Use Classification System (HILUCS) classes from the INSPIRE DSs on LU. To some extent, the LUA block has been modified according to conceptual consideration compared to the original structure of HILUCS. The LU types are mutually exclusive in

TABLE 6.2

Land Use Attributes of the EAGLE Matrix, Hierarchically Structured from Left to Right

Primary production sector	Agriculture	Commercial crop production
		Agricultural facilities
		Production for own consumption
	Forestry	Short rotation
		Interim or long rotation
		Continuous cover, selective logging
	Mining and quarrying extraction sites	Surface mining
		Underground mining
		Under water mining
		Salines
	Aquaculture and fishing	
	Other primary production	
Secondary production Sector/ Industries	Manufacturing/producing industry	
	Energy production	
Tertiary production sector/ services	Commerce, Finances	
	Communication, information services	
	Accommodation, gastronomy	
	Community services	Public administration, defense, military, security
		Science, research, education
		Health and social services
		Religious facility
		Other community services
	Culture, entertainment, recreational	

(Continued)

TABLE 6.2 (*Continued*)

Land Use Attributes of the EAGLE Matrix, Hierarchically Structured from Left to Right

Transport networks, logistics, Utilities	Transportation Logistics Utilities	
Residential	Permanent residential	
	Residential–commercial mixed	
	Other residential	
Other nonsocioeconomic functions	Inland water functions	Drinking water
		Irrigation
		Fire-fighting
		Reservoir for artificial snow
		Nature protection
		No specific function
	Flood protection (water retention area)	
	Nature protected land	
	Renaturation	
	Abandoned	
	No use, not known, not relevant	

their meaning, but not referring to the occurrence in space or time; in contrary, they can be overlapping either at the same time, or occur sequentially along a season for a particular land unit.

For better readability, it is showed here also horizontally instead of columns and also reduced in hierarchical depth.

With matrix elements from the third matrix part, the CH block (Table 6.3), LCCs and LU types can be further described by attaching more detailed characteristics. Here, qualitative properties such as land management forms or crop types/mining products are listed, as well as spatial patterns (e.g., homogenous, mosaic, scattered) or quantitative parametric properties (e.g., soil sealing degree, crown cover density). The CH block is the most flexible part of the matrix, and is also under elaboration.

With the combination of model elements from all three matrix blocks, a land surface unit or a specific class definition can be described in its pure LC aspects, the present LU(s), and further characterizing properties.

These combinations attached to a certain LC class of one classification system can be compared with the componential description of a similar class of another classification system. In the User Manual of FAO's LCCS (Di Gregorio & Jansen 2000), the term "independent diagnostic criteria" is used

TABLE 6.3

Further Landscape Characteristics of the EAGLE Matrix, Hierarchically Structured from Left to Right

Land management	Agricultural cultivation type	Arable crop land
		Permanent crop land
		Permanent grass land
	Cultivation practices	Crop rotation
		No crop rotation
		Plantation (intensive)
		Orchards (extensive)
		Agroforestry
		Shifting cultivation
	Cultivation measures	Fertilization
		Irrigation
		Drainage
		Mowing
		Grazing
		Shrub clearance
	Forest management type	Intensive monoculture
		Regular
		Extensive (selective logging)
	Forest history type	Endemic, primary
		Reforestation
		Afforestation
Spatial patterns	Texture patterns	Homogenous
		Mosaic
		Scattered
		Mixed, heterogeneous
	Linear patterns	Hedge rows
		Tree rows
		Stone walls
		Ditches
	Built-up patterns	Single houses
		Single blocks
		Row houses
		City street blocks
		Large complexes
Crop type	Arable crops	
	Permanent crops	
	Grass	
Species type	*Open for any kind of list*	
Mining product type	Energy producing materials	
	Metal ores	
	Salt	

(*Continued*)

TABLE 6.3 (*Continued*)

Further Landscape Characteristics of the EAGLE Matrix, Hierarchically Structured from Left to Right

	Peat	
	Others	
Habitat/ecosystem Types	*For example, EUNIS classes*	
(Bio)physical characteristics	Abiotic characteristics	Soil sealing degree
	Vegetation characteristics	Leaf type
		Crown cover density
		Phenology
	Water characteristics	Water regime
		Tidal influence
		Water salinity
Status	Under construction	
	Clear-cut	
	Damaged	
	Contaminated	
	Out of use	
Temporal parameters	Instant event	
	Alteration frequency	
	General duration	
	Determined period	
General parameters	Height	
	Width	

in the sense of a collection of conditions that leads to the categorization of a certain LCU. The EAGLE concept works in principle in a similar way, but is able to distinguish the LCC, LUA, and CH in a more consistent and logical way. The elements that are here described and distinguished as LCC, LUA, and CH are usually in classification part of the class definitions in a mixed-up manner without any clear distinction between them. As the CH block now can be handled in separation from LC and LU information, it enables a more flexible combination of all model elements.

Taking cultivation practices or cultivation measures under land management as an example, it is at first sight arguable whether to consider, for example, the activity of mowing or grazing as a kind of (agricultural) LU; however, when sticking to the definition of LU as a "socioeconomic function or purpose," it becomes clearer where to draw the line between generic LU and land management measures. Like that, the model is more flexible with keeping LU separate from land management, so the generic LU type "agriculture" can be combined with all kinds of land management activities (fertilizing, irrigating, grazing, mowing, etc.) and further spatial patterns (mosaic) and status (burnt area) as additional attributes.

Also, the EAGLE model can be set up in a modular configuration, so that only a selection of elements that is relevant for a certain thematic field of work (e.g., agriculture, forestry, biomass estimations) is included.

Once some model elements—for example, in the LUA or CH block—have to be added or changed due to new requirements, the LCC block stays stable and can be kept, whereas the LUA or CH are flexible to be modified.

6.4.2.2 Structure of Application Schema and Unified Modeling Language Chart (EAGLE Data Model)

The UML is the most well known and used modeling language in the field of software engineering (UML 1998). It is a graphical language for visualizing, specifying, constructing, and documenting a system. It describes its conceptual model and provides a standard way to write a system's blueprints.

The EAGLE data model is available as a UML class diagram that is following the ISO standard 19109 (Geographic information—Rules for application schema).

6.4.2.2.1 General Structure

The object-oriented data model is based on the INSPIRE Directive Annex II—LC DS, where an LC data set consists of a collection of LCUs. At each LCU, the LC has been observed on one or more observation dates. The extension of the EAGLE data model starts at the LCU level, where each unit contains one to many LCCs. Figure 6.2 shows the basic structure of the EAGLE data model.

The LCU is described as a discrete geometric feature, whereas the LCC is described as a parametric observation; also several LCCs can occur in combination with one another inside the containing LCU. Coming from the extension of the INSPIRE LC specification the parametric observation is meant to store either the presence of a particular LC type, a countable parameter (i.e., number of trees), or a percentage value indicating the covered area within the enclosed geometry. In terms of the EAGLE UML model—being an object-oriented data model—the LCC itself is handled as an abstract UML-class* and cannot be instantiated. It means that only the inherited UML classes can be instantiated. A discrete LCC must therefore either be Abiotic/Nonvegetated, Biotic/Vegetation, Water, or any of their subclasses. Figure 6.3 shows a reduced excerpt from the EAGLE data model (version 1.21) and shows the Biotic/Vegetation LCCs, including attributes and custom types, which have been transformed from the Characteristics block from the EAGLE matrix into the syntax of the UML application schema. Here, the landscape characteristics are now distributed among the Biotic LCC where applicable, with the related code lists also displayed aside in the UML chart.

---

* Not to be confused with land cover "class" within a legend or a classification system.

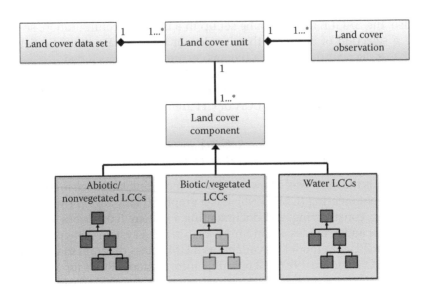

FIGURE 6.2
Simplified structure of the EAGLE data model.

Attributes defined in the parent UML class are inherited to the subclasses, therefore they do not need to be repeated on the lower sublevel. Like this, attributes from the subclasses add up with the attributes from the superior classes. For example, the parent LCC "Biotic/Vegetation" contains all attributes defined in the abstract generic LCC UML class plus the attributes displayed in Figure 6.3. The LCC "Herbaceous" adds another attribute (Mowing) to that list.

6.4.2.2.2 *Level of Detail*

The EAGLE data model contains a method to represent LC in different levels of detail (LoD). This concept was adopted from CityGML, an application schema for the Geography Markup Language (GML) standardized by the Open Geospatial Consortium (OGC)* that describes three-dimensional objects in variable complexity depending on the scale. The LoD is implemented at the LC data set level, meaning that each LC data set can be associated with another LC data set on a higher LoD. The best results will be carried out by combining this method with the grid approach. Grid cells can be upscaled easily to match the spatial constraints (in UML terminology) for different reporting obligations.

---

* URL: http://www.opengeospatial.org/standards/citygml.

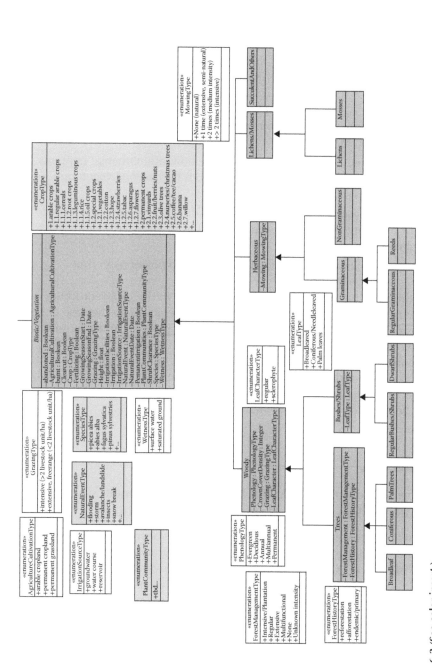

FIGURE 6.3 (See color insert.)

Biotic/Vegetation block in the EAGLE data model in detail—green boxes represent classes, white boxes represent custom types (= characteristics/attributes) with a limited number of options (enumerations), gray boxes represent custom types with an unlimited number of options (code lists).

6.4.2.2.3 Temporal Aspect

Modeling the temporal aspect of LC allows the description of LC change. The EAGLE data model uses ISO 19108 (Geographic information—Temporal schema) and extends the concepts described there. Each LCC holds besides other attributes an attribute of the type TimeDimensionType. This attribute is designed to allow four different kinds of temporal variations of the mapped objects as follows:

- Instant changes for sudden events without temporal duration (e.g., landslide on May 8).
- Seasonal frequency (alteration) to store how often change happens per year (e.g., change twice per year from herbaceous vegetation to bare soil).
- Seasonal duration to store the temporal duration of state of land surface by using coded values (e.g., duration of snow cover half year, quarter of a year).
- Period of a state which is determined by two positions in time.

In addition to just using one of the types above, the different kinds can be combined with each other, allowing for example to map the case of seasonal crops that have a frequency of two changes per year and a duration of 3 months each.

Attributes can be also modeled using this method. An example for such a use case is a building that is under construction for a particular period. The object is still a building, but after a determined time span the attribute UnderConstruction changes its value.

6.5 Application of the EAGLE Matrix: Enhancement of CORINE Land Cover Nomenclature Guidelines

As a first operative use case, the EAGLE concept has been applied for enhancing the nomenclature guidelines of the CLC classification system (Heymann et al. 1994). Since its initiation in 1986, the CLC inventory has served as a principle source of information on the state of environment for Europe by providing data on LC and LU and their changes in a comparable form across countries. With its geographical coverage (39 countries), a timeline of data sets since 1990 (reference years 1990, 2000, 2006, and 2012 updates are in progress) (EEA 2004, 2009), and a common nomenclature (Bossard et al. 2000) and methodology (Büttner et al. 2004; Büttner & Kosztra 2007), the CLC concept has established itself as a "quasi-standard" for land monitoring in Europe in a broad field of applications (e.g., Sifakis et al. 2004; Stathopoulou & Cartalis

2007; Gallego & Bamps 2008; Janssen et al. 2008; EEA 2010; Feranec et al. 2010; Suau-Sanchez et al. 2014).

However, during the past 30 years, user requirements as well as technical circumstances of CLC production have considerably changed, triggering a need for revision of nomenclature and concept. Driven by intention to reduce high-labor costs of photointerpretation, to improve repeatability, and to achieve consistency with national data sets, an increasing number of countries are moving from conventional visual photointerpretation toward a semiautomatic approach of CLC production (Büttner et al. 2013). Semiautomated generation of data according to the CLC legend from national data sets by using GIS generalization and data merging requires a systematic decompository description of CLC classes. The CLC nomenclature guidelines in its current form (Bossard et al. 2000) is designed to assist and guide visual photointerpretation of satellite images, therefore shows some weakness in describing classes as needed for the abovementioned semiautomated approaches.

Inherent logical inconsistencies of the nomenclature were also revealed, when in a number of cases the common guidelines were interpreted by neighboring data producers differently, which is traceable in the CLC mapping results, as for example at the border of Norway and Sweden with CLC classes "322 moors and heathland" and "333 sparsely vegetated areas" (Hallin-Pihlatie & Valcarcel 2013). As both sides applied the mapping rules according to the guidelines correctly, it was the overlap in class definitions themselves that allowed different interpretation of the same type of landscape.

In response to this arising need and to provide a more coherent systematic description of the classes' thematic content, the CLC nomenclature has been analyzed using the EAGLE matrix as a decompository descriptive tool to

- Reveal inherent semantic inconsistencies, gaps, and overlaps within and between class definitions.
- Clearly distinguish between mandatory as well as optional landscape features for assignment of a given class, as well as their characteristics and features that are excluded by definition.
- Characterize classes from both an LC and LU viewpoint.

In addition, the typical landscape situations where each class is applicable were revised. Through the refinement of class definitions, this use case showed the EAGLE concept's capability to enhance semantic consistency for a harmonized understanding of thematic content of the CLC nomenclature.

As an applicable outcome, the guidelines' enhancement is made available in the form of an EEA internal report (Kosztra & Arnold 2013) to support a more consistent CLC production (by both traditional photointerpretation and bottom-up/semiautomated methods) during forthcoming inventories.

A foreseen next stage of implementing the concept is to provide a machine-readable description of CLC classes with help of the EAGLE model, which can directly be used in deriving CLC data by generalization of national data sets with higher thematic and spatial resolution and precision.

6.6 Relation of the EAGLE Concept to Other Existing Standards

The EAGLE concept aims at harmonization and comparability of LC/LU classification approaches. During the development phase, it was attempted not to be in substantial contradiction with existing standards. The EAGLE concept was therefore aligned with existing standards where possible and diverted from them where necessary.

This paragraph explains briefly the relation between the EAGLE concept and other existing approaches. Similar to the motivation for the development of the FAO's LCCS (LC Classification System) and the following submission of the LCCS to ISO Technical Committee 211 on Geographic Information (Di Gregorio & Jansen 2000; Herold & Di Gregorio 2012), the development of the EAGLE concept was as a response to the need for semantic interoperability between LC classification systems. Although the LCCS development had a geographical focus on the African continent, for EAGLE the focus is on the specific European needs for the better integration and harmonization of national mapping activities with European land monitoring initiatives, among others, the CLC. The LCCS follow-up ISO standard 19144-2 LCML (ISO 2012), representing an inspiring approach for the EAGLE concept toward a more elementary categorization of land, still mixes LC with LU terms in its taxonomy, in some cases leading to semantic overlap within the theme LC. According to Ahlqvist (2008), the LCCS was assessed for its applicability from the scientific point and some enhancing modifications were made. Jansen, in Chapter 4, also addresses categorization of land, the questionable seek for that one-fits-all classification system, and how it is possible/impossible in general to draw clear lines between classes. Several other studies were initiated on, for example, how CLC can be described and translated using FAO's LCCS nomenclature (Herold et al. 2009), where the primary target was to express all CLC classes from an object-oriented perspective.

During the course of its development, the EAGLE matrix was also first tested against the CLC legend to express its classes by LCCs, LUAs, and further descriptive characteristics. Taking into account ongoing European developments such as the Copernicus HRLs, the INSPIRE DSs, as well as new emerging user requirements from the LC and environmental monitoring community, the EAGLE data model was further extended by additional attributes to characterize LCUs. To comply with the INSPIRE DSs in technical modeling terms, the EAGLE UML model (see also Figure 6.2) is structured in a way to match to the UML structure of the INSPIRE LC data model. In terms of thematic content, there is also a strong coherence between the INSPIRE *Land Cover* DSs (INSPIRE 2013a)* and the EAGLE LCC. Furthermore, a direct link to the INSPIRE DS on *Land Use* (INSPIRE 2013c)[†] has been established by inclusion of the HILUCS into the EAGLE matrix/data model as LUAs, although with some enhancing modifications. With the arrangement of buildings and surface water bodies in the EAGLE matrix/data model, two additional links to the INSPIRE DSs on *Buildings* (INSPIRE 2013b)[‡] and on *Hydrology*[§] have been made. Also the INSPIRE specifications of Habitats and Biotopes (INSPIRE 2013e)[¶] and Biogeographical regions are to some extent integrated as a code list under the model characteristics. In these specifications the EUNIS[**] habitat classification system is seen as primus inter pares. However, the habitat types from the Annex I of the Habitats Directive have also gained an overall importance in Europe due to reporting obligations, and the level of detail of these Annex I habitat types is in parts finer than those covered by the EUNIS habitats classification. Further, for the Status "Damages" and the Characteristics in the EAGLE model, a cross-reference to the hazard code list from the DS of the INSPIRE theme Natural Risk Zones (INSPIRE 2013d)[††] seems reasonable.

When comparing the terms as used in the EAGLE model with the INSPIRE terminology, some wordings may be varying, but still are comparable in their meaning.

Although the development of LCML was recognized as a reference structure for LC classification systems, it was decided to proceed with the EAGLE model independently, as the review on the ISO 19144-2 was still ongoing, and proposing adaption/extensions to the European conditions were considered as too impractical at this stage. Moreover, the ISO 19144-2 copyright issues hindered the free distribution to the involved experts and the use of the LCML as reference. The EAGLE data model has initially evolved independently from LCML, but at a later stage many similarities between those two approaches have been identified. Because of that affinity, the EAGLE concept has the potential to adapt the LCML to European conditions, respectively, to support in removing some of the remaining shortcomings in LCML like

---

* INSPIRE, 2013. D2.8.II.2 *Data Specification on Land cover*—Draft Technical Guidelines version 3 (Document identifier D2.8.II.2_v3.0rc3).
† INSPIRE, 2013. D2.8.III.4 *Data Specification on Land use*—Draft Technical Guidelines version 3 (Document identifier D2.8.III.4_v3.0rc3).
‡ INSPIRE, 2013. D2.8.III.2 *Data Specification on Buildings*—Draft Technical Guidelines version 3 (Document identifier D2.8.III.2_v3.0rc3).
§ INSPIRE, 2013. D2.8.I.8 *Data Specification on Hydrology*—Draft Technical Guidelines version 3 (Document identifier D2.8.I.8_v3.0rc3).
¶ INSPIRE, 2013. D2.8.III.18 Data Specification on *Habitats and Biotops*—Draft Technical Guidelines version 3 (Document identifier D2.8.III.18_v3.0rc3).
** EUNIS—European Nature Information System, URL: http://eunis.eea.europa.eu.
†† INSPIRE, 2013. D2.8.III.12 Data Specification on *Natural Risk Zones*—Draft Technical Guidelines version 3 (Document identifier D2.8.III.12_v3.0rc3).

- The LCML includes a precise description of the horizontal and vertical structure and strata of the LC up to the leaf shape, but still incorporates some elements that are defined by LU (e.g., roads and railways).

The list of LCML element characteristics related to LU and management appears quite selective and focuses mainly on crop cultivations. Moreover, the Boolean separation of cultivated, seminatural, and natural land as characteristics has been proven difficult in practical applications, as in CLC, for example. Instead, there should be a way to respectively describe the naturalness and the intensity of LU.

As the EAGLE concept was first focused on the remote sensing point of view, it considers the earth's surface summing up to a maximum of 100% without spatially overlapping LCCs, whereas LCML does includes layers (or "strata" in modeling terms) that can add up to more than 100% area coverage (e.g., grassland under tree crowns). The EAGLE group is aware of the vertical aspects of landscape, but for the moment is still under discussion whether to include it already in the data model or leave it for future modifications. In contrary to the LCML, which was already tested by the scientific community and applied to some extent, the EAGLE model is still in a development stage. As a particular topic within the broad discussion about taxonomy and semantic issues, the question of how to address the vertical aspects of overlapping LC features is also discussed by Milenov in Chapter 11.

Beside EEA's CLC as the most important LC mapping initiative, Eurostat has since 2006 regularly (3-year cycle) organized the LUCAS field survey (Land Use and Cover Area Frame Survey), which has its own classification system. In The EAGLE concept was designed to be , it has been taken care of the capable ility of to expressing the LUCAS classes also with the presented taxonomy heremodel elements; almost all of the LUCAS LC classes can be expressed by EAGLE LCC, only whereas the LUCAS crop types have their corresponding elements are placed under the Characteristics block of the EAGLE model.

6.7 Database Merging and Grid Approach

Land Monitoring in European countries is currently changing from ad hoc monolithic projects toward integrated production using national spatial data infrastructures (SDIs). This trend follows from a technical transition from strictly visual interpretation of satellite images toward automated analysis of combined and complex data sources (aerial images, multitemporal and multisensor satellite images, laser scanning data, etc.). One characteristic of this development is the increased use of ancillary databases. Traditionally, the land monitoring community has used primarily topographic maps and orthophotos as ancillary data, and only for the purpose of visual inspection. With the implementation of the INSPIRE directive and the move toward more open data policies, new possibilities of accessing additional ancillary databases (e.g., land parcel identification system, building registries, road databases, national forest inventories, geocoded statistical socioeconomic data) emerge. The compilation of land monitoring data from *existing* data sources is referred to as database merging (Strand 2013). This approach is already used for example, in Austria, Finland, Netherlands, Norway, Sweden, and the United Kingdom. In cases where more than one ancillary database is used, countries apply a "data integration phase" (phase 1), where the ancillary databases are merged together into a single, integrated data set with common regular spatial units (mostly 25 m raster cells). In a subsequent "data aggregation phase" (phase 2) various attributes are retrieved, generalized, and merged into the national or regional land monitoring code lists (*classification*).

The process of integrating various databases with their varying geometric representations into a standardized spatial unit can also be taken one step further. This is defined as the "grid approach." A *grid* is—by abstract explanation—a spatial model falling somewhere between the vector model and the raster model. The grid is a spatial partition of the earth's surface into nonoverlapping and nonempty parts. Similar to the raster model, the grid consists of regular spatial units (cells) of uniform size and shape. The most common identification of grid cells are either the coordinates of the central point or the upper left/lower left corner. However, they can be thought of as a spatial representation of the land unit in the form of a determined extent of quadrangles rather than only the identifying point coordinates of pixels. The objective of the grid approach is not simply to *classify* each single grid cell and give it a single coded value, but to *characterize* these grid cells with several properties, thus following the approach where spatial units are described using a set of independent diagnostic criteria* (Jansen & Di Gregorio 2002; Gomarasca 2009). In terms of the EAGLE concept, these diagnostic criteria can be the LCCs, LUAs, and further Characteristics (CH).

The outlines of grid cells are not mapped in the traditional way, but they represent an *a priori* determined subdivision of space in the form of regular units (objects) that can be "populated" and "characterized" using, for example, the EAGLE data model elements. Grid cells can be directly populated (attributed) with remote sensing information, but this process can also involve counting of frequency or density, or results from geometrical or statistical processing of various input data that represent observations made

---

* "Criteria" here in the meaning of landscape properties; not to be mixed up with the required criteria for a data model.

inside the area of each grid cell. Common methods used to calculate spatial coverage inside a grid cell are as follows:

1. To use a geometric intersection between the grid and various polygon data sets and attribute the grid cell with the covered percentage of each (e.g., LC) class that falls inside the grid cell
2. To register the class that is dominating (maximum cover percentage) the grid cell
3. To register unweighted presence/absence of classes, or register only the class that is observed at a particular location, for example in the center of the cell

Other attribution methods are counting number of objects (e.g., buildings) or calculating the length of objects (e.g., roads) or performing statistical aggregations of observations in the grid cell (e.g., average height of objects). The EAGLE data model is capable of logically structuring the various attributes within each grid cell and expressing their relation to each other through distribution and spatial patterns. When applied to all relevant source data sets, the data model also enables a European aggregation of data sources.

To assess the usability of the grid approach, a number of challenges and advantages can be listed. The main challenges and drawbacks are as follows:

1. A conglomerate of different data qualities and precisions among data sources.
2. Thematic overlap from different data sources with partly redundant information content and/or partly differing meaning with respect to class definitions.
3. An unfamiliar format (abstract squares) for users.
4. An arbitrary partition of the real world without any direct relation to the spatial extent of landscape features, whereas the size setting of the grid cell still can indirectly indicate a scale.
5. Internal changes inside single grid cells may not be visible, because only overall net changes are recorded for each cell.

The advantages of the grid approach are as follows:

1. Several parameters/attributes from various data sources and of different data types (not only spatial thematic data but also statistical data) can be handled.
2. Combinations and queries of the information can be used to reveal spatial relationships or associations.
3. Improved statistical analysis is possible due to the fixed (unchanging) spatial structure.

Besides these technical aspects, the grid represents an additional advantage regarding data access restrictions due to confidential information content or property rights. National data that are not accessible in their original form because they are considered too detailed can be made available in a generalized and thus anonymous form (e.g., LPIS*).

Regular grids are already used in several settings: After receiving CLC data in polygon vector format from its MS, the EEA provides compiled pan-European CLC data not only as vector data set but also as a rasterized data set (EEA 2009) within the standardized INSPIRE Grid with 100×100 m resolution using a ETRS89 Lambert Azimuthal Equal Area 52N 10E projection; the Copernicus HRLs populate a subdivision of this grid with 20×20 m; Eurostat has launched the GEOSTAT[†] project in 2010, which was carried out by the European Forum for GeoStatistics (EFGS), where grids are used for spatial integration and harmonization of socioeconomic data (e.g., representing national census data in a European population grid) (EFGS 2012).

The idea for a standardized future European land monitoring is to transpose the grid approach to the LC/LU domain and apply it consistently throughout Europe. The European grid can be populated (characterized) according to the EAGLE data model in a combined bottom-up and top-down approach using national land monitoring data when available, together with centrally produced pan-European data such as the Copernicus HRL. The grid approach was described by G. H. Strand in an unpublished paper from the EU-FP7 project HELM. A brief summary of the content is found in Ben-Asher (2013, pp. 68–71). The model is based on ideas previously published and discussed in Strand & Bloch (2009) and Strand (2011).

6.8 Outlook and Next Steps

In this chapter, the EAGLE concept and individual aspects of its implementation are described and discussed. Most of the important parts of the EAGLE concept are already drafted, but the final implementation the EAGLE concept as a whole still has to be fine-tuned, consolidated, and documented to serve as an operational framework. Future work of the EAGLE group is therefore concentrated on the following activities to foster the concept's operational use.

---

* LPIS, Land Parcel Information System as it is applied for the Integrated Administration and Control System (IACS) in distributing EU subsidy to farmers.
† URL: http://epp.eurostat.ec.europa.eu/portal/page/portal/gisco_Geographical_information_maps/geostat_project.

6.8.1 Explanatory Documentation of the EAGLE Concept and Unified Modeling Language Application Schema

Thorough documentation is a fundamental prerequisite for a concept's operational implementation and integration within various applications. Therefore, an explanatory documentation of the EAGLE concept is being created. It will be available for all users without any access restrictions. The document will introduce and explain the scope of the EAGLE concept, the overall structure of the matrix/data model formulation, and the definitions of the matrix and model elements. It will also provide a user guideline on how to apply the concept for practical purposes, for example, particular data set description or semantic translation between classification systems. In addition to that, a machine-readable ISO-conform application schema of the data model is created based on the UML by using special software (e.g., Rational Rose, Enterprise Architect). It is planned to also use XMI (XML Metadata Interchange) ISO markup language for exchanging UML models and GML (OGC Geography Markup Language) (see also Section 6.4.2.2). Regarding the close relation with INSPIRE, the UML model will also be in line with the DS of the theme LC (from the INSPIRE Directive Annex II), but also will take into account other INSPIRE themes—where necessary or appropriate—such as LU; Buildings, Habitats, and Ecosystem types; and Hydrology that describes the same, similar, or connected geographic objects.

In close connection with the documentation, a parallel phase of fine-tuning the data model is carried out. That is to implement all collected comments from experts as well as integrate identified requirements from thematic fields of work related to land monitoring activities.

Besides the fact that such documentation is a living document, a first release is available from 2015 on. This version will then be published in the EAGLE website.*

6.8.2 The EAGLE Model Population and Comparison Tool

The EAGLE concept introduces a synoptic view into land monitoring and enables the interoperability of various land monitoring products by opening new possibilities to describe various LC/LU data sets and explore their semantic relationships, both within and between nomenclatures. Nevertheless, a practical tool for model/matrix population and comparison supporting the easy application of the concept is still missing and is planned to be developed. The goal is therefore to further promote and support the use of the EAGLE data model by development of such a tool that facilitates the application of the data model and its population to semantically describe class definitions according to the specifications determined by different

---

* http://sia.eionet.europa.eu/EAGLE.

nomenclature or classification systems. This kind of tool is the precondition to systematically and machine-readably compare the definitions of similar LC classes to assess their semantic matching for data integration procedures from various data sources.

The main aim is to further foster the multinational communication and collaboration by setting up a knowledge base on semantic aspects of land monitoring in Europe. For that purpose, the setting up of an interactive web-based platform for semantic description and comparison of national land monitoring nomenclatures is the final target.

Another field of work that gains more and more significance for operational initiatives is the collection of volunteered geographic information: the process of crowd sourcing of spatial data, like open street mapping, gives a spectacular example for topographic and route mapping. Also for the domain of LC data some innovative approaches are developing quite rapidly, like with the Geo-Wiki platform* to engage citizens in the environmental monitoring (Perger et al. 2012).

6.8.3 Common Integrated Generalization and Aggregation Rules Set for Land Monitoring Data

Recent developments show that more and more European countries implement methods to derive LC/LU data from national sources to the pan-European level and scale. Meanwhile, all of them are practicing their own methods that differ among countries due to varying starting situations. During the first (still ongoing) episode of content-wise harmonization of pan-European data of CLC, the data production follows uniform mapping guidelines, which are valid for all countries and explain how to map the landscape through visual interpretation of satellite imagery. In recent times—as an increasing number of countries use (semi)automated procedures to derive European data from existing national data (Valcarcel et al. 2008; Arnold 2009; Aune-Lundberg & Strand 2010; Hazeu et al. 2011; Törmä et al. 2011; Banko et al. 2013; Steinmeier 2013; Manakos & Braun 2014)—it will also be necessary to harmonize the generalization practices used in bottom-up CLC production to guarantee the comparability of the final result across borders.

The EAGLE concept is a basis for a semantic translations tool in support of the derivation of land monitoring data for European LC classifications from national sources. Nevertheless, the entire process of data transformation has to deal with spatial properties of geometrical feature and their generalization. Here, the aim is to develop a CIGARS agreed upon by stakeholders across Europe. Aspects like similarity between classes need to be assessed, guidelines on how to aggregate and geometrically generalize from the usually smaller spatial units of national data to the larger spatial units of European data need to be specified, and the relations between source feature

---

* Geo-Wiki URL: http://www.geo-wiki.org/.

types and target feature classes have to be agreed upon. This way it can be guaranteed that, at least up to a certain hierarchical level, the results of bottom-up data derivation remain comparable across borders.

In terms of standardization of spatial units, the grid approach can be considered as a promising approach that will make statistical analysis and assessment of spatial data easier and transparent.

Again, the EAGLE concept will help implement the solution by providing conceptually consistent elements to describe the landscape by diagnostic criteria instead of classifying it.

6.9 Summary and Conclusions

Current demands for landscape information and the necessity to reuse and share data has resulted in a need for a land data model that is flexible, extendable, and applicable to a range of spatial and temporal scales on national as well as on European levels. The demands and the development of such a model has required a paradigm shift away from simple legend lists of class names toward an attribute-rich characterization of objects to increase the information value and usefulness of land data sets.

The EAGLE concept incorporates a semantic model, either represented as an UML model or as a matrix in the form of spreadsheet, both having the same thematic content. Being structured in the three blocks of LCCs, LUAs, and further Characteristics (CH), the model allows it to capture, expose, and visualize a wide range of important land properties while remaining connected and compliant with other established nomenclatures and standards (e.g., CLC, INSPIRE, ISO TC 211, LCML, LUCAS). It is by its structure also flexible enough to react to the needs of upcoming new thematic fields of work and to modifications of existing standards or future emerging activities.

As key message, the EAGLE concept can be as follows:

- A useful framework for the integration of LC/LU information from various data sets in one single data model.
- A tool for standardizing the componential description of LC/LU classes to provide added value beyond a class label.
- A tool for semantic analysis of class definitions to identify inconsistencies (semantic overlaps and gaps) within a nomenclature.
- A vehicle for comparison and semantic translation between different LC/LU nomenclatures, and that facilitates data exchange.
- Implemented as a LC/LU data collection standard and mapping guidelines for national land monitoring initiatives.
- A coherent common data framework for several single Copernicus products (CLC, HRLs, Urban Atlas).

The EAGLE concept was explored within the context of the HELM project, which aimed to provide a roadmap to make European land monitoring more productive by increasing the alignment of national and subnational land monitoring endeavors and by enabling their integration to a coherent European data system. A common European framework for land monitoring, which is able to connect semantically several data sets of LC and foster information interchange, was seen as a key criterion. Such a development can be of benefit only for the societal impacts of Earth Observation–based land monitoring, monitoring of valuable natural-protected sites, food supply and harvest monitoring, quick reaction on disasters, and so forth.

The target of this framework is supporting the land monitoring harmonization process and addressing the proposed synergy of a decentralized (bottom-up) and centralized (top-down) production initiatives. Only after being established as fully operational, the EAGLE concept will reach its final goal to increase the overall maturity and effectivity of integrated European land monitoring, when national land monitoring initiatives share compatible concepts and data models and they provide the technical solution toward data integration on the European level. Further in time, the static classification approach of CLC is envisaged to be gradually replaced by the EAGLE concept's object-oriented and parameterized descriptive approach of storing information about the land as resource, its status and use. As a first step of this paradigm shift, the enrichment of CLC databases by populating its polygons with quantitative and qualitative descriptors is proposed. The EAGLE concept now embarks on further testing phases against national and European data sets, fine-tuning of the data model, and tool development. The overall long-term aim and vision of this concept is to serve as the conceptual basis for the operational implementation of a future European Land Monitoring System.

6.10 Annex: Who Is EAGLE?

6.10.1 Forming of the Group

The EAGLE group is an open assembly of experts in the field of land monitoring. It was formed in 2009 after the invitation of IGN Spain to discuss object-oriented issues in connection with land monitoring initiatives. Many of the group members have the role as an NRC for LC, being the national contact point within EEA's EIONET. The group works and meets on a voluntary basis, so far not financed by any allocated budget, except the experts' home institution seconding their employees to the annual working group meetings. Besides, some travel budget from external sources has helped to cover at least travel expenses. The meetings themselves have partially been supported by the FP7 geoland2 project that has provided resources for travel expenses of working group members.

The most active core members of EAGLE group are the following:

- Germany (represented by Stephan Arnold, DeSTATIS, and Michael Bock, DLR)
- Austria (represented by Gebhard Banko, UBA, and Christoph Perger, IIASA)
- Hungary (represented by Barbara Kosztra and Gergely Maucha, FÖMI)
- Finland (represented by Markus Törmä and Elise Järvenpää, SYKE)
- Norway (represented by Geir-Harald Strand, NFLI)
- Spain (represented by Nuria Valcarcel, Julian Delgado, and Guillermo Villa, IGN; Roger Milego and Cesar Martinez, UAB; Emanuele Mancosu, UMA)
- Switzerland (represented by Charlotte Steinmeier, WSL)
- United Kingdom (represented by Geoff Smith, Specto Natura)
- Czech Republic (represented by Tomas Soukup, GISAT)
- The Netherlands (represented by Gerard Hazeu, Alterra)
- Luxemburg (represented by Stefan Kleeschulte, Geoville Environmental Services)

Within EAGLE, an editing committee (B. Kosztra, N. Valcarcel, M. Bock, and T. Soukup under the lead of S. Arnold) was entrusted to collect the group's conceptual input and develop working material including the data model and its explanatory documentation. The European Topic Centre on Spatial Information and Analysis (ETC SIA) provides secretariat webmaster host service.

6.10.2 Connections with Other Committees and Stakeholders

EAGLE is supported by EEA, but is independent from a political and technical point of view. The group tries to bring together the knowledge from the existing EU and national LC and LU classifications and initiatives. EAGLE members are also engaged in a broad range of European land monitoring-related activities as follows:

- Several EAGLE members have participated in the INSPIRE process as experts in the TWG LC and Land Use (TWG LU), which resulted in a productive interchange of expertise concerning the transboundary context and the conceptual working out of the INSPIRE DSs for LC and for LU.
- Consultations with the CLC Technical Team (also members in EAGLE) regarding the data model content and its compatibility to CLC are part of the process.

- Overlapping membership between EAGLE and the FP7 project "Harmonized European Land Monitoring" (HELM), which embraced a broader circle of the European MS land monitoring community, including many NRCs (project closed by end of 2013).

- Involvement in the former validation of HRLs concept, which was elaborated under FP7 project geoland2.

- Contacts to GMES Initial Operations (GIO) staff in the field of administration and industry on national and European level are maintained on a regular basis.

- Experiences and best practices from already existing object-oriented and/or national bottom-up approaches or from those being under development have been brought together from predecessor projects. Some examples are LCCS (FAO)/LCML (ISO Standard 19144-2), SIOSE (Spain), DLM-DE (Germany), LISA (Austria), AR50 (Norway), LCM (UK), and SLICES (Finland).

- Close cooperation with EEA for the further development of the EAGLE concept, its documentation, and future implementation.

- Eurostat considers integrating the ideas behind the EAGLE concept into the long-term strategy of its LUCAS survey; continuous exchange exists between Eurostat and the group.

References

Ahlqvist, O. 2005. Using uncertain conceptual spaces to translate between land cover categories. *International Journal of Geographic Information Science* 19: 831–857.

Ahlqvist, O. 2008. In search for classification that supports the dynamics of science—The FAO Land Cover Classification System and proposed modifications. *Environment and Planning B* 35(1): 169–186.

Arnold, S. 2009. Digital Landscape Model DLM-DE—Deriving Land Cover Information by Integration of Topographic Reference Data with Remote Sensing Data. Proceedings ISPRS Workshop Hannover. Volume XXXVIII-1-4-7/ W5. Federal Agency for Cartography and Geodesy (BKG), Frankfurt am Main, Germany, accessed November 24, 2014, http://www.isprs.org/proceedings /XXXVIII/1_4_7-W5/paper/Arnold-167.pdf.

Arnold, S., & Kosztra, B. 2013. Final Report on Task 4.3: Enhancing the European Land Monitoring System—collection of criteria for a future data mode. In *Prerequisites and Criteria for Aligning National/Sub-National Land Monitoring Activities.* EU FP7 HELM Deliverable 4.1, 57–84, Part 3 Nomenclature and data models, accessed August 22, 2014, http://www.fp7helm.eu/fileadmin/site/ fp7helm/HELM_4_1_Aligning_national-land_monitoring_activities.pdf. and In Ben-Asher, Z., (Ed.), *HELM—Harmonised European Land Monitoring: Findings and Recommendations of the HELM Project*, 56–59. Tel-Aviv, Israel. http://www .fp7helm.eu/fileadmin/site/fp7helm/HELM_Book_2nd_Edition.pdf.

Arnold, S., Kosztra, B., Banko, G., Smith, G., Hazeu, G., Bock, M., & Valcarcel Sanz, N. 2013. The EAGLE concept—A vision of a future European Land Monitoring Framework. In Lasaponara, R., Masini, L., Biscione, M., (Eds.), *Towards Horizon 2020: Earth Observation and Social Perspectives*, 551–568. 33th EARSeL Symposium Proceedings. Matera, Italy: EARSeL and CNR.

Aune-Lundberg, L., & Strand, G. H. 2010. *CORINE Land Cover 2006*. The Norwegian CLC2006 Project. Report 11/2010 from Norwegian Forest and Landscape Institute. Ås, Norway.

Banko, G., Franzen, M., Ressl, C., Riedl, M., Mansberger, R., & Grillmayer, R. 2013. *Bodenbedeckung und Landnut-zung in Österreich—Umsetzung des Projektes LISA zur Schaffung einer nationalen Geodateninfrastruktur für Landmonitoring*, 14–19. Strobl, Blaschke, Griesebner, Zagl (Hrsg.): Beiträge zum 25. AGIT Symposium Salzburg.

Ben-Asher Z., (Ed.). 2013. HELM—Harmonised European Land Monitoring: Findings and recommendations of the HELM Project. Tel-Aviv, Israel, accessed April 10, 2015, http://www.fp7helm.eu/fileadmin/site/fp7helm/HELM_Book_2nd _Edition.pdf.

Blanes Guardía, N., Green, T., & Simón, A. 2014. The users' role in the European land monitoring context. In Manakos, I., Braun, M., (Eds.), *Land Use and Land Cover Mapping in Europe*. Practices & Trends. Dordrecht, the Netherlands: Springer.

Büttner G., Feranec, J., Jaffrain, G., Mari L., Maucha G., & Soukup, T. 2004. The CORINE Land Cover 2000 Project. In Reuter, R., (Ed.), *EARSeL eProceedings*. Volume 3EARSeL, Paris, pp. 331–346.

Büttner, G., & Kosztra, B., 2007. *CLC2006 Technical Guidelines*. EEA Technical Report 17/2007.

Büttner, G., Kosztra, B., & Maucha, G. 2013. Final Report on Task 3.1 Operational Working Procedures. In *Commonalities, Differences and Gaps in National and Sub-National Land Monitoring Systems*. EU FP7 HELM Deliverable 4.1, 6–39. Part 3 Operational working procedures, accessed November 21, 2014, http:// www.fp7helm.eu/fileadmin/site/fp7helm/HELM_3_1_Commonalities_and _Differences.pdf.

Comber, A., Lear, A., & Wadswort, R. 2010. *A Comparison of Different Semantic Methods for Integrating Thematic Geographical Information: The Example of Land Cover*. 13th AGILE International Conference on Geographic Information Science 2010. Guimarães, Portugal.

Comber, A. J. 2008. The separation of land cover from land use using data primitives. *Journal of Land Use Science* 3(4): 215–229.

Comber, A. J., Fisher, P. F., & Wadsworth, R. A. 2007. Land cover: to standardise or not to standardise? Comment on "Evolving standards in land-cover characterisation" by Herold et al. *Journal of Land Use Science* 2(4): 283–287.

Di Gregorio, A., & O'Brien, D. 2012. Overview of land-cover classifications and their interoperability. In Giri, P. C., (Ed.), *Remote Sensing of Land Use and Land Cover: Principles and Applications*, 37–48. Boca Raton, FL: CRC Press.

Di Gregorio, A., & Jansen, L. J. M. 2000. *Land Cover Classification System (LCCS)— Classification Concepts and User Manual*, 177. Environment and Natural Resources Service, GCP/RAF/287/ITA Africover—East Africa Project and Soil Resources, Management and Conservation Service. FAO/UNEP/Cooperazione italiana, Rome, Italy.

EC—European Commission. 2014. 7th European Union's Research and Innovation Funding Programme (FP7). http://ec.europa.eu/research/fp7/index_en.cfm.

EC—European Commission. 2011. Communication from the Commission, Roadmap to a Resource Efficient Europe, COM/2011/0571.

EEA—European Environment Agency. 2004. CORINE Land Cover 2000. Mapping the decade of change. EEA, OPOCE, accessed November 21, 2014, http://www.eea.europa.eu/publications/brochure_2006_0306_103624.

EEA—European Environment Agency. 2009. CORINE Land Cover 2006 (CLC2006) 100m, version 12/2009. European Environment Agency, Copenhagen, Denmark, accessed August 22, 2014, http://www.eea.europa.eu/data-and-maps/data/corine-land-cover-2006-clc2006-100-m-version-12-2009.

EEA—European Environment Agency. 2010. The European Environment, State and Outlook, Land Use 2010, ISBN 978-92-9213-160-9, EEA, Copenhagen.

EEA—European Environment Agency. 2013. GIO land (GMES/Copernicus initial operations land) High Resolution Layers (HRLs)—summary of product specifications. Version 7, accesssed May 7, 2015, http://land.copernicus.eu/user-corner/technical-library/gio-land-high-resolution-layers-hrls-2013-summary-of-product-specifications-1.

EFGS—European Forum for GeoStatistics. 2012. GEOSTAT 1A—Representing Census data in a European population grid. Final Report.

Eurostat. 2014. LUCAS—Land Use/Cover Area frame statistical Survey, accessed May 07, 2015, http://ec.europa.eu/eurostat/statistics-explained/index.php/LUCAS_-_Land_use_and_land_cover_survey.

Feranec, J., Jaffrain, G., Soukup, T., & Hazeu, G. 2010. Determining changes and flows in European landscapes 1990–2000 using CORINE land cover data. *Applied Geography* 30(1): 19–35.

Gallego, J., & Bamps, C. 2008. Using CORINE land cover and the point survey LUCAS for area estimation. *International Journal of Applied Earth Observation and Geoinformation* 10(4): 467–475.

Gomarasca, M. A. 2009. *Basics of Geomatics*. Netherlands: Springer.

Hallin-Pihlatie, L., & Valcarcel, N. 2013. Final Report on Task 2.4 Transfer experiences from already achieved coordination between member States. In *Practical Experiences with Bottom-Up Land Monitoring Approaches*. EU FP7 HELM Deliverable 2.3, Part 1, accessed November 21, 2014, http://www.fp7helm.eu/fileadmin/site/fp7helm/HELM_2_3_Botton-Up_Experiences.pdf.

Hazeu, G. W., Bregt, A. K., de Wit, A. J. W., & Clevers, J. G. P. W. 2011. A Dutch multi-date land use database: Identification of real and methodological changes. *International Journal of Applied Earth Observation and Geoinformation* 13: 682–689.

Herold, M., & Di Gregorio, A. 2012. Evaluating land-cover legends using the UN land-cover classification system. In Giri, P. C., (Ed.), *Remote Sensing of Land Use and Land Cover: Principles and Applications*, 65–88. Boca Raton, FL: CRC Press.

Herold, M., Hubald, R., & Di Gregorio, A. 2009. *Translating and Evaluating Land Cover Legends Using the UN Land Cover Classification System (LCCS)*. GOFC-GOLD report no. 43. Land Cover Project Office, Jena, Germany, accesssed May 7, 2015, http://nofc.cfs.nrcan.gc.ca/gofc-gold/Report%20Series/GOLD_43.pdf.

Herold, M., Latham, J. S., Di Gregorio, A., & Schmullius, C. C. 2006b. Evolving standards in land cover characterization. *Journal of Land Use Science* 1(2–4): 157–168.

Heymann, Y., Steenmans, C., Croissille, G., & Bossard, M. 1994. *Corine Land Cover Technical Guide*. EUR12585 Luxembourg, Office for Official Publications of the European Communities.

INSPIRE. 2013a. D2.8.II.2 Data Specification on Land cover—Draft Technical Guidelines version 3 (Identifier D2.8.II.2_v3.0rc3), accesssed May 7, 2015, http://inspire.ec.europa.eu/documents/Data_Specifications/INSPIRE_Data Specification_LC_v3.0.pdf.

INSPIRE. 2013b. D2.8.III.2 Data Specification on Buildings—Draft Technical Guidelines version 3 (Identifier D2.8.III.2_v3.0rc3), accesssed May 7, 2015, http://inspire.ec.europa.eu/documents/Data_Specifications/INSPIRE _DataSpecification_BU_v3.0.pdf.

INSPIRE. 2013c. D2.8.III.4 Data Specification on Land use—Draft Technical Guidelines version 3 (Identifier D2.8.III.4_v3.0rc3), accesssed May 7, 2015, http://inspire.ec.europa.eu/documents/Data_Specifications/INSPIRE_Data Specification_LU_v3.0.pdf.

INSPIRE. 2013d. D2.8.III.12 Data Specification on Natural Risk Zones—Draft Technical Guidelines version 3 (Identifier D2.8.III.12_v3.0rc3), accesssed May 7, 2015, http://inspire.ec.europa.eu/documents/Data_Specifications/INSPIRE _DataSpecification_NZ_v3.0.pdf.

INSPIRE. 2013e. D2.8.III.18 Data Specification on Habitats and Biotops—Draft Technical Guidelines version 3 (Identifier D2.8.III.18_v3.0rc3), accesssed May 7, 2015, http://inspire.ec.europa.eu/documents/Data_Specifications/INSPIRE _DataSpecification_HB_v3.0.pdf.

ISO 19144-2. 2012. Geographic information—Classification systems—Part 2: Land Cover Meta Language (LCML), accesssed May 7, 2015, http://www.iso.org /iso/home/store/catalogue_tc/catalogue_detail.htm?csnumber=44342.

Jansen, L. J. M, & Di Gregorio, A. 2002. Parametric land-cover and land-use classifications as tools for environmental change detection. *Agriculture, Ecosystems & Environment* 91(1–3): 89–100. http://www.sciencedirect.com/science/article /pii/S0167880901002432.

Janssen, S., Dumont, G., Fierens, F., & Mensink, C. 2008. Spatial interpolation of air pollution measurements using CORINE land cover data. *Atmospheric Environment* 42(20): 4884–4903.

Kosztra, B., & Arnold, S. 2013. Proposal for Enhancement of CLC Nomenclature Guidelines. ETC/SIA deliverable EEA subvention 2013 WA1 Task 261_1_1: Applying EAGLE concept to CLC guidelines enhancement. EEA internal report, Copenhagen.

Manakos, I., & Braun, M., (Eds.). 2014. *Land Use and Land Cover Mapping in Europe. Practices & Trends*. Dordrecht, The Netherlands: Springer.

Neumann, K., Herold, M., Hartley, A., & Schmullius, C. 2007. Comparative assessment of CORINE2000 and GLC2000: Spatial analysis of land cover data for Europe. *International Journal of Applied Earth Observation and Geoinformation* 9: 425–437.

Partzsch, D. 1970. Daseinsgrundfunktionen, I. Die Raumansprüche der Funktionsgesellschaft. In *Handwörterbuch der Raumforschung + Raumordnung*, 424–430. Akademie für Raumforschung und Landesplanung. Jänecke Verlag, Hannover.

Perger, C., Fritz, S., See, L., Schill, C., van der Velde, M., McCallum, I., Obersteiner, M. 2012. A campaign to collect volunteered geographic information on land cover and human impact. In Jekel, T., Car, A., Strobl, J., and Griesebner, G., (Eds.), *GI_Forum 2012: Geovizualisation, Society and Learning*, 83–91. (Wichmann, Berlin/ Offenbach, Germany). Salzburg, Austria.

Sifakis, N., Paronis, D., & Keramitsoglou, I. 2004. Combining AVHRR imagery with CORINEL and Cover data to observe forest fires and to assess their consequences. *International Journal of Applied Earth Observation and Geoinformation* 5(4): 263–274.

Stathopoulou, M., & Cartalis, C. 2007. Daytime urban heat islands from Landsat ETM+ and Corine land cover data: An application to major cities in Greece. *Solar Energy* 81(3): 358–368.

Steinmeier, C. 2013. CORINE Land Cover 2000/2006 Switzerland, 30. Final Report. Birmensdorf, Swiss Federal Institute for Forest, Snow and Landscape Research WSL.

Strand, G. H., & Bloch, V. V. H. 2009. Statistical grids for Norway. Documentation of national grids for analysis and visualisation of spatial data in Norway. Document 2009/9. Statistics Norway, Oslo, Norway.

Strand, G. H. 2013. Final Report on Task 4.1 Explore database merging and grid approaches. In *Prerequisites and Criteria for Aligning National/Sub-National Land Monitoring Activities*. EU FP7 HELM Deliverable 4.1, Part 1 Database merging and grid approaches, 5–44, accessed August 22, 2014, http://www.fp7helm.eu /fileadmin/site/fp7helm/HELM_4_1_Aligning_national_land_monitoring _activities.pdf.

Strand, G. H. 2011. Uncertainty in classification and delineation of landscapes: A probabilistic approach to landscape modeling. *Environmental Modelling and Software* 26: 1150–1157.

Suau-Sanchez, P., Burghouwt, G., & Pallares-Barbera, M. 2014. An appraisal of the CORINE land cover database in airport catchment area analysis using a GIS approach. *Journal of Air Transport Management* 34: 12–16.

Törmä, M., Härmä, P., Hatunen, S., Teiniranta, R., Kallio, M., & Järvenpää, E. 2011. Change Detection for Finnish CORINE Land Cover Classification. Proceedings SPIE 8181. Earth Resources and Environmental Remote Sensing/GIS Applications II, 81810Q, October 26.

UML (Unified Modeling Language Specification). 1998. Object Management Group. Framingham, MA, accesssed May 7, 2015, http://www.omg.org.

UNEP—United Nations Environment Programme. 1992. Convention on Biological Diversity 1760 UNTS 79; 31 ILM 818.

UNFCCC—United Nations Framework Convention on Climate Change. 1997. Kyoto Protocol to the United Nations Framework Convention on Climate Change adopted at COP3 in Kyoto, Japan, on December 11.

Villa, G., Valcarcel, N., Arozarena, A., Garcia-Asensio, L., Caballero, M. E., Porcuna, A., Domenech, E., & Pece, J. J. 2008. Land cover classifications: An obsolete paradigm. *The International Archives of the Photogrammetry, Remote Sensing and Spatial Information Sciences*. Proceedings Beijing, China. Volume XXXVII. Part B4.

Valcarcel, N., Villa, G., Arozarena, A., Garcia-Asensio, L., Caballlero, M. E., Porcuna, A., Domenech, E., & Peces, J. J. 2008. SIOSE—a successful test bench towards harmonization and integration of land cover/use information as environmental reference data. The International Archives of the Photogrammetry, Remote Sensing and Spatial Information Sciences. Proceedings Beijing. Volume XXXVII. Part B8.

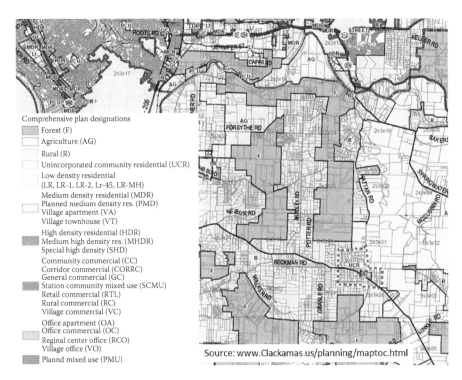

Comprehensive plan designations

Forest (F)

Agriculture (AG)

Rural (R)

Unincorporated community residential (UCR)

Low density residential
(LR, LR-1, LR-2, Lr-45, LR-MH)

Medium density residential (MDR)
Planned medium density res. (PMD)
Village apartment (VA)
Village townhouse (VT)

High density residential (HDR)
Medium high density res. (MHDR)
Special high density (SHD)

Community commercial (CC)
Corridor commercial (CORRC)
General commercial (GC)
Station community mixed use (SCMU)
Retail commercial (RTL)
Rural commercial (RC)
Village commercial (VC)

Office apartment (OA)
Office commercial (OC)
Reginal center office (RCO)
Village office (VO)

Plannd mixed use (PMU)

Source: www.Clackamas.us/planning/maptoc.html

FIGURE 1.5

Clackamas County, Oregon: A portion of land use plan map.

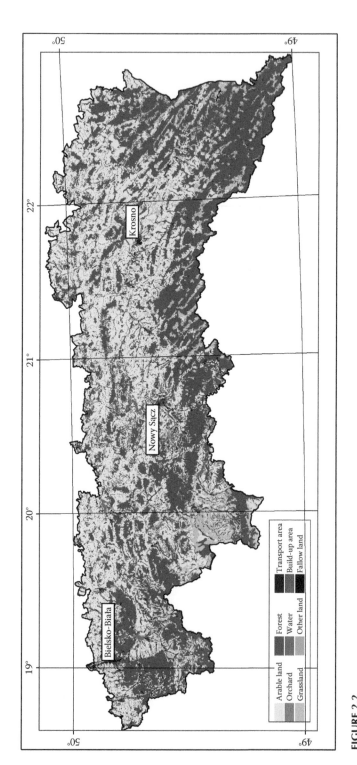

FIGURE 2.2
LULC tessellation in the Polish Carpathians in 2006.

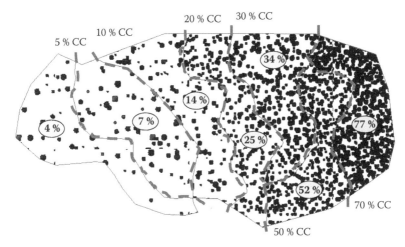

FIGURE 3.1
The effect of thresholds for forest mapping on gradual CC in a hypothetical landscape. (From Skånes, H.M., and Andersson, A., Flygbildstolkningsmanual för Uppföljningsprojektet Natura 2000 version 4.0. UF 19, *Naturvårdsverket*, p. 73, [in Swedish, unpublished], 2010, working document.)

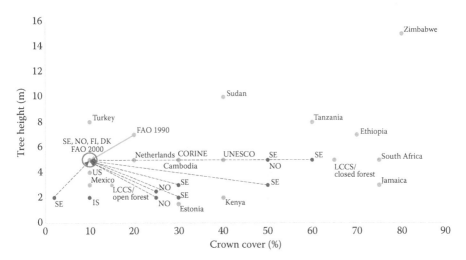

FIGURE 3.2
An illustration of the semantic plasticity of the forest concept, both in a global (blue dots) and in a Nordic (red dots) context. Forest classifications vary greatly on a global and national scale. The dashed lines indicate the general and conceptual shifts from previous Nordic definitions toward a harmonized use of the FAO 2000 definition. However, some of the other definitions are still in use in their respective fields of application. (Modified after Ahlqvist, O., *Environment and Planning B: Planning and Design*, 35, 169–186, 2008a, based on data from Lund, H.G., *Definitions of Forest, Deforestation, Afforestation, and Reforestation*, Forest Information Services, Gainesville, VA, 2014, and the Internet-based query for this article.) SE, Sweden; NO, Norway; FI, Finland; DK, Denmark; IS, Iceland.

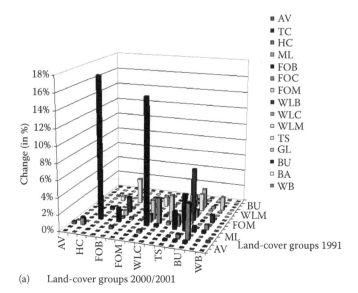

(a)   Land-cover groups 2000/2001

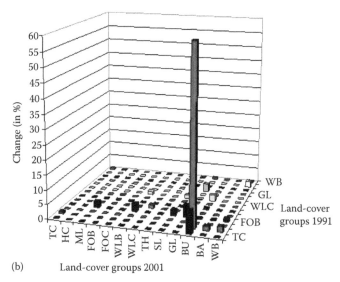

(b)   Land-cover groups 2001

FIGURE 4.4

Land-cover changes 1991–2001. (a) At National level. (b) In Tirane District. AV, Aquatic vegetation; TC, Tree & shrub crops; HC, Herbaceous crops; ML, Managed lands; FOB, Broadleaved forests; FOC, Coniferous forests; FOM, Mixed forests; WLB, Broadleaved woodlands; WLC, Coniferous woodlands, WLM, Mixed woodlands; TS, Thickets and shrublands; TH, Thickets; SL, Shrublands; GL, Grasslands; BU, Built-up areas; BA, Bare areas; WB, Water bodies.

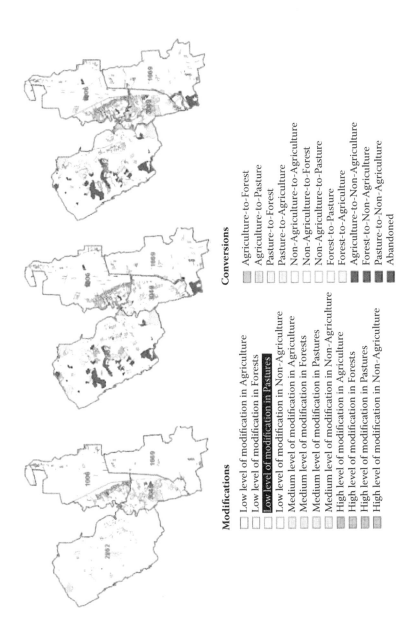

Modifications

☐ Low level of modification in Agriculture
☐ Low level of modification in Forests
☐ Low level of modification in Pastures
☐ Low level of modification in Non-Agriculture
▦ Medium level of modification in Agriculture
▦ Medium level of modification in Forests
▦ Medium level of modification in Pastures
▦ Medium level of modification in Non-Agriculture
▦ High level of modification in Agriculture
▦ High level of modification in Forests
▦ High level of modification in Pastures
▦ High level of modification in Non-Agriculture

Conversions

☐ Agriculture-to-Forest
☐ Agriculture-to-Pasture
☐ Pasture-to-Forest
☐ Pasture-to-Agriculture
☐ Non-Agriculture-to-Agriculture
☐ Non-Agriculture-to-Forest
☐ Non-Agriculture-to-Pasture
☐ Forest-to-Pasture
☐ Forest-to-Agriculture
▦ Agriculture-to-Non-Agriculture
▦ Forest-to-Non-Agriculture
▦ Pasture-to-Non-Agriculture
▦ Abandoned

FIGURE 4.5

Land-use change in Preza Commune in Albania: from left to right in the period 1991–1996, 1996–2003, and 1991–2003 (numbers refer to the four cadastral zones).

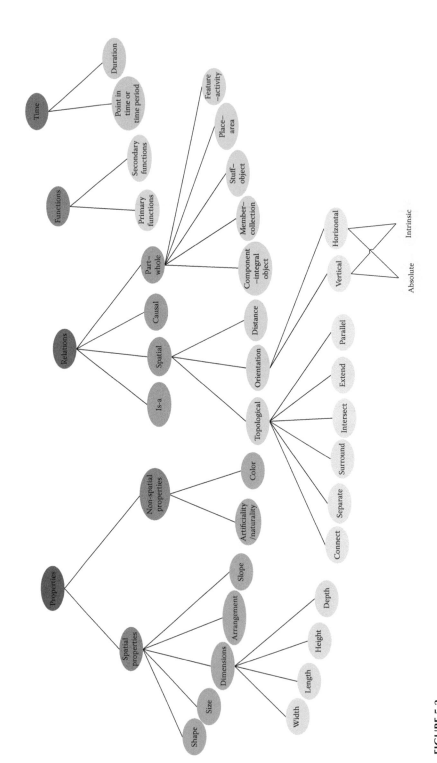

FIGURE 5.2
The taxonomy of semantic elements.

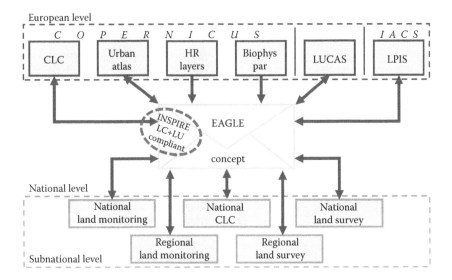

FIGURE 6.1
The integration scheme of the European Land Monitoring Framework as envisioned by the EAGLE concept.

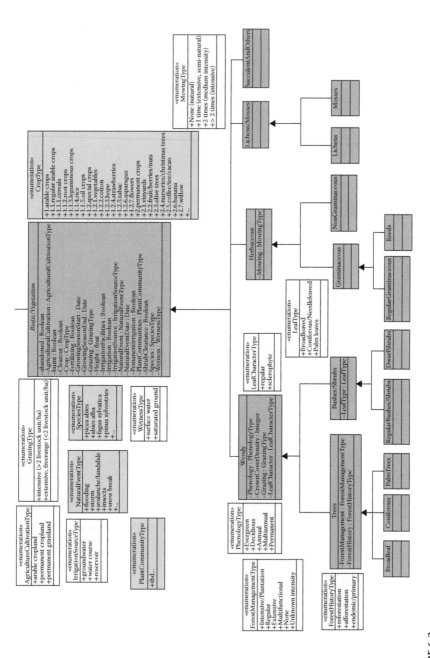

FIGURE 6.3

Biotic/Vegetation block in the EAGLE data model in detail—green boxes represent classes, white boxes represent custom types (= characteristics/attributes) with a limited number of options (enumerations), gray boxes represent custom types with an unlimited number of options (code lists).

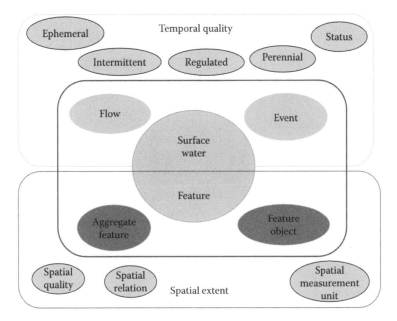

FIGURE 7.2
Top-level surface water ontology classes for land cover.

FIGURE 10.5
Example of a split process. Background map from OSM (2014).

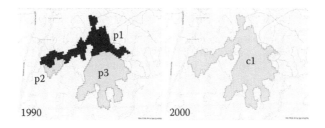

FIGURE 10.6
Example of a merge process. Background map from OSM (2014).

FIGURE 10.7
Example of an increase of fire risk process. Background map from OSM (2014).

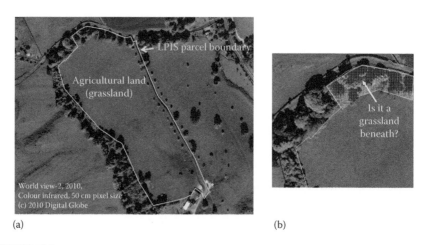

(a)   (b)

FIGURE 11.5
Overview image (a) Example of a reference parcel, enclosing "grazed" land; note the extension of the boundary inside the eastern part of the tree-covered area. Inset (b) Enlargement of the northern part. Includes material © DigitalGlobe (2010), all rights reserved.

(a)

(b)

FIGURE 11.6
(a) Aerial view of the field observation viewpoint with five arrows pointing the direction of the photographs taken on site (1, 2, and 3) or obtained from © Google Street View (4 and 5). Includes material © BING Maps (2010), all rights reserved. (b) The photographs capturing the substrate of the land cover phenomenon beneath the "solid top face." The abbreviations T1–T6 indicate the footprint location of a particular tegon Ti found on place. Includes material Street View © 2010 Google, all rights reserved.

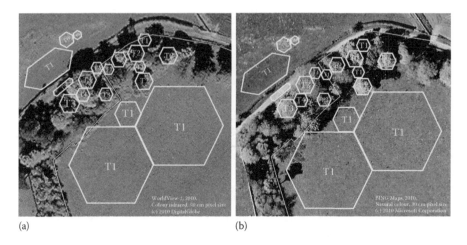

(a)   (b)

FIGURE 11.8
(a) Footprints of the tegons from Figures 11.6 and 11.7 presented on the area of interest over
WorldView-2 satellite image. Includes material © DigitalGlobe (2010), all rights reserved. (b)
The same footprints over an aerial image. The difference of visual appearance of the tegons
is due to the different fields of view of the two acquisitions, Includes material © BING Maps
(2010), all rights reserved. For simplicity and compliance with Figure 11.7, tegon footprints are
drawn as hexagons.

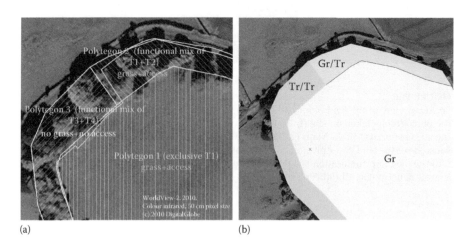

(a)   (b)

FIGURE 11.9
(a) The resulted polytegons on the area of interest. (b) The resulted spatial features (polygons)
on the area of interest stored in the GIS with their correspondent map codes. Includes material
© DigitalGlobe (2010), all rights reserved.

1. CONTINUOUS MOSAIC URBAN FABRIC – ПЪЛНА МОЗАЕЧНА УРБАНИЗИРАНА СТРУКТУРА

Description	Functional entity of three Tegons with abiotic substrate of artificial built-up material. One type of Tegons contains vegetation in its upper stratum. The distribution of the Tegon within the polytegon creates a repeated pattern of linear and non-linear structures.

Continuous urban fabric → Horizontal pattern 1 → Vegetation 1 · Abiotic surface 1 · Abiotic surface 2 → Vegetation · Litoral surface · Non litoral surface · Cultivated and managed vegetation

Horizontal pattern 2

Map code			UBC
Tegon 1			Artificial abiotic built-up material-network
Cover %			Artificial abiotic built-up Area
Characteristic	Artificial Linear Surface-Streets	Stratum 1- Material	Abiotic-Artificial bare Area
		Apperance	Built-up surface – network
		Life cycle	Permanent
Tegon 2			Artificial abiotic built-up material
Cover %		Stratum 1- Material	Abiotic-Artificial Bare Area
Characteristic	Artificial Non-Linear Surface-Building	Apperance	Built-up surface - base
		Life cycle	Permanent
		Stratum 2- Material	Abiotic-Artificial Bare Area
		Apperance	Built-up surface - building
		Life cycle	Permanent
Tegon 3			Artificial abiotic built-up material-network, with vegetation on top
Cover %		Stratum 1- Material	Abiotic-Artificial Bare Area
Characteristic	Artificial Linear Surface (Streets) with Cultivated and Managed Vegetation (line of Trees)	Apperance	Built up surface – network
		Life cycle	Permanent
		Stratum 2- Material	Biotic-Vegetation
		Apperance	Woody
		Life cycle	Perennial

INTERPRETATION KEYS (EXTRACT)

Key 1	Located inside the boundary of the administrative unit
Key 2	Intrinsic non-interrupted mix of streets network and clusters of low or tall buildings
Key 3	Sparse vegetation along the streets and the inner yards might occur
Key 4	Usually the center part of the city

(a)

(b)

(c)

FIGURE 11.12

Illustration of the land cover class "Continuous urban fabric," used in the project SPATIAL. (a) Land cover profile: application of the tegon following the classification concepts of ISO 19144-2 (LCML). (b) As seen from above (National orthophoto map, 2011, provided by the Bulgarian Ministry of Agriculture and Food). (c) As seen from the ground (includes material © ASDE [2014], all rights reserved).

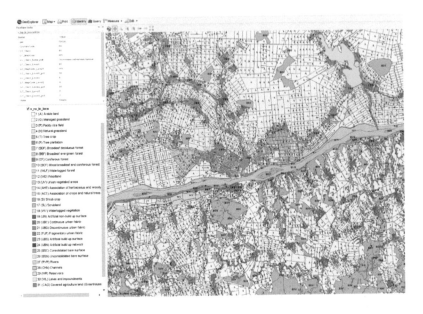

FIGURE 11.13
Extract and legend of the reference transborder land cover data set between Bulgaria and Romania, derived from the Geoportal of ASDE. Note that the agricultural lands are subdivided into LPIS production blocks, whose stable borders are consistent with polytegon edges. Contains material © ASDE (2014), all rights reserved.

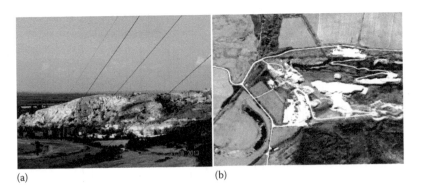

(a)   (b)

FIGURE 11.14
(a) Example of class "Artificial non-built-up surface" as seen from the ground. Includes material © ASDE (2014), all rights reserved. (b) Example of class "Artificial non-built-up surface" as seen from above. (COPERNICUS CORE 03 Image Dataset 2011). © European Union, 2014, all rights reserved. Includes material provided under http://gmesdata.esa.int/web/gsc/ terms_and_conditions, SPOT 5 © CNES (2010–2013); distribution Astrium Services/Spot Image S.A., all rights reserved.

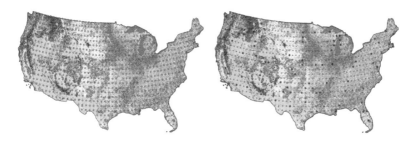

FIGURE 13.1
The NLCD 2006 overlaid by confluence points (left). Stratified random sampled confluence points, 77 total sampled, 7 in each land cover class (right).

FIGURE 13.2
Screenshot of the CatScan interface of an ongoing mock-up experiment.

	BA	CC	dL	dO	EW	FO	GS	OW	PH	SS	WW
BA	46.43	2.86	0	0	7.14	0.71	2.86	0	2.86	36.43	0.71
CC	9.29	37.14	0	2.14	2.86	0	34.29	0.71	11.43	2.14	0
dL	0	0	57.86	30	0	0	7.86	0	2.86	1.43	0
dO	0.71	0	46.43	35.71	0.71	0	2.86	0	7.86	5.71	0
EW	6.43	5	0	0	2.14	33.57	12.14	0.71	17.86	16.43	5.71
FO	0.71	0.71	0	0	2.86	72.14	0	0	1.43	20	2.14
GS	23.57	13.57	0.71	0.71	0.71	0	35	0	15	10	0.71
OW	0	0	0	0	0.71	0.71	0.71	92.14	0	0	5.71
PH	14.29	2.14	2.86	3.57	3.57	12.14	34.29	0	9.29	14.29	3.57
SS	45	0.71	0.71	0	0.71	0.71	10	0	3.57	37.14	1.43
WW	0	1.43	1.43	0	1.43	71.43	0	0	0	7.14	17.14
Total	13.31	5.78	10	6.56	2.08	17.4	12.73	8.51	6.56	13.7	3.38

FIGURE 13.3
Confusion matrix for Experiment 1 (lay participants with no intervention) showing percentages of correct (diagonal) and misclassified landscape images (rows). Misclassified classes between 5% and 25% are indicated by light pink, misclassifications between 25% and 50% are light orange, and misclassifications above 50% are red. The "Total" row indicates the percentage of classification choices made in each class.

FIGURE 13.4
An example of what the laypeople see before and during the experiment.

	BA	CC	dL	dO	EW	FO	GS	OW	PH	SS	WW
BA	42.14	0.71	0.71	0	7.14	0	0.71	2.14	0.71	45.71	0
CC	8.57	44.29	0	0.71	3.57	0	30	1.43	10	0.71	0.71
dL	0	0	47.86	47.14	0	0	3.57	0	1.43	0	0
dO	0	0	37.86	57.86	0.71	0	2.14	0	1.43	0	0
EW	3.57	1.43	0	0	9.29	34.29	6.43	0	20.71	17.14	7.14
FO	1.43	0	2.86	0.71	4.29	79.29	0	0	1.43	5.71	4.29
GS	15.71	14.29	0	0	0.71	0	41.43	0	13.57	13.57	0.71
OW	0	0	0	0	0	0	0	93.57	0	0	6.43
PH	12.14	4.29	2.86	1.43	6.43	10.71	32.86	0	11.43	17.86	0
SS	35.71	0.71	0.71	0	5	2.14	6.43	0	3.57	45.71	0
WW	0	0	0.71	0.71	3.57	77.86	0	0	0	2.14	15
Total	10.84	5.97	8.5	9.87	3.7	18.57	11.23	8.83	5.84	13.5	3.11

FIGURE 13.5
Confusion matrix for Experiment 2 (lay participants with intervention).

	BA	CC	dL	dO	EW	FO	GS	OW	PH	SS	WW
BA	14.29	3.57	0	0	14.29	3.57	0	0	0	64.29	0
CC	0	67.86	0	0	0	0	0	0	25	7.14	0
dL	0	3.57	78.57	10.71	0	0	3.57	0	3.57	0	0
dO	0	0	46.43	46.43	0	0	0	0	7.14	0	0
EW	0	17.86	0	0	17.86	42.86	3.57	0	3.57	10.71	3.57
FO	0	0	0	0	0	78.57	0	0	3.57	17.86	0
GS	0	14.29	0	0	14.29	0	21.43	0	21.43	28.57	0
OW	0	0	0	0	0	0	0	100	0	0	0
PH	0	21.43	0	0	0	0	25	0	46.43	7.14	0
SS	17.86	0	0	0	0	3.57	17.86	0	0	60.71	0
WW	0	0	0	0	0	92.86	0	0	0	7.14	0
Total	2.92	11.69	11.36	5.19	4.22	20.13	6.46	9.09	10.06	18.51	0.32

FIGURE 13.8
Confusion matrix for Experiment 3 (experts).

7

An Applied Ontology for Semantics Associated with Surface Water Features

Dalia E. Varanka and E. Lynn Usery

CONTENTS

ABSTRACT Surface water land cover plays a major role in a range of geographic studies, including climate cycles, landform generation, and human settlement and natural resource use. Extensive surface water data resources exist from geographic information systems (GIS), remote sensing, and real-time hydrologic monitoring technologies. An applied ontology for surface water was designed to create an information framework to relate data in disparate formats. The objective for this project was to test whether concepts derived from a GIS hydrographic data model based on cartographic relational table attribute data can be formalized for semantic technology and to examine the differences evident using the ontology for database semantic specification. The surface water ontology was initially derived from the National Hydrography Dataset (NHD) GIS data model. The hypothesis is that ontology semantics can be consistent with a long-term empirically collected database. An automated conversion of classes and properties was then manually refined with the support of an upper ontology. The results were tested for reliable class associations, inferred information, and queries using SPARQL Protocol and RDF Query Language (SPARQL). The ontology reflects studies of the physical environment, the objectives of the supporting institution, the reuse of GIS, and the adaptation of semantic technology. The results contribute to the development of an ontology model that leverages large data volumes with information user access.

KEY WORDS: *geospatial ontology, hydrography, semantic technology.*

7.1 Introduction

Surface water accumulates in depressions on the earth's surface at geographic scales, persists for periods of time, and flows or recedes over the surface as a function of elevation. Surface water is a primary category of human environmental interest; its study and representation as land cover has a long history. The charting of surface water crosses cultures, technologies, and symbolic languages. The recognition of surface water features results in part from the direct experience of the environment, such as from overland travel, but because features at the geographical scale can become too broad to easily see, then ideas of landscape often result from the study of geographical texts and maps. Varying criteria are possible for categorizing and labeling surface water entities in texts. For example, water bodies may persist over a period, or appear periodically or intermittently due to movement through seepage, replenishment by precipitation, or loss through evaporation. In many cases, the specific meanings of categories become ambiguous when separated from their context or defined by different groups of users.

Ontology is the study of what exists, and findings from this branch of philosophy can be applied to guide the design of data models. A central objective of applied ontology is to specify semantic information about data that usually remain within a broader context of knowledge and experience of users, or are represented in texts such as writing or graphic sources. Such knowledge is not encoded as part of the data but provided cognitively by the user during database interaction. Such contextual semantics are difficult to include as coherently reasoned media because they are technically incompatible with geographic information system (GIS) databases, the conceptual developments of which are based on expanding the capabilities of mapping by manipulating related data attribute tables.

The National Hydrography Dataset (NHD), the surface water component of *The National Map* of the U.S. Geological Survey (USGS), is one such GIS database (USGS 2014a, 2014b). The NHD is the digital version of the surface water theme appearing on topographic maps of the United States since the late nineteenth century. The data were collected according to surveying instructions, both from field and aerial photography sources, and converted from maps to digital vector data in the late twentieth century. The NHD is centrally maintained with information edits from state partners, resulting in a complex technical design that has been developed over 25 years by an extensive user community. An ontology design for surface water data and its integration with empirical data as a semantic technology system are expected to improve the clarity of surface water data such as the NHD. In turn, empirical data are needed to validate ontological surface water concepts.

The objective of this study is to present the development of a surface water ontology for semantic technology that reflects information about real-world entities and leverages legacy databases aligned with a different technical data model. The vision for the ontology is that its future application by users will aid accessibility to the data. The approach is to build semantic concepts with ontology modeling practices on a foundation of NHD data as they were developed through extensive hydrographic modeling practice, and to test whether this specific surface water ontology, which will be called SWO NHD, can be used to clarify the NHD semantics that are not supported or are often confusing in GIS. Typical of GIS, the NHD data model consists of numerous tables defining feature classes in various forms: as points, lines, and areas; as feature domains (types); events; the Watershed Boundary Dataset (WBD); attribute tables; metadata; and processing domains. The hypothesis is that a semantic approach will clarify these tables, making them more categorically aligned with the expectations of users. This will be achieved by reorganizing the geometrically constrained data categories, clarifying codes, and relating similar concepts to reduce redundancy while still supporting semantic detail.

The sections of this study are organized as follows. Section 7.2 is a review of significant literature on applied surface water ontology, and the approach is briefly summarized in Section 7.3. Section 7.4 details the development of the SWO NHD. The steps include the automated conversion of NHD data from GIS to Resource Description Framework (RDF) triples that result in an ontology called GIS NHD and the manual refinement of the GIS NHD as the SWO NHD (Cyganiak et al. 2014). The SWO NHD follows top-level knowledge models, including upper ontology and surface water science. The SWO NHD has an instance database component organized as gazetteer. Section 7.5 describes testing the SWO NHD by applying reasoning to the ontology for inferred triple statements, and Section 7.6 describes information retrieval using use case queries with competency questions and SPARQL graph patterns. The ontology application is followed by discussion and conclusion. The digital ontology file is available on the Internet (Varanka 2014).

7.2 Literature

An ontology design is abstracted from the context of a subject at varying levels, including the physical world; cultural abstractions represented through language; quantitative, scientific, and logic models; upper ontology concepts; and technical implementations. Research contributions have been made toward these aspects of surface water ontology, though toward different objectives and with varying parameters. The results of some key studies are summarized in this section.

The study of semantics normally begins with natural language. In a major systematic linguistic analysis, the lexical term "body of water" was parsed into English-language synsets by the WordNet project (Princeton University 2014). Body of water was assigned to domain categories of river, lake, and ocean, and related with two predominant properties, type and part, to broader or narrower classes. This synset provides a basic level of surface water semantics, but excludes important spatial and temporal relations, and provides no other context for each term other than a natural language definition (gloss). Synsets are designed for computational linguistics and natural language processing, related to semantic technology, but different in their focus on informal terms rather than formal variables and relations; terms in semantic technology are arbitrarily assigned labels.

Although language is an important source for ontological analysis and resolution, linguistically derived ontology will lead to several inconsistencies because terms vary for reasons such as cultural and geographic difference, geographic scale, or technological approach. Research in multilingual categorization indicates the complexity of drawing equivalent or related classes for data integration or interoperability of multilanguage spatial data infrastructures (Duce and Janowicz 2010; Feng and Sorokine 2014). Although these studies confirm the variability in the concepts used to distinguish water features between languages and cultures, some qualities, such as shape and size differentiation, may be widely recognized across cultures.

A hydrology ontology published by the British Ordnance Survey (OS) is rooted in national topographic data sources similar to SWO NHD. The files list extensive geospatial feature types as primitive classes with spatial relation properties, Web Ontology Language (OWL) axioms, and annotations to help clarify the semantics (OS 2008; Hart and Dolbear 2012; W3C OWL Working Group 2012). The ontologies are supported by reasoning software. Most terms, however, rely on information derived from natural language with few defined classes that specify class criteria based on ontological analysis. Because a large number of information queries are satisfied by identifying the taxonomic type of a geospatial feature, hierarchy and subsumption play a central role in ontology development and function. Taxonomic specification is limited, however, with a single property between primitive, meaning basic, terms. Primitive terms alone are insufficient in specifying the relations forming a complex proposition formed by multiple related properties. Without a formally defined framework involving properties such as parts or specifically identified properties for the application, an ontology composed of predominantly natural language terms lacks sufficient specificity and equivalence for the operation of inference.

Ontologies aim to resolve semantic variability by creating restrictions on category criteria that reflect complex relations. Among these may be aspects of physical reality based on direct observation or experience of the world, such as size, shape, and material. Property restrictions to include spatial semantics may be functions such as navigation, force dynamics such as water flow, or metric values such as hydrographic shape or size. For example, an ontology of Cree hydrography specified geospatial feature pairs, such as big brother/little brother lakes (Wellen and Sieber 2013). Quantitative methods have been applied for surface water ontology design, including artificial neural net processing (Li et al. 2012). Santos and Bennett (2005) used formal concept analysis to create a concept lattice of object attribute ranges for the water domain: shape, size, flow, depth, and origin. Supervaluation semantics are applied to model threshold-value variability (Bennett 2001). This approach differs from the development of ontology from cognitive or experiential-derived observation, where specifics can be applied at the instance level. The automatic classification of quantitative data helps build ontology by identifying salient qualities from reoccurring instances of a preselected object.

Hydrographic ontology requires further logical restrictions based on systematically organized science principles concerning surface water features. For example, the objective of EnvO is the formalization of environmental ontology (EnvO 2013). In the EnvO ontology, surface water is a subclass of water and environmental material. EnvO has a class called Hydrographic Feature, defined as "a geographical feature associated with water" with 22 subclasses. Unfortunately, variability, even among scientists, persists. Synonyms for Hydrographic Feature include Fluvial Feature, Marine Feature, Tidal Rip, Upwelling, Eddy, and Overfalls; these classes are not synonyms with one another. Some of these terms could arguably be called superclasses of feature events; others could be events rather than features of an enduring type. Some sibling classes include mixed surface water/terrain features types, such as island, inlet, coast, and harbor, but also include biological elements to surface water, such as algal bloom, or causes, such as beaver dam; and engineered features such as wells, which are subsurface water.

The extension of spatial representation to other science ontologies is an important function of a surface water ontology. The realm HydroBody module of the Semantic Web for Earth and Environmental Terminology (SWEET) ontologies has mostly hydrologic classes and properties, such as MethaneIce, with some included hydrographic features, such as Floodbank (SWEET 2013). The class Coastal, sharing the EquivalentTo property with CoastalRegion, for example, has sibling classes consisting of mechanical and chemical hydrology, imported from other separate modules. Unlike SWEET, the SWO NHD aims to clearly define spatial elements while supporting hydrologic modeling.

Hahmann and Brodaric (2012) clarified aspects of hydro-ontology by formalizing spatial voids, primarily holes and gaps that help define the integration of surface and subsurface parts of hydrogeology. Voids define areas within the earth's surface or other physical materials that host surface water. A top-level ontology was used to establish rules for earth/water spatial properties within voids. The demonstrated research of the study specifically focuses on groundwater formalizations, but that can also apply to the creation and persistence of surface water areas or features within their terrain hosts for surface water. Upper ontology also guided a surface hydrologic ontology developed with the Basic Formal Ontology (BFO) for the design of a hydro-ontology (Feng et al. 2004).

A surface water ontology pattern published by Sinha et al. (2014) is composed of two essential modules, one representing earth surface terrain that supports the accumulation and flow of water, called a dry model, and the second representing surface water and its properties, called a wet model. The central focus of the ontology is that the dry model influences the shape of the water bodies and water courses in the wet model, but water flow and pooling, and flow direction, is modeled in the wet model. The nature of a pattern is that as a small ontology, reasoning may be complete within the pattern, but is incomplete when expanded to specific applied situations (Gangemi and Presutti 2010). For example, though channels need incline to be on a path of greatest descent, elevation and slope are neither implied in the Dry Model, nor are obstructions and natural or artificial diversions such as dams or rapids. These exclusions are partially because other inputs for determining flow and pooling are possible, such as groundwater rise and rainfall. Instances of Fluence, an object class defined in the ontology pattern roughly representing surface water flow, would normally include "micro" features, such as water turbulence, mixing of water qualities such as temperature, and so forth, or the extension of a feature into a topologically joined feature, such as the movement of a river beyond the ocean coastline. Such microfeatures are neither accounted for by the pattern, nor are events such as flood conditions. Also, there are no prescriptive directions for feature geometry, for example, whether a channel should be represented as a line or linear feature with width. However, the presence of features may be scale independent, so the basic ontology model is not affected greatly this way.

Surface water ontologies have contributed linguistic propositions, quantitatively measured morphology, earth science dynamics, and formal logic designs to surface water studies. The SWO NHD allows for these ontological sources, and adds the benefit of technical integration with GIS and a large empirical database. An approach to creating a stable ontology that systematically organizes extensive data must allow repeated application with changing empirical detail and is sufficiently abstract so that inference relations produce intuitively true statements. These goals, used in the approach to develop the SWO NHD, are detailed later.

7.3 Approach

Classes and properties for the SWO NHD were initially converted directly from the GIS data model of NHD to enable the capture of all concepts considered to be relevant to the database users and to capture all legacy data. This initial version of the SWO NHD is called GIS NHD. In addition to classes and properties, many domain and range sets were identified based on the GIS attribute table. GIS NHD was manually aligned with top-level concepts, particularly upper ontologies, geographic theory, and RDF data model design. The SWO NHD is characteristic of descriptive logic, involving classes, instances, and properties, and first-order logic, such as domain and range classes (Pease 2011). Restrictions were applied to surface water domain–level classes and properties, such as hydrographic feature types, surface water flow processes, and spatial and temporal constraints.

The resulting version of the ontology was validated by producing inferred triples using SPARQL Inference Notation (SPIN) and examining the results to see if they seem reasonable. Three use cases and corresponding competency questions and SPARQL queries were developed to demonstrate capabilities for retrieving data that could be particularly challenging using GIS. These include "What types of waterbodies are subject to inundation?" and "What is the temporality of surface water flow associated with particular terrain feature types?" Lastly, the project is discussed and conclusions drawn.

7.4 Surface Water Ontology

7.4.1 Geographic Information Systems National Hydrography Dataset

The initial trial triple data were converted directly from GIS relational tables to the RDF triple data model by a custom designed program creating subjects from unique identifiers of rows, properties from column headings, and objects from cell values (Mattli 2013). Output triples of data from *The National Map* use Resource Description Framework Schema (RDFS) and OWL vocabulary terms in addition to RDF (Brickley and Guha 2014). Universal Resource Identifiers (URIs) are assigned to each resource and can be found in the header of the RDF document. The relational data model of NHD stores segments of the spatial geometry of features as unique rows in a database table, but the conversion program creates geometry objects in Well Known Text (WKT) format for GeoSPARQL standard compatability. The sample data set includes almost all NHD classes and properties, but is not an exact replica of NHD data at any specific time or version. The NHD data model changes and inconsistencies may occur between the data model and data set documentation.

After the sample data set was converted, it served as a starting point for further ontology development (Viers 2012). No URIs were created for table row groupings of geometry feature classes—point, line, and polygon—because the instance triples in the ontology, which is not constrained by geometry, were reorganized into topographic feature classes. In addition to feature instances created by the conversion program, however, tables specifying the column formats were manually converted to domain and range classes for NHD properties as part of the ontology. This allowed all instances, that is, rows with unique identifiers that share the same generated attribute values, to be part of the domain class that restricts the instances the property can draw upon to serve as the subject. For example, the NHD table called NHD VerticalRelationship describes three column headings available to any instance that participates in a vertical relationship, where one feature crosses over another feature. The relationship itself has an ID (Permanent_Identifier), the feature above has an ID (Above_Permanent_Identifier) and the feature below has an ID (Below_Permanent_Identifier). Those three attributes were converted to properties to connect subjects to the possible or allowed object values, for example belowPermanentIdentifier. By establishing NHDVerticalRelationship as the domain class for belowPermanentIdentifier, only members of NHDVertical_Relationship are useable subjects for that property.

Because the conversion resulted in the creation of a very large number of properties, a subset of data triples was selected to focus on the specific question of surface water feature types. Though the design and recognition of feature types and classes are highly cognitive, implementations to support geospatial data analysis involve technical specification as well. Classes and properties without geospatial qualities, such as source data identification, were not considered. Much of the information that was unspecified in SWO NHD was moved to other modules where they could be linked to other major ontologies used within the semantic technology community, such as one of several well-established provenance, metadata, or business systems ontologies (Figure 7.1). Important linkages exist for dimension and measurement units, such as the OGC Observations & Measurements ontology to provenance ontologies such as PROV-O and others (Cox 2011; Lebo et al. 2013). No software is known for ontology-driven mapping, but data can be exported to the Geography Markup Language (GML) to be digitally mapped.

New classes and properties were created only when essential and missing from the many column headings that were converted to properties from the GIS NHD model. The need for new triple resources occurred because of unspecified assumptions in the database or the lack of properties due to the tabular design of GIS rather than graphs.

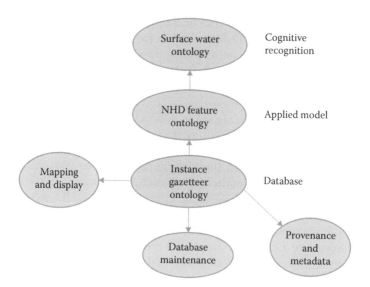

FIGURE 7.1
Surface water ontology layers and ancillary modules.

7.4.2 Top-Level Concepts

Though top-level principles are relatively independent of more specific subject domains, guidelines from upper ontology, geographic theory, and general database design principles provided insight to more specific SWO NHD classes and properties. Upper ontologies were used to provide guidance for forming the taxonomic order. Geographic concepts, such as elevation measurement, provided insight to interrelations between entities. Database design aligned the ontology to the instance gazetteer.

Upper ontology formalizations specify the relations between material objects and nonmaterial concepts and their attributes, such as qualities, roles, and the processes within which they engage. These general rules then apply to subject-specific subclasses and subproperties through inference, the inheritance of relations through the transitive property. Two documented upper ontologies were used: the BFO and the Suggested Upper Merged Ontology (SUMO) (Pease 2011; Smith 2014). For more intuitive understanding of the ontology described in this study, some upper ontology concepts were renamed to more specifically indicate spatial surface water land cover concepts. The *natural language* term is rather arbitrary because the ontology resource is defined by the formal logic.

Every triple resource (subject, property, or object) representing an entity takes the conventional form of a qualified name, meaning a prefix to indicate the URI separated from the class, property, or instance name by a colon. For example, BFO uses bfo as the prefix for its qualified names, so an example of a class name from that ontology is bfo:Entity. Class and property names of the SWO NHD described in this study omit the prefix of the namespace and use just the colon before the resource name, as in:flow, to indicate that SWO NHD is the default ontology being referenced. Class names begin with uppercase letters and property names begin in lowercase letters.

The surface water feature concept is defined by two general parts: topography, meaning the solid earth, and surface waterflow. When a drop of rainwater falls on the land, it flows downslope toward a singular water feature accommodated on and within the terrain, such as a stream. Surface water then flows downstream; no matter what juncture it comes to, the stream continues along the most straightforward channel. SWO NHD accommodates feature classes at this general level of the NHD and the included WBD. A characteristic of the NHD is that it includes many earth surface-type classes, such as :Diversion, a channel. The WBD centers data on nested hydrologic unit, such as a basin, subbasin, or watershed. Modifications of the earth surface that affect the collection of water as NHD features are indicated by a class called :HU_Mod, indicating a type of modification to natural overland flow such as :UrbanArea or :SpecialCondition subclasses such as :Glacier or :Karst. The terrain features described in this study will be the NHD surface features, and not those of the WBD.

BFO class definitions were used to reorganize the surface water concepts along ontology principles (Figure 7.2). The results were subgroups that encompass a large number of hydrologic feature types and properties. These classes include the earth surface formations indicated as :Feature (equivalent to bfo:MaterialEntity) with subtypes :Object and :ObjectAggregate. Surface water is indicated as :Flow (equivalent to bfo:Process), including standing water and hydrological events such as damming. The class :SpatialExtent (related to bfo:SpatialRegion) includes :SpatialQuality, :SpatialRelation, and :SpatialMeasurementUnit subclasses. :Temporality (equivalent to bfo:TemporalRegion) has subclasses :Ephemeral, :Intermittent, :Perennial, :Regulated, and :Status. SWO NHD includes bfo:Function, a class for socially defined areas serving a role by virtue of their dispositions (not depicted in Figure 7.2). This class was included to link to separate but related graphs, such as for land use or the role of surface water in other ontologies. These superclasses include many subclasses in the digital file that are too numerous to include in this article, though some specific examples are discussed in the following sections.

Figure 7.2 indicates the solid material components that are characterized by form and spatial extents (continuants) and fluid materials that are characterized by processes and temporal change (occurrents). This distinction is not completely disjoint, in that solid materials that interface with water are not completely static. Debris flows, landslides, and glaciers are examples of solid earth change affected by surface water. This specific interaction is not described in the SWO NHD.

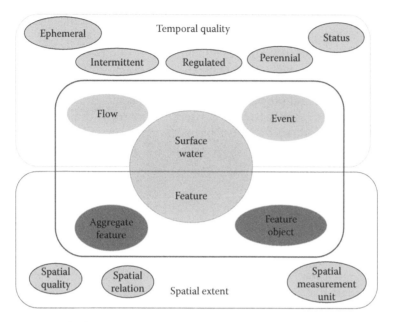

FIGURE 7.2 (See color insert.)
Top-level surface water ontology classes for land cover.

Ontology models allow for greater specification of feature qualities, roles, and relative spatial positions. In determining these specifics, a common problem was identifying the primitive terms that were combined in NHD attribute names to specify two or more classes at once, such as, areaAcres (area in acres) or DEDEM10 (Drainage-enforced 10 m Digital Elevation Model). A balance was struck between splitting such terms whenever possible to increase the reuse of classes and to reduce database redundancy and maintain attribute names for linking to NHD data (USGS 2014c).

7.4.2.1 *Feature*

Geographical or geospatial feature type is a term that is widely used in geospatial analysis literature, standards, and database design (Usery 2015). A feature is a relatively stable entity and so for the SWO NHD, the term is classified as equivalent to the structures that support the collection and flow of water, such as terrain or engineered channels and basins. The class :Feature is a subclass of bfo:Material Entity and conceptualizes a real-world material object in time and space, but infers additional semantics from the geographical literature. Feature type class semantic specifications apply to its subclasses :Object and :ObjectAggregate, a distinction that also appears in SUMO. These Feature subclasses allow for distinctions to identify material objects that are normally separated by spatial gaps, such as one single stream channel from others, from aggregates of objects, such as rocks of a reef or an area of complex channels. Object instances have cardinality normally restricted to one. :ObjectAggregate instances could have a restriction allowing for one group or many members. This distinction between an object and aggregated object allows, for example, the differentiation between a single-dredged channel and other nondredged channels of a braided stream river. Though bfo:FiatObjectPart was not used, the meronymy property :partOf, which allows objects such as a bay or inlet with a bona fide or fiat separation from an otherwise singular entity such as a sea, was added to the class of object properties.

7.4.2.2 Spatial Extent

Surface water often involves spatial extent as a criterion for classification. The class :SpatialExtent is related to bfo:SpatialRegion, but bfo:SpatialRegion is represented by spatial coordinates, and :SpatialExtent includes relative and qualitative spatial representation; the SWO NHD class for spatial coordinates is called :Geometry within the instance gazetteer. Upper ontologies lack broad guidelines of spatial and geographic theory for spatial extents that can be found in geographic information science literature. The subclasses devised for :SpatialExtent are :SpatialQuality, such as :Area or :Length; :SpatialRelation, such as :Elevation; and :SpatialMeasurementUnit, such as :Acre.

A distinction was drawn between spatial qualities of objects and spatial measurements. If a term was a spatial dimension of an object, such as length, this class or property was treated as a quality. If a spatial relation exists between objects whose computation is based on spatial coordinates, such as distance, then that entity was classified as a spatial relation. The actual measurement is a specific value for each instance and is documented in the gazetteer. The :SpatialQuality class includes geometric dimension classes, :Length and :Area, that are applied to features in general. :SpatialMeasurementUnit includes :Acre, :Kilometer, :SquareKilometer, :Meter, and :SquareMeter. These subclasses are available in commonly used ontologies and can serve as links to broader and widely used ontology modules. :SpatialRelation subclasses, indicating certain vertical and horizontal relations between features and representations as real-world entities and as measurements, such as a :SoundingDatumLine, include :Direction, :Elevation, :RelationshipToSurface, :Route, :SoundingDatumLine, :Route, :Stage, and :VerticalRelationship. :RelationshipToSurface and :Stage have several subclasses, such as :Underground or :AboveWater, and :FloodElevation or :NormalPool (Figure 7.3). Other topological relations are defined by the GeoSPARQL standard and applied to geospatial feature geometry objects in the instance gazetteer (Perry and Herring 2012).

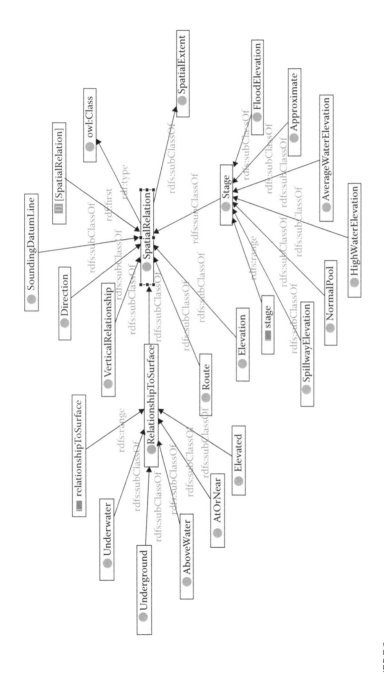

FIGURE 7.3
Subclasses of :SpatialRelation.

Surface water features have physical qualities that can lead to socially defined functions and roles, perhaps particularly true for engineered features designed and built for a purpose. A class called :Function is related to bfo:RealizeableEntity, with criteria that if a particular feature bearing a quality, role, disposition, or function is removed, the feature may be changed, but continues to exist. The SWO NHD :Function class links to classes such as NHD :HazardZone or :SpecialUseZone, found in a separate graph.

:Feature and :SpatialExtent classes focus on hydrographic entities of temporal endurance, relative to the more changeable temporality of surface water. The qualities of surface water and temporality are discussed in the following section.

7.4.2.3 Flow and Temporality

The :Flow class was designed separately from the :Feature class for modeling temporal processes such as :Waterbody and :Event. BFO defines bfo:Process as a bfo:Occurrent, an entity with temporal parameters that for some time is dependent on some material-entity participant to play itself out, that in this case is the water. Processes are weakly modeled in GIS relative to continuant entities defined primarily by their spatial ranges, so relations between these classes were drawn more from surface water science domain knowledge.

:Flow is the class of features consisting of water and flow dynamics. :Event is a subclass of :Flow consisting of hydrological monitoring types associated with particular features, such as :Dam or :Divergence. :Event is a class to integrate with possible hydrology ontologies. :Waterbody has subclasses for spatial parameters with regard to the terrain, such as :Rapids, :SinkOrRise, :SpringOrSeep, or :Waterfall. A much smaller number of such features are named compared to the number of :Feature subclasses. :Temporality, representing the temporal aspects of processes as defined in BFO, includes :Ephemeral, :Intermittent, :Perennial, :Regulated, and :Status, meaning a state of being.

7.4.3 Gazetteer Ontology

The gazetteer, or database, of the ontology consists of classes, but differs from the feature type taxonomy because categories are sets of instances and not subtypes. For example, the class :Name is a collection of instances of names, not a taxonomy of types of names. One characteristic of this difference is that subtypes of a parent class must be mutually exclusive, but instances may be members of more than one subclass.

A gazetteer consists of traditional categories: :Names (toponyms), :Geometry (spatial coordinates), and :Identifier, but added to these in the SWO NHD is the :Hydro_Net class, which is the entire coordinate geometry network created by all the combined geospatial features in the selected data set when a subset of the NHD is downloaded from the national database. Gazetteer classes, being sets of instances, and properties, relating to instances, were mostly taken directly from the GIS NHD ontology. The taxomony of surface water land cover required ontological reorganization that could be modeled as a graph, but once those classes were specified, sets of instances fell in place along the ontology design. This is a benefit of building "bottom-up," that is, starting the ontology with the GIS database. The feature type and gazetteer instance modules are interconnected using properties between classes. The gazetteer includes a great number of properties for instances. Most of the feature ontology properties are object properties, drawing relations between continuant entities, but the majority of properties in the gazetteer are datatype properties, storing specific values for instances. Though many triple model object resources of instances in the gazetteer ontology take the form of literals, the creation of an object class in the feature ontology is required to define them as instances of sets. For example, the :Geometry class contains the objects of the :hasGeometry property.

A class within the Hydro_Net called :HydroNetJunction is a set of NHD vector nodes forming junctions of different features in the geometry network. These junctions support surface water flow modeling. Flow modeling and watershed boundaries, forming nested hydrological units, have transitive properties that are compatible with inference. According to the Strahler Stream Order, if a first-order stream feeds to a second-order stream, and if the second-order stream flows to a third-order stream, then the first-order stream flows to the third-order stream (Strahler 1952). Within the WBD, subwatersheds are units contained within watersheds, and watersheds are contained within basins, then subwatershed are contained within basins. The inferred data from the Hydro_Net can be queried to trace a route along multiple stream segments and linkages from one point on the network to points downstream. These relations are calculated "on-the-fly" using GeoSPARQL topological relation analysis.

For the SWO NHD as a whole, the more specific the subclasses of those aligned with upper ontology, the more semantic specification is required. In addition to asserted classes, the effective use of inference is a key objective for the surface water domain ontology. Different methods are available for specifying semantics and inference; among these are formal proofs (Hitzler et al. 2010), graphic representations (Allemang and Hendler 2011), and an expressive language such as ISO Common Logic (ISO/IEC 2007). Although logical proofs capture the details of the algorithms and graphic representations do not, graphics were used for this study, as in examples shown below, because of their clarity for anticipating inference processes. Formalizations were left to the ontology, triplestore, and reasoning software.

7.5 Inference

Inference can be executed using the subsumption relation between owl:Class and rdfs:subClassOf, setting domain and range classes for properties, OWL axioms, defined classes using the property owl:equivalentTo, and using other restrictions such as cardinality. The top-level classes described in this study so far form a taxonomic hierarchy of primitive or asserted classes. A primitive class, using the subsumption (type-of) relation between parent and child classes, is defined in ontology as having necessary, but not sufficient conditions to support inference. Defined classes have necessary and sufficient conditions. This is indicated by specifying an equivalent-to relation between triple resources. For example, the :Flow subclass :Waterbody was converted to a defined class equivalent to the intersection of :Flow and one of the :Waterbody subtypes (The list of subtypes appearing below includes only a few of the eleven possible.)

> :Waterbody owl:equivalentTo :Flow and (:Rapids or :SinkOrRise or :SpringOrSeep or :Waterfall)

After applying reasoning software to the ontology, new triples were defined, indicating class membership through the transitive property. The following triple for :Waterbody is inferred:

> :Waterbody rdfs:subClassOf (:Waterbody or :Event)

Rather than adding taxonomic classes to the ontology to expand perceived distinctions, for example, engineered from natural feature types, the goal is to specify the formal semantics of each defined class to indicate the criteria by which subclasses vary. Feature types should cluster in the graph according to restrictions rather than additional taxonomic definitions. The number of classes was kept as small as possible to focus on key ontology properties. Nevertheless, the :FeatureObject class is particularly large, including engineered objects with operational parts, such as :LockChamber; natural objects with complex criteria, such as :SwampOrMarsh; and simple objects consisting of a single type of matter, such as :EarthenMaterial. The specifications for various defined classes are not fully established yet for the ontology as a whole, but some individual examples are described later. These limited semantics are partially to quickly complete initial drafts and will be addressed in later edits, and partially to facilitate sharing mutual natural language semantics with other hydrography data sets. As a result, many of the classes are simply terms for named entities and require further logic specification.

7.5.1 Feature Class Semantics

The SWO NHD has a greater number of triple resources to model than can be described in this study. This section presents models for two specific feature type classes, :InundationArea (Figure 7.4), defined by the NHD as "An area of land subject to flooding" and :AreaToBeSubmerged (Figure 7.5), defined as "The known extent of the intended lake that will be created behind a dam under construction" (USGS 2014c). These two classes are chosen because of their similar but slightly different semantics for :spatialExtent and :flow.

To model :InundationArea and :AreaToBeSubmerged, both were first identified as features. Features have certain dispositions based on internal physical qualities of the entities in question, as is so with :InundationArea and :AreaToBeSubmerged; flooding is possible only if the surface water height exceeds flood elevation. The two types of features differ in their external influences, which are uncontrolled natural forces or controlled human decisions and actions. With:InundationArea and with :AreaToBeSubmerged, a consequence is assumed, but for one site,

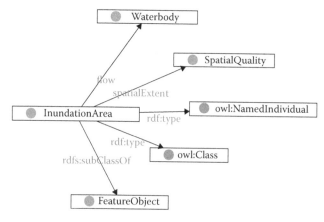

FIGURE 7.4
A semantic model for the class :InundationArea.

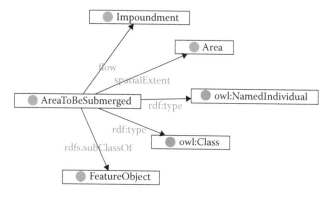

FIGURE 7.5
A semantic model for the class :AreaToBeSubmerged.

intermittent inundation from variable water flow, and for the other, permanent submersion from damming. For both models, the potential presence of surface water is assumed.

The two features have the same properties of rdf:type; rdfs:subClassOf; and :flow, meaning they are associated with surface water; and have :spatialExtent of their physical formation in common, though the objects of these properties are different. :Feature has a :flow property that is fulfilled by :Waterbody; :Waterbody has properties of :stage, :temporality, and :event (Table 7.1). For the class :InundationArea, which has a simpler set of criteria than :AreaToBeSubmerged, the property :flow has a wider range of possible object values and thus a more general range class. The range class of :AreaToBeSubmerged is a subclass of :Waterbody.

7.5.2 Inference on Asserted Classes

The inference engine executed using SWO NHD was SPIN. SPIN is a RDF vocabulary that formalizes constraints using SPARQL. SPIN is an expressive way to formalize rules that will apply to classes (Knublauch 2011). The results in Table 7.1 are inferred triples produced from asserted classes. Inferred triples based on the semantic graphs for :InundationArea (Figure 7.4) and :AreaToBeSubmerged (Figure 7.5) are included in the results listed in Figure 7.6, together with sibling and other classes of the SWO NHD. Inferencing at this step of ontology development demonstrates that some restrictions are declared by the RDF and RDFS vocabulary. For example, by declaring a domain and a range class for a property, several inferences are invoked. The subject of the statement will be inferred to be an instance of the class in the domain of the property, and the object of the statement will be inferred to be an instance of the class in the range of the property. However, if a property has more than one domain or range, the resource will be inferred to be an instance of both. As a result, the ontology will probably be more correct if fewer general classes are declared for domain and range than several

TABLE 7.1

Domain and Range Classes for Selected SWO NHD Properties

Property	Domain	Range
:flow	:Feature	:Waterbody
:stage	:Waterbody	:Stage
:temporality	:Waterbody	:Temporality
:event	:Waterbody	:Event
:spatialExtent	[none]	[none]

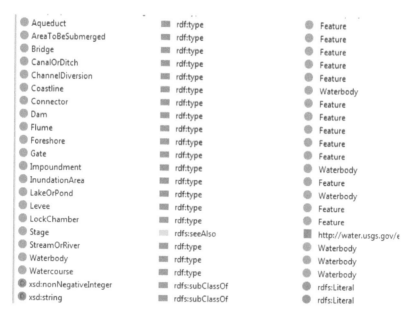

Aqueduct	rdf:type	Feature
AreaToBeSubmerged	rdf:type	Feature
Bridge	rdf:type	Feature
CanalOrDitch	rdf:type	Feature
ChannelDiversion	rdf:type	Feature
Coastline	rdf:type	Waterbody
Connector	rdf:type	Feature
Dam	rdf:type	Feature
Flume	rdf:type	Feature
Foreshore	rdf:type	Feature
Gate	rdf:type	Feature
Impoundment	rdf:type	Waterbody
InundationArea	rdf:type	Feature
LakeOrPond	rdf:type	Waterbody
Levee	rdf:type	Feature
LockChamber	rdf:type	Feature
Stage	rdfs:seeAlso	http://water.usgs.gov/e
StreamOrRiver	rdf:type	Waterbody
Waterbody	rdf:type	Waterbody
Watercourse	rdf:type	Waterbody
xsd:nonNegativeInteger	rdfs:subClassOf	rdfs:Literal
xsd:string	rdfs:subClassOf	rdfs:Literal

FIGURE 7.6
Triples derived from :InundationArea and :AreaToBeSubmerged semantics.

specific classes. Declaring rdfs:domain and rdfs:range classes accomplishes one stage of creating an expanded graph of inferred triples.

The inferred triples in Figure 7.6 highlight two particular inference rules. Subclasses acquire the type relation to their parent class in addition to the subclass relation that was asserted in the class hierarchy. The property rdfs:subClassOf is used to state that all the instances of one class are instances of another class. The property rdf:type is used to state that a single instance of a class is an instance of another class. Second, a class is reflexive, meaning a class is a type of itself.

The transitive property of inference applies to properties as well. In the Dublin Core Metadata Initiative (DCMI) vocabulary terms, which uses the prefix "dcterms," the dcterms:partOf property is a subproperty of dcterms:relation (Dublin Core Metadata Initiative 2012). Through inference, a triple such as :BayOrInlet dcterms:partOf :SeaOrOcean will also lead to the creation of the triple :BayOrInlet dcterms:relation :SeaOrOcean. Subproperties have domain and range classes whose parent classes will be inferred for the parent property (TopBraid Composer 2011). If the parent property has domain and range classes, then additional triples, such as :BayOrInlet dc:terms:Relation :Waterbody, will result. Such inference expands the range of associated category types for a triple and supports information retrieval.

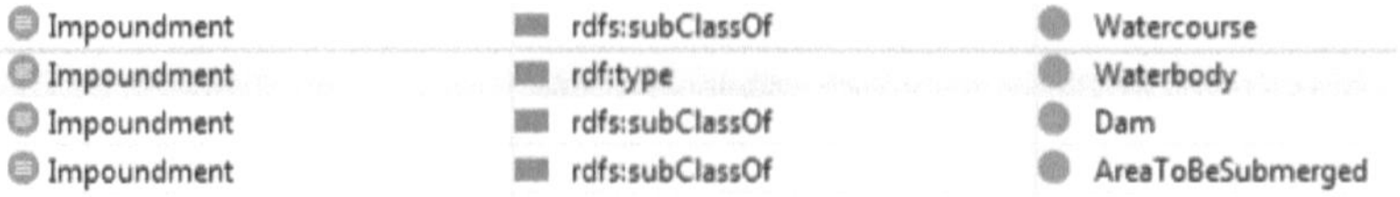

Impoundment	rdfs:subClassOf	Watercourse
Impoundment	rdf:type	Waterbody
Impoundment	rdfs:subClassOf	Dam
Impoundment	rdfs:subClassOf	AreaToBeSubmerged

FIGURE 7.7
Triples derived from the defined class :Impoundment.

7.5.3 Inference on Equivalent Classes

Subsumption or taxonomic relations, indicating types or subsets, are frequently not sufficient to establish criteria for membership in a class. The application of restrictions, meaning conditions to which specific instances must adhere, creates defined classes that are considered to establish necessary and sufficient conditions for a class. The primary property for establishing restrictions is owl:equivalentTo. The property owl:equivalentClass converts an asserted class to a defined class. The owl:equivalentClass property exists in addition to rdf:subClassOf, not in place of it.

The class :Impoundment, defined as "A body of water formed by impoundment," was defined to be equivalent to :Watercourse and :Dam and :AreaToBeSubmerged. The conjunction "and" indicates the intersection of the three classes, one a :Waterbody (:Watercourse), an :Event (:Dam), and a :Feature (:AreaToBeSubmerged). Certain inference rules are invoked by these semantics for the defined class :Impoundment. The triples that result from running the inference engine, shown in Figure 7.7, indicate that the defined class :Impoundment is a subclass of each of the members of its equivalent class, meaning that members of the set of :Impoundment may be a member of the class :Watercourse, :Dam, or :AreaToBeSubmerged, but :Impoundment is not a type of these equivalent sets.

Establishing taxonomic classes, domain and range classes, properties and subproperties, and defined classes are basic ways of building semantics in graph databases. Other possible restrictions support other new inferred triples. The graph with the original and inferred triples from asserted and defined classes formed the basis of a triplestore for SPARQL queries.

7.6 Information Retrieval

An objective for the design and development of the SWO NHD is to see whether semantic technology eases information access. To explore this question, the use case method, which assumes the perspective of a system user, was selected for information retrieval executed with SPARQL queries (Wiegers 2003; Fox and McGuinness 2008). Three use cases are described in this section. The use cases have corresponding competency questions designed to demonstrate queries that would otherwise be complex to retrieve in GIS. GIS primarily uses Structured Query Language (SQL) for queries. SPARQL is similar to SQL, but the potential expression of potential SPARQL queries on RDF data is limited at this stage of its technical development (Patroumpas 2014).

The use case/competency question method involves scoping capabilities of the system for particular objectives. Parts of the scoping process are to ask questions and assess resources for relevant and acceptable results. The competency question method originated in human interviewing techniques to answer criteria-based questions and thus has a greater focus on the cognitive semantics. Competency questions are an important part of the use case approach because ontology formalizations are mediated with psycholinguistic semantics by users.

7.6.1 Use Case 1

Use Case 1 poses the task: retrieve classes of different types that are related to each other, such as surface water and terrain. Use Case 1 is designed to seek specific information given a general set of parameters. The competency question is stated as, What types of waterbodies are subject to inundation? The question must be reformulated to work with SPARQL. The following SPARQL Query specifies a variable to select called ?wb to stand for waterbody. The WHERE clause, which specifies the triple pattern to match against the data, is at the point of the query process at which the natural language question is formalized as a logic statement, reversing the order of the subject and object. :InundationArea represents the subject and "has type" is the predicate (the rdf:type property) and "waterbodies" represent the object. The subject is modified as "are subject to inundation" by virtue of the :InundationArea class definition.

```
Query:

SELECT ?wb                      Selects and displays all
                                waterbodies that match the
WHERE {                         constraints of the WHERE clause
:InundationArea :flow ?wb.      Restricts triple results to
                                waterbodies that have the
}                               subject :InundationArea and
                                property :flow
```

```
The results of this query submitted to the triplestore are
copied below.
  ?wb
  LakeOrPond
  Reservoir
  StreamOrRiver
  SwampOrMarsh
```

TABLE 7.2

Inundation Area and Waterbody Associations as
Retrieved Using a GIS Attribute Table

Feature	Waterbody
:InundationArea	:LakeOrPond
:InundationArea	:StreamOrRiver
:InundationArea	:StreamOrRiver
:InundationArea	:LakeOrPond
:InundationArea	:SwampOrMarsh
:InundationArea	:Reservoir
:InundationArea	:StreamOrRiver
:InundationArea	:StreamOrRiver
:InundationArea	:StreamOrRiver
:InundationArea	:LakeOrPond
:InundationArea	:Reservoir
:InundationArea	:SwampOrMarsh
:InundationArea	:Reservoir
:InundationArea	:SwampOrMarsh

A similar query using GIS would filter the data first by one variable and then the second. The data retrieval results shown in Table 7.2 would return the entire columns of both. Though some software offers the additional option of identifying just the unique values, that step is not the basic way the tables function. SPARQL supports data retrieval as subgraphs of the graph being queried, but results such as those shown in Use Case 1 suggest that triplestores can also be used as a knowledge base of statements that answer information questions.

7.6.2 Use Case 2

Use Case 2 poses the following task: retrieve values from a category not directly related to a feature type; for example, to model the relation of objects to their temporal qualities. The competency question is, What is the temporality of surface water flow associated with particular terrain feature types? The SPARQL Query has a variable ?F for any type of feature class and ?T for any type of flow temporality, given that temporality is associated with processes, not objects. The query and results appear below.

```
Query:

SELECT ?F ?T               Select and display all values for
                           variables F and T
WHERE {                    that match the constraints of the
                           WHERE clause
    ?F :temporality ?T.    the variable F has temporality T
}
```

```
Results:

[F]                    T
AreaToBeSubmerged      Regulated
InundationArea         Intermittent
```

Though one-to-one relationships are easily modeled in GIS, a query such as this one will return the possible options within the database, not just a list of values within the cells of selected rows.

7.6.3 Use Case 3

Use Case 3 poses the following task: get more information about a concept. In this use case, the competency question could be, How can I get more information about the term :Stage in surface water studies? This query builds toward the development of triple data linkages to other information about a single entity, instead of an entire metadata document, as is common in GIS. ?I is the variable representing additional information.

```
SELECT ?I
WHERE {
Stage rdfs:seeAlso ?I.
}

Results:

?I
http://water.usgs.gov/edu/dictionary.html
```

Though ideally, URIs link to a specific gloss associated with the NHD class :Stage, in this instance, http://water.usgs.gov/edu/dictionary.html is a document with multiple glosses for an entire vocabulary; the specific gloss for :Stage must be manually sought.

The use cases demonstrate that triples can contribute semantic detail to any number of primitive entities or complex concepts without duplication that increases file size and visual complexity for the user. The implication of this is that the added semantic detail does not need to be specified for every instance because classes work as sets of instances.

7.7 Discussion/Conclusion

A surface water ontology was developed from an empirical base, organized in accordance with top-level ontology models, and formalized for basic inference using asserted, domain, range, and defined classes. Parts of the SWO NHD were validated through inferring new triples and querying the triplesstore within the parameters of use cases. The NHD data model structure was regrouped around related concepts, creating semantically a similar context for complex parts of the GIS database. For example, terrain categories were grouped together distinct from water flow processes and spatial and temporal qualities. The GIS data that were captured from the automated conversion aligned within classes and properties with identical URIs. The legacy data can be managed with minimal change to the SWO NHD because of the flexibility of the graph-based data model.

By developing the SWO NHD, feature classification was no longer based on geometric constraints of layer-based GIS, but on relations between concepts made more intuitive to the user through natural language. For example, GIS data layers were organized by feature geometry, which constrained water feature and flow modeling along Flowlines, modeled as linear features in one layer, through water bodies, modeled as polygons, formed as a separate layer. A class of objects called Artificial Paths was required to resolve the discrepancy between lines and polygon disconnect in layer-based NHD. With the SWO NHD, water flow is easily modeled along the surface water network in a way that more closely resembles the real world because coordinate geometry constraints are removed. Feature types, processes, and qualities were reorganized in semantic technology along guidelines consistent with cognitive understanding of real-world entities. The conclusion of this study is that though the ontology requires further refinement, it demonstrates the potential of semantic technology for advancing surface water data use.

References

Allemang, D. and Hendler, J. 2011. *Semantic Web for the Working Ontologist. Effective Modeling in RDFS and OWL*, 2nd edition. Burlington, MA: Morgan Kaufman.

Bennett, B. 2001. Application of supervaluation semantics to vaguely defined spatial concepts. In D. R. Montello, ed., *Spatial Information Theory: Foundations of Geographic Information Science: Proceedings of COSIT01*, vol. 2205 of LNCS, pp. 108–123. Morro Bay, CA: Springer.

Brickley, D. and Guha, R. V. 2014. RDF Schema 1.1. W3C, accessed, April 16, 2015, http://www.w3.org/TR/rdf-schema/.

Cox, S. J. D. 2011. OWL representation of ISO 19156 (Observation model), accessed July 21, 2014, http://def.seegrid.csiro.au/isotc211/iso19156/2011 / observation.

Cyganiak, R., Wood, D., and Lanthaler, M. 2014. RDF 1.1 Concepts and Abstract Syntax. W3C, accessed, April 16, 2015, http://www.w3.org/TR/rdf11-concepts/.

Dublin Core Metadata Initiative. 2012. DCMI Metadata Terms, accessed August 28, 2014, http://dublincore.org/documents/dcmi-terms/.

Duce, S., and Janowicz, K. 2010. Microtheories for Spatial data infrastructures — accounting for diversity of local conceptualizations at a global level. In

Fabrikant, S. I., Reichenbacher, T., van Kreveld, M. J.., and Schlieder, C., eds., *6th International Conference on Geographic Information Science* (GIScience 2010), vol. 6292 of *Lecture Notes in Computer Science*, pp. 27–41. Springer, Berlin.

EnvO. 2013. EnvO, accessed August 12, 2014, http://environmentontology.org/.

Feng, C-C., Bittner, T., and Flewelling, D. 2004. Modeling surface hydrology concepts with endurance and perdurance. In Egenhofer, M., Freksa, C., and Miller, H. J. eds., *Geographic Information Science: Proceedings Lecture Notes in Computer Science*, 67–80. Berlin, Heidelberg: Springer.

Feng, C.-C., and Sorokine. A. 2014. Comparing English, Mandarin, and Russian Hydrographic and Terrain Categories. *International Journal of Geographic Information Science* 28:1294–1315.

Fox, P., and McGuinness, D. L. 2008. TWC Semantic Web Technology, accessed November 27, 2013, http://tw.rpi.edu/web/doc/TWC_SemanticWebMethodology.

Gangemi, A., and Presutti, V. 2010. Towards a pattern science for the semantic web. *Semantic Web* 1: 61–68.

Hahmann, T., and Brodaric, B. 2012. The void in hydro ontology. In Donnelly, M., and Guizzardi, G., eds., *Frontiers in Artificial Intelligence and Applications*, 239, 45–58. Formal Ontology in Information Systems, Proceedings of the 7th International Conference. Amsterdam, The Netherlands: IOS Press.

Hart, G., and Dolbear, C. 2012. Ordnance Survey Hudrology Ontology V2.0. Ordnance Survey, accessed October 12, 2012, http://www.ordnancesurvey .co.uk/oswebsite/ontology/Hydrology/v2.0/Hydrologyv2.0.mht.

Hitzler, P., Krötzsch, M., and Rudolph, S. 2010. *Foundations of Semantic Web Technologies*. Boca Raton, FL: CRC Press.

ISO/IEC. 2007. Information technology—Common Logic (CL): A framework for a family of logic-based languages. First edition. Reference number ISO/IEC 24707:2007. Geneva, Switzerland.

Knublauch, H. 2011. SPIN—SPARQL Syntax. W3C, accessed August 26, 2014, http:// www.w3.org/Submission/2011/SUBM-spin-sparql-20110222/.

Lebo, T., Sahoo, S., and McGuinness, D. 2013. PROV-O: The PROV Ontology, accessed July 21, 2014, http://www.w3.org/TR/prov-o/.

Li, W., Raskin, R., and Goodchild, M. 2012. Semantic similarity measurement based on knowledge mining: An artificial neural net approach. *International Journal of Geographical Information Science* 26(8):1415–1435.

Mattli, D. 2013. Geospatial Semantics and Ontology. NationalMap2rdf-new.py. Computer program, accessed August 12, 2014, http://cegis.usgs.gov/ontology .html.

Ordnance Survey. 2008. *Hydrology*. Ordnance Survey. Written Communication (Original document no longer available over the Internet, but copied and in possession of the author).

Patroumpas, K., Giannopoulos, G., and Athansiou, S. 2014. Towards GeoSpatial Semantic Data Management: Strengths, Weaknesses, and Challenges Ahead. ACM-SIGSPATIAL 2014. November 4–7. Dallas, TX.

Pease, A. 2011. *Ontology, A Practical Guide*. Angwin, CA: Articulate Software Press.

Perry, M., and Herring, J. 2012. OGC GeoSPARQL—A Geographic Query Language for RDF Data. Open Geospatial Consortium project document OGC 11-052r4, v. 1.0 Wayland, Mass.

Princeton University. 2014. WordNet, A lexical database for English. Princeton University, accessed January 21, 2014, http://wordnet.princeton.edu/.

Santos, P., and Bennett, B. 2005. Supervaluation semantics for an inland water feature ontology. IJCAI-05. *Proceedings of the Nineteenth International Joint Conference on Artificial Intelligence,* 564–569. Edinburgh, Scotland, July 30–August 5, 2005. Morgan Kaufmann, San Francisco, CA.

Sinha, G., Mark, D., Kolas, D., Varanka, D., Romero, B. E., Feng, C., Usery, E. L., Liebermann, J., and Sorokine, A. 2014. *An Ontology Design Pattern for Surface Water Features.* GIScience.

Smith, B. 2014. *Basic Formal Ontology.* 2.0. Draft Specification and User's Guide, accessed July 20, 2014, http://www.ifomis.org/bfo/.

Strahler, A. N. 1952. Hypsometric (area-altitude) analysis of erosional topology. *Geological Society of America Bulletin* 63:1117–1142.

SWEET Ontologies 2.2. 2013. realmHydroBody.owl. NASA Raleigh, North Carolina, accessed November 27, 2013, http://sweet.jpl.nasa.gov/ontology/.

TopBraid Composer. 2011. Set domains and ranges. TopBraid Composer Maestro Edition, v. 3.6.1.v20120622-1546R. TopQuadrant.

U.S. Geological Survey. 2014a. Hydrography; National Hydrography Dataset, Watershed Boundary Dataset. U.S. Geological Survey, accessed December 18, 2013, http://nhd.usgs.gov/.

U.S. Geological Survey. 2014b. *The National Map.* U.S. Geological Survey, accessed December 18, 2013, http://nationalmap.gov/.

U.S. Geological Survey. 2014c. NHD User Guide. U.S. Geological Survey, accessed July 22, 2014, http://nhd.usgs.gov/userguide.html.

Usery, E. L. 2015. Spatial Feature Classes. In D. Richardson, N. Castree, M. F. Goodchild, A. L. Kobayashi, W. Liu, and R. Marston, eds., *The International Encyclopedia of Geography; People, the Earth, Environment, and Technology.* John Wiley, Hoboken, New Jersey.

Varanka, D. 2014. Surface Water Ontology. U.S. Geological Survey, accessed November 28, 2014, http://cegis.usgs.gov/ontology.html.

Viers, W. 2012. GIS NHD Ontology. U.S.Geological Survey, accessed August 12, 2014, http://cegis.usgs.gov/ontology.html.

W3C OWL Working Group. 2012. OWL 2 Web Ontology Language Document Overview (Second edition), accessed April 16, 2015 http://www.w3.org/TR/owl2-overview/.

Wellen, C. C., and Sieber, R. E. 2013. Toward an inclusive semantic interoperability: The case of Cree hydrographic features. *International Journal of Geographical Information Science* 27:168–191.

Wiegers, K. E. 2003. *Software Requirements.* Redmond, WA: Microsoft Press.

8

Land Type Categories as a Complement to Land Use and Land Cover Attributes in Landscape Mapping and Monitoring

Anders Glimskär and Helle Skånes

CONTENTS

ABSTRACT The purpose of land cover and land use descriptions varies, and this influences how these concepts are perceived in different contexts. The increasing need for spatial data for multipurpose monitoring and modeling also increases the demands for compatibility, repeatability, detail, and well-documented criteria.

We suggest that threshold values along a continuous scale can be used to create nominal classes for a common conceptual framework. However, the exact values of these thresholds need to be based on well-defined functional and systematic criteria. Ecological and environmental gradients are often mosaic and complex, and several types of land use may coexist at the same site. In reality, land use can be seen as a "shifting cloud" of activities varying in both time and space. We advocate the use of strict definitions of land cover as physical structures and land use as human activities, which raises the need for a complementary concept, which we call "land type," with stable threshold values based on mutually exclusive functional criteria. Such functional criteria often put clear limits to what spatial resolution is appropriate, since the suitability for a certain purpose (e.g., agriculture or forestry) is determined by the user of the land, rather than by the independent observer.

Our example of land type categories comprises a two-level hierarchical classification with seven main types and altogether 28 subtypes. As an example, we discuss the overlapping Swedish definitions of forest and arable land. The criteria that define our main land types are less dependent on how the area is managed at a specific moment in time, and they are therefore less sensitive to short-term variation. The land types define the limits for what land cover and land use can be expected at a certain site, given, for example, ground conditions, water, or artificial structures. Since such land types need to incorporate functional and qualitative understanding and interpretation, human visual interpretation is needed, whereas automated remote sensing methods are suitable mainly for the structural aspects of land cover.

KEY WORDS: *Land type classification; functional criteria; arable land; forest land; overlapping definitions; compatible concepts; ownership types.*

8.1 Background

In environmental monitoring of biodiversity and landscape, both stable criteria and flexibility in classification are crucial for the ability to confidently show changes through time and, at the same time, adapt the analyses according to different ecological patterns, change trajectories, natural values, and geographical scales (Käyhkö and Skånes 2006, Ståhl et al. 2011, Normander et al. 2012). When mapping for describing landscape changes through time, it is not sufficient to use only delimitation criteria based on arbitrary thresholds along a quantitative scale, since it could always be argued that some other threshold would give other results.

In our work with monitoring and mapping of the Swedish landscape, we have for a long time acknowledged the usefulness of quantitative data (Skånes, Glimskär, and Allard 2011; Ståhl et al. 2011), but also felt an increasing need for a more coherent and operatively well-defined set of criteria for classifying contents of the landscape that can be consistently applied in the field and by remote sensing. This set of criteria should ensure stability and transparency in the application of major classes of land and make the classification less sensitive to temporary or short-term changes or fluctuations. For example, the open conditions of a clear-cut area should not render a change from forest to grassland, since within a cycle of 10–50 years, the area will be again covered by forest. Correspondingly, the dynamics in tree cover on grassland depending on long-term grazing pressure fluctuations should not have to indicate a break in grassland continuity (Käyhkö and Skånes 2006).

Land cover and land use are of fundamental importance for any description and classification of the environment and landscape. In the literature, the importance of separating between land cover and land use has been repeatedly stated, but also the apparent problems relating the two concepts to each other have been acknowledged (Cihlar and Jansen 2001; Comber, Wadsworth, and Fisher 2008; Bakker and Veldkamp 2008). Comber, Wadsworth, and Fisher (2008) emphasize physical characteristics as a main component of land cover, and actual human activities as that of land use. This is an important distinction, but these components are not enough for a full-cover map or a complete description of landscape structure, since the functional aspects that include the interaction between land use and land cover are not included, nor are the ecological and other environmental processes that form the landscape.

There is not only one, but a large number of aspects that are generally overlooked in the debate on classification criteria. A complete land and landscape description may contain several layers of information, of which land cover and land use are only a subset, as exemplified below:

- Physical structures: areas with or without a closed tree canopy, artificial structures, water, etc.
- Human activities: active cultivation on arable land, forestry measures, recreation, etc.
- Land surface: rocky outcrops, soil texture, peat, topographical conditions, etc.
- Ecological/environmental processes: Harsh climate (wind erosion, frost), water-saturation/flooding, etc.
- Ownership type (or other legal/administrative criteria)

In a landscape perspective, human land use is often a strongly influential factor, but the character and the intensity of the land use impact varies considerably (Cihlar and Jansen 2001). Land use also interacts with ecological, geomorphological, hydrological, and climatic factors in a complex relationship. We need to incorporate many such factors in a common context, not only purely structural and descriptive but also including functions and processes. Our main question is therefore, "What is the most influential factor determining the conditions at a certain place?" This simultaneously puts the limit to what priorities and attributes are relevant in each case. These are principles that should characterize an instrumental and effective system for landscape classification:

- Relative stability in space and time, allowing small-scale variation in land use and land cover within each class.

- Ensure mutually exclusive and spatially exhaustive classes suitable for general-purpose mapping, defined in such a way that physical borders are relatively sharp and well-defined and possible to extract in a similar way from several data sources (in the field or by remote sensing). Class limits defined as a certain point along a continuous gradient should have a clear functional motivation.

- Simple and straightforward classification criteria based on functionally motivated and unambiguous limits, preferably also with clear links to administrative needs and prerequisites for land use and land cover, for example, *productive forest* or *arable land*.

- Nomenclature that explicitly and unambiguously states what constitutes the classification criteria and the priorities between them. This may lead to class legends, that is, metadata that are relatively long and explanatory, but less prone to alternative interpretations.

- The criteria should be comprehensible and generally applicable in various contexts and geographical regions.

8.2 Bridging the Land Use/Land Cover Dilemma Through a Land Type Classification: A Swedish Example

To overcome and bridge some of the well-known issues with the land cover and land use concepts and to meet the demands of a more stable and consistent broad classification system, we have developed a generally applicable classification system for the mapping and monitoring of landscapes in Sweden. Our suggested classification system consists of 7 main land types and altogether 28 subtypes (Table 8.1). The criteria for the 7 main types aim at identifying the main structuring conditions of land. To justify our focus on only certain aspects of land cover and land use, and to incorporate both in the same classification system, we return to our initial question, as formulated above: What is the most influential factor determining the conditions at a certain place? This factor, to large extent, also determines what other attributes are relevant and what patterns in space and changes through time can be expected.

One way of dealing with the fact that there is no one-to-one relationship between land use and land cover is to acknowledge the multifunctionality of land cover (Cihlar and Jansen 2001). Bakker and Veldkamp (2008) introduce the terms *primary land use* and *secondary land use*. The primary land use has a strong impact on land cover and claims a certain area, whereas secondary land use often does not. Several types of secondary land use can

TABLE 8.1

Division of Proposed Main Land Types with Their Respective Primary Criterion and Subtypes

Main Land Type	Constructed or Artificial Land	Arable Land and Former Arable Land	Seminatural Pasture or Meadow	Land with Closed Forest or Active Forestry	Other Terrestrial Land with Human or Natural Disturbance	Semiaquatic Land without Closed Forest	Aquatic Area
Primary criterion for main land type	The area is constructed, with artificial surface or sown/planted vegetation (excl. arable land).	The area is influenced by plowing and is or has been suitable for cultivation of arable crops.	The area has grassland vegetation suitable for grazing or mowing.	The potential for forest production is high enough for forestry. No other dominant land use.	Cover of trees and shrubs is limited by shallow soil, harsh climate, natural/human disturbance, or other land use.	The vegetation is influenced by water saturation or frequent flooding (excl. closed forest).	The area has permanent water cover.
Subordinate land types	1. Transportation area 2. Residential or commercial area 3. Agricultural area (except arable land) 4. Industrial area 5. Recreational area	1. Arable land used for arable crops 2. Arable land with permanent grazing or mowing 3. Unused arable land 4. Former arable land with permanent grazing or mowing 5. Unused former arable land	1. Managed pasture or meadow 2. Unmanaged pasture or meadow 3. Rocky outcrop influenced by grazing	1. Terrestrial land influenced by forestry 2. Terrestrial closed forest without forestry 3. Terrestrial land with successional closed forest 4. Semiaquatic land influenced by forestry 5. Semiaquatic closed forest without forestry 6. Semiaquatic land with successional closed forest	1. Natural rocky outcrop 2. Terrestrial land influenced by harsh climate or natural disturbance 3. Terrestrial land influenced by other land use or human disturbance	1. Peat-forming land (mire) not by shore 2. Peat-forming land (mire) by shore 3. Other semiaquatic land not by shore 4. Other semiaquatic land by shore	1. Aquatic area not in mire mosaic 2. Aquatic area within mire mosaic

occur, in different combinations and with different intensity, and are often not restricted to a specific land area. Secondary land use can seldom be reliably described even at a single field visit, but requires detailed administrative information, interviews, or various census data. As Cihlar and Jansen (2001) and Bakker and Veldkamp (2008) emphasize, primary land use interacts strongly with land cover, and it is therefore reasonable to include both aspects in a spatially exhaustive and mutually exclusive classification system for landscape mapping. Such a system would clearly acknowledge that human land use is the dominant formative agent in some areas, but not in others.

In addition to the distinction between primary and secondary land use, we suggest that the qualitative *functional land cover* is treated separately from the more descriptive *structural land cover*. Functional land cover is based on the ecological or anthropogenic functions that determine and characterize the land cover (e.g., wetland, arable field, road, and urban area), typically requiring a component of human interpretation. The criteria for delimiting such functional land cover types could be based on quantitative criteria, but only if they constitute well-motivated and transparent functional criteria. In contrast, structural land cover is strictly descriptive, in terms of physical characteristics, for example, tree height, tree cover, exposed substrate, buildings, topography, and open water surface, and includes the features and properties that often can be quantitatively described by automated or semiautomated remote sensing methods. Such information should preferably be included either as quantitative attributes or as pixels in complementary raster maps. A somewhat similar distinction between function and attributes for describing land use/land cover features has been used by Feng and Flewelling (2004).

For practical classification and mapping purposes, each decision preferably should be based on a single, quite straightforward criterion (Table 8.1). The criteria for identifying subtypes in our classification system are unique for each main land type, based on the factor that is most characteristic for each main type, but are of no or minor importance in other main types. For forest, it is the forestry practices, for arable land it is the crops and cultivation practices, and for semiaquatic land it is the influence of water saturation or flooding.

To avoid conflict between criteria for a land type classification with mutually exclusive classes, we therefore need to decide which factor is the most formative. For example, if an open wetland area is also used for grazing or mowing, we suggest that the influence of water is the dominating factor and management subordinate. The logic is that the influence of water often is enough to keep the land open, even in the absence of grazing or mowing, whereas in the terrestrial seminatural pastures and meadows abandoned land will, in a few decades, typically develop into forest. Our aim is also that the subtypes within or between main types should be possible to combine to yield aggregated classes that correspond to existing definitions of, for example, forest land, wetland, or agricultural land. The possibility to use the subtypes as such building blocks increases transparency and flexibility, in that the underlying priorities and the conflicting definitions of different established systems are highlighted and pin pointed. It should be emphasized that the most formative factors can differ depending on the location on earth and hence this system below might look different in different parts of the world. Still, on a conceptual level, it is valid in an international context.

8.2.1 Constructed or Artificial Land

This main type most often contains a mosaic of different structures in land constructed for a specific purpose, such as buildings, paved ground and lawns for living or trade in urban areas, for transport along roads or in airports, for recreation in golf courses, for agriculture in farmsteads, or for industry in industrial sites. Since this mosaic is an integrated part of the function of such sites, it may not always be justified or practical to subdivide it in detail. Such a principle also accords with the common notions of for example industrial sites. Since the function is what determines the way the sites are used and how they are constructed, it also makes sense to assign the main intended purpose as the main criterion for division into subtypes. We have chosen to include all constructed/artificial land with a closed tree canopy in this land type, not in the forest type, because they mostly have distinct land use and function within the particular context, for example, rows of tree in parks or planted trees in gardens. The vegetated areas in constructed sites are often reshaped or strongly modified by addition or replacement of soil and by sowing and planting of introduced or exotic species of trees, shrubs, grasses, or decorative plants. Some constructed land may also develop ruderal vegetation in early successional stages, for example, in recently abandoned extraction pits or in disturbed areas along railways.

The intended delimitation of constructed or artificial land agrees closely with the corresponding definitions of the global Land Cover Classification System (LCCS) criteria at the highest hierarchical level (Di Gregorio 2005): *Cultivated and managed terrestrial areas* and *Artificial surfaces and associated areas*, which include strongly modified vegetation or artificial cover. We have chosen to put arable land in a separate land type class, which of course again could be combined with constructed or artificial land for certain comparisons.

The separation between different structures within a constructed area is mainly a topic for complementary land cover attributes. Similarly, a more detailed description of the use of individual parts may be included as land use attributes. Also, interspersed with such constructed areas, there may be remnants of natural vegetation, that essentially look as they would do in some other context and should be classified as other land types (e.g., *terrestrial land influenced by other land use or human disturbance*) (Table 8.1).

8.2.2 Arable Land and Former Arable Land

The Swedish official definition of arable land states that the land should be suitable for plowing and growing crops, but does not say much about its actual use at a specific time (Table 8.2). The most common reasons for arable land to cease being suitable for plowing is the colonization of woody plants to the unused area, and sometimes a rise in groundwater level or insufficient draining. This means that arable land according to the accepted Swedish definition also includes both unused and permanently grazed land (Table 8.1). The semi-open former arable land in extensively used landscapes contains large areas that have grassland vegetation but are still strongly influenced by plowing. We believe that this increases the internal consistency in the land type system, but also attracts attention to a structurally distinctive feature in the landscape. This means that the sharp boundary between plowed land and other land can be maintained also at later stages of succession, which otherwise would be replaced by a sometimes variable and heterogeneous frontier of tree and shrub colonization.

Apart from the cultivated arable land with crops, the four other suggested subtypes are the factorial combination of *arable land* or *former arable land* with either permanent grazing/mowing or abandonment. This subdivision is admittedly based on the active land use, but we argue that this is justified in this case, since it allows the reclassification of abandoned former arable land to *forest land*, and former arable land with permanent grazing to *pasture* if necessary, according to established definitions (e.g., Table 8.2).

TABLE 8.2

Official Swedish Definitions of Ownership Types (Sw. *ägoslag*)

Official Swedish Term	Official Swedish Definition
Arable land (Sw. *Åkermark*)	Land that is suitable for plowing and used or suitable for use in crop cultivation or grazing.
Permanent pasture (Sw. *Betesmark*)	Land that is used or can suitably be used for grazing and that is not suitable for plowing.
Forest land (Sw. *Skogsmark*)	Land that is suitable for timber production and that is not to an appreciable extent used for other purposes, or land where there should be forest for protection against sand or soil erosion or the lowering of the Subarctic tree line. Fully or partly unused land shall not be considered as forest land if, due to certain conditions, it should not be used for forest production. Land shall be considered useful for forest production if it, according to generally accepted criteria, can produce on average at least one cubic meter of timber per year and hectare.
Other land (Sw. *Annan mark*)	Land unsuitable for crop cultivation, grazing, or timber production.

Source: Statistics Sweden 1981. *Svensk standard för ägoslagsklassificering av mark för jordbruk och skogsbruk* [Swedish standard classification of land use in agriculture and forestry]. Statistiska Centralbyrån, Meddelanden i samordningsfrågor [Statistics Sweden, Reports on Statistical Co-ordination] 1981:4. Stockholm, Sweden. [in Swedish]

8.2.3 Seminatural Pasture or Meadow

The official Swedish definition of pasture, including meadows, which have small areal extent in Sweden, is comparable to arable land, in that it includes land suitable for grazing or mowing. For comparability within the agricultural sector, this definition excludes land suitable for plowing, that is, fulfilling the definition of arable land. The definition of "suitable for grazing and mowing" requires that the field layer has a forage value, which means that it should have a grass sward dense enough and containing plant species that can be eaten and digested by domestic grazing animals. How the suitability and the forage value influences actual use is affected by issues such as market prices for milk or meat or subsidies for landscape conservation. This is one reason why more and more grazing is taking place on arable land rather than seminatural grassland.

A relevant conceptual analysis was made by Halvorsen et al. (2009). They describe two aspects of land use: *fundamental management intensity*, which is the long-term, formative impact of, for example, grazing in grasslands, and *current land use intensity*, which refers to the short-term impact. Along this gradient of intensity, Halvorsen et al. (2009) distinguished among three levels of impact: natural land (*No: naturmark*), seminatural land (*No: kulturmark*) and constructed/artificial land (*No: kunstmark*). Whereas low intensity of management gives only minor relative changes in vegetation structure and composition, and high-intensity management leads to a highly artificial state (including arable land), management in seminatural land gives a quasi-stable state with distinctive vegetation properties but still interacting strongly with the natural environment. These are the particular conditions forming this land type also in our Swedish approach.

In Sweden, pastures and meadows frequently contain considerable amounts of trees and shrubs. In the shaded conditions under a canopy, the grass sward suitable for grazing and mowing is loosened up between 50% and 70% canopy coverage, and the grassland plants are successively replaced by shade-tolerant plant species of less value for fodder (A. Glimskär personal observation based on monitoring data; cf. Ståhl et al. 2011). Above this cover, we therefore consider the land type to be forest, rather than pasture, in accordance with the intentions of the official definition of pasture. This also crudely corresponds to the cover of *productive forest* with an established, closed tree canopy. We have also chosen to include a third subtype, which is areas with thin soil layer (including rocky outcrops) strongly influenced by grazing (Table 8.1). These are comparable to *natural rocky outcrops*, but are often included in mosaics with seminatural grassland swards and contribute strongly to the natural values of such sites. One example is the Alvar vegetation of the Baltic islands of Öland and Gotland, which is very important to include in this context, even if the vegetation sometimes is quite sparse.

8.2.4 Land with Closed Forest or Active Forestry

There is a well-established Swedish official definition of productive forest, which is the potential to produce more than 1 m³ of wood per hectare and year (Table 8.2). According to Swedish legislation, forestry is not supposed to occur on land that does not fulfill this production criterion (Svensson 2006). We have therefore chosen a definition based on two main criteria, either a closed tree canopy or obvious use of the land for forestry (or both). The strictest procedure for determining the production potential, for example, as implemented in the Swedish National Forest Inventory (NFI) (SLU 2011), is quite complex. A preliminary suggestion for Swedish conditions, and based on earlier studies (Rafstedt and Andersson 1981) and on our experience from regional and national grassland inventories, is that this threshold in general corresponds to a closed canopy with cover of woody plants that exceeds 60%. This threshold does not correspond with the international Food and Agricultural Organization (FAO) or European Union Nature Information System (EUNIS) definitions of forest, at 10% tree cover (Di Gregorio 2005; Davies, Moss, and Hill 2004). However, it is very close to the definition of *closed forest* (60%–70%), as defined by Di Gregorio (2005). For Swedish conditions, a threshold at 10% is rather arbitrary and of minor practical use, as there are large areas of land with 10%–60% tree cover that will not be considered suitable for forestry, but rather be appreciated for their natural, historical, and recreational values as mires, mountain birch forest, or grasslands suitable for grazing or mowing.

The division into subtypes includes a separation between terrestrial and semiaquatic forested land on one hand and successional stage and forestry impact on the other. Water saturation can be expected to lower the production potential, so many production forests are intentionally drained. However, we assume that any wet forests influenced by forestry are important to distinguish, also for the internal consistency of the classification. We have chosen to separate between (1) land clearly influenced by forestry, in which actual canopy cover is not the main criterion; (2) successional land with closed canopy; and (3) forest that has reached a more stable canopy structure, which may be more or less natural. The semiaquatic forest types can be aggregated with other semiaquatic land, if that is required, and the distinction between successional and more stable forest allows an even more flexible description of the gradual changes from open land to mature forest.

8.2.5 Other Terrestrial Land with Human or Natural Disturbance

The common denominator for this group is that it is not a part of the other main land types, and as such it contains a wide range of land covers and uses. As mentioned above, land with dominant use other than forestry or agriculture is not included in the definitions of forest or arable land, nor is land with lower productivity that cannot support commercial forestry or agriculture. In some contexts it is obvious that intensive human land use dominates as the formative factor, for example, in power-line corridors, but in other contexts, the human and the natural disturbances interact. For example, in the Scandinavian mountains, given the climatic conditions and topography, with frost, strong winds, landslides, etc., natural processes are obviously a strong factor. But the influence of domestic reindeer also has a strong influence on both woody plants (e.g., willows) and ground vegetation, as they clearly interact with the natural conditions and modify their impact on vegetation and other environmental conditions. A similar distinction between human and natural disturbance was made by Halvorsen et al. (2009), with exploitation as a third category. This third category will in our classification belong to constructed or artificial land.

This land type may often occur in mosaics with constructed/artificial land, for example, within areas used for recreational or residential purposes. This means that in a site set aside for recreation, for example, a camping site, constructed and nonconstructed areas have different land types, but sometimes the same or similar land use. For the subtype classification, we treat separately areas with human use or disturbance as the dominating factors, because abandonment from human use would most often lead to a succession to forest. We also distinguish rocky outcrops with no or shallow soil layers as a third subtype, as they are not really dependent on any frequent natural or human disturbance to keep open and nonproductive.

8.2.6 Semiaquatic Land without Closed Forest

Land can be kept open by the influence of water in several ways. The most important of these in Sweden are the disturbing influence of flooding and fluctuating water levels along shores, and the formation of peat during oxygen-deficient conditions on water logged ground. A dense tree canopy can develop only when the groundwater is lower or when the flooding is less intense. Typically, under reasonably nutrient-poor conditions, the accumulating peat layer eventually is taken over by peat mosses (*Sphagnum* spp.) that can form a very thick layer of water-absorbing peat, in its most developed state (in bogs) only supplied by nutrients from precipitation, that is, ombrotrophic conditions (Rydin and Jeglum 2006).

Mires and other wetlands can to a certain extent be used for grazing or mowing. For a land type classification with mutually exclusive classes, we therefore need to decide which factor is the most formative, the water or the management impact. In this case, we suggest that the influence of water is the dominating factor, since the water in itself often is enough to keep the land open, even in the absence of grazing or mowing. This can be exemplified by open fens in northern Sweden that have been mown extensively for hay in earlier times, but in which mowing could be reinstated immediately, still after 50–100 years of abandonment. We suggest a factorial division into four subtypes, with peat formation as one factor and influence by flooding and so on by shores as another. Peat-forming mires close to shores are often more influenced by nutrients from stream or lake water than other mires, and can, therefore, be described as limnogenous mire (Rydin and Jeglum 2006). Wetlands without peat formation can also be wet heaths or wet grasslands in the inland, where the wetland conditions are caused by high or fluctuating ground water or a temporary/seasonal inflow of rainwater in poorly drained conditions.

For integration into an international framework, a comparison to the Ramsar Convention on Wetlands is important. In the Ramsar classification criteria (Ramsar Convention 2002), *Non-forested peatlands* and *Forested peatlands* are distinguished as two major inland wetland categories, which agree to a large extent with our corresponding land-type classes. It is acknowledged that peat soils also occur in other inland or coastal wetland categories, in which they have specific, important ecological and hydrological functions (Ramsar Convention 2002). This is why we include the limnogenous mires (Rydin and Jeglum 2006) in the subordinate land type category *Peat forming land by shore* within the main type *Semiaquatic land without closed forest* (Table 8.1).

8.2.7 Aquatic Area

Of course, permanently water-covered areas in rivers, lakes, and seas have other characteristics than terrestrial land or many wetlands. However, for mapping and monitoring, the decision of what is permanently water-covered sometimes requires experience and consideration. Along shores with fluctuating water levels, the border between terrestrial and semiaquatic land may be fairly easy to distinguish as marked changes in vegetation cover and plant species composition at the high-water level. At the lower level of the shore, however, between semiaquatic and aquatic areas, it is more difficult to draw out differences. This means that the theoretically straightforward criterion of permanent water cover can sometimes be difficult to apply in practice. A number of visits or remote sensing images from different periods during the year may be required to decide where the low-water level actually is. In agreements with the European system EUNIS (Davies, Moss, and Hill 2004), the classification of aquatic areas that are constructed but contain a seminatural aquatic fauna and flora may be ascribed to this class, whereas areas with no or unnaturally restricted species lists or domination by exotic species (Davies, Moss, and Hill 2004) should be classified as *Constructed or artificial land*.

In mires, depressions in a thick peat layer may form water-filled pools or flark pools, either because of peat accumulation between pools or secondary deepening of flarks to become flark pools (Rydin and Jeglum 2006). We consider these as an integrated part of the mire mosaic and the dynamics of mire ecology, and therefore we suggest that these form a separate subtype—aquatic areas within a mire mosaic. This subtype can be easily identified and treated in the same context as peat-forming semiaquatic land, with which it is closely functionally linked.

8.3 Examples of Conflicting Criteria in Sweden

In this section, we elaborate on the existing official definitions and the data sources used for Swedish national statistics. Even if the availability of national, administrative data and full-cover maps in Sweden is relatively good and has a long tradition, there are some fundamental problems that restrain our attempts to give a complete picture for areal statistics and landscape. Each of the dominating sectors has definitions adapted to the requirements of its respective administrative systems. Environments that have no obvious use for forestry, agriculture, or urban development are often not taken into account in a relevant and reasonable way, or they are claimed by more than one sector through definitions that are not mutually exclusive.

The main official categories of agricultural land in Sweden are arable land (Sw: *åkermark*) and pasture (Sw: *betesmark*), which also includes grassland with mowing (Table 8.2). The main criteria for these categories are "suitable for plowing and growing arable crops" and "suitable for grazing or mowing, but not suitable for plowing." This means that there is no need for active management at the time. This is reflected by the Swedish word for these categories, which in a crude translation is "ownership types" (Sw: *ägoslag*), that is, an administrative or prescriptive concept, rather than a description of actual land use activities. However, in the Swedish forestry legislation, land with enough capacity for forest production (>1 m^3 timber/ha and year) should be considered as forest if there is no other dominating land use (Statistics Sweden 1981; Svensson 2006). In the Swedish NFI, this is interpreted as 3 years of abandonment from active agricultural use (SLU 2011). This means that there is a serious overlap between the official definitions of arable land and forest land (Table 8.3), and that the forestry sector and the agricultural sector both tend to define their own interests in terms of potential use and all other interests in terms of actual use.

Among the ownership types of the Swedish NFI, the only wetland type included is mire, which comprises only wetland with peat-forming vegetation that does not fulfill the requirements for productive forest (SLU 2011). To incorporate all wetland in the national statistics, Statistics Sweden (2013) has considered the wetland mask from Swedish topographical and land cover maps, in which the tree cover threshold for non forested wetland is 30%. However, the classification criterion included in the Swedish NFI is productive forest, which in effect corresponds to a much higher tree cover, and the international FAO threshold for forest is 10% tree cover (Di Gregorio 2005; Davies, Moss, and Hill 2004). Statistics Sweden (2013) tried to combine these two datasets, but the result was an area sum that was considerably larger than the total land area of Sweden. The Swedish NFI includes total tree cover as a continuous variable in their data collection, so a more sophisticated analysis could combine the classes and the attributes to give more useful and comparable results. Unfortunately, the definitions of what trees should be included in the cover value (based on height or on species) are also different.

TABLE 8.3

Comparison between Swedish Official Ownership Types and Our Second-Level (Subordinate) Land Types

Swedish Ownership Types	Subordinate Land Types
Arable land	Arable land used for arable crops
	Arable land with permanent grazing or mowing
	Unused arable land
Permanent pasture	Managed pasture or meadow
	Unmanaged pasture or meadow
	Former arable land with permanent grazing or mowing
Forest land	Terrestrial land influenced by forestry
	Terrestrial closed forest without forestry
	Terrestrial land with successional closed forest
	Semiaquatic land influenced by forestry
	Semiaquatic closed forest without forestry
	Semiaquatic land with successional closed forest
	Unused arable land
	Unused former arable land
	Unmanaged pasture or meadow

Source: Statistics Sweden 1981. *Svensk standard för ägoslagsklassificering av mark för jordbruk och skogsbruk* [Swedish standard classification of land use in agriculture and forestry]. Statistiska Centralbyrån, Meddelanden i samordningsfrågor [Statistics Sweden, Reports on Statistical Co-ordination] 1981:4. Stockholm, Sweden. [in Swedish]

Furthermore, the application of the definitions for the administration of the agricultural sector is not consistent, and there is a tendency that actively used agricultural land is better represented than unused land, making the statistics and other information incomplete and inconsistent. In practice, the administrative system is well updated for arable fields included in agricultural payment schemes, but not for arable land where the farmer for reasons unknown has not applied for such subsidies. This tendency is even more accentuated for seminatural pastures than for arable land, because they are often managed more extensively, and the EU agri-environmental payments have even stricter requirements for being granted, not least as restrictions in the allowed number of trees per hectare. In Sweden, pastures and meadows frequently contain considerable amounts of trees and shrubs, partly as a consequence of more extensive land use over larger areas, but also as a consequence of multifunctional utilization of land that has historical roots in Sweden in the form of wooded pastures (Ihse and Skånes 2008). This is currently debated within the EU, since the pastures of most other parts of Europe are by tradition much more open. It is also a primary reason for the ambiguity in the classification of forest versus pasture. However, in environmental monitoring and mapping, where data collection should be independent of administrative systems, the prerequisites are rather different. The actual use of the land may change during one season, and sometimes it is the intention of the farmer that defines the use. This may be difficult to infer by an independent observer, especially from remote sensing. In this case, the suitability for grazing or cultivation for crops is rather more straightforward than the actual momentary use.

8.4 Links to FAO's Land Cover Classification System

It is no coincidence that the structure of the suggested land type classification is similar to the structure of the LCCS (Di Gregorio 2005). An initial, strictly defined hierarchical classification followed by a large number of optional attributes is an effective way to assure transparency, comparability, and operability, and at the same time allow maximal flexibility in mapping and monitoring (Di Gregorio 2005; Ahlqvist 2008; Jepsen and Levin 2013). To this, we have added some criteria to increase the usefulness and comparability with the Swedish official *ownership types*, which mainly concern suitability for forestry and agriculture (Table 8.2). We have included some aspects of actual land use, to allow for the fact that the Swedish ownership types defined by each sector are inconsistent and overlapping in this respect. For the comparability with Swedish ownership types, we have added some criteria concerning the suitability for or the influence by forestry and agriculture according to Swedish practices and regulation, with a functionally motivated tree cover threshold as one criterion.

Concerning the hierarchical module of LCCS and the suggested land type classification, the overall logic is almost identical, as are the definitions of artificial/constructed land (with arable land included or as a separate class), and semiaquatic/flooded land. The difference in view may be a consequence of LCCS's broader scope as an international system, but the arguments valid from a Swedish point of view may also be useful in other contexts. For example, our emphasis in this chapter on functional criteria as a main basis for classification applies also on vegetated vs. nonvegetated land (Table 8.4). We have chosen to include only rocky outcrops with shallow soil layer as a separate subtype, because such rocky outcrops are permanent features, mostly defined by their lack of soil. Most other types of nonvegetated land must be kept nonvegetated by repeated disturbance, either by man, by wild or domesticated animals, or by some natural disturbance. Otherwise they will soon be colonized by denser vegetation. Our conclusion is that a certain land area can be nonvegetated for a large number of reasons, and different types should therefore be treated in different contexts, not as one group separate from vegetated areas.

With the addition of some simple structural land cover attributes to our land type system and some simple secondary land use attributes to the LCCS, the two systems are compatible, at least for use in a Swedish context. Regarding scale-independence, the functional criteria we advocate should put the limits for what resolution for mapping and monitoring is appropriate. For example, the suitability for forestry and agriculture is not dependent

TABLE 8.4

Comparison between LCCS Classes and Our Land Types at the First Hierarchical
Levels

LCCS-1	LCCS-2	LCCS-3	Main Land Types
Primarily vegetated area	Terrestrial	Cultivated and managed terrestrial area	Arable land and former arable land
			Constructed or artificial land
		Natural and seminatural terrestrial area	Seminatural pasture or meadow
			Land with closed forest or active forestry
			Other terrestrial land with human or natural disturbance
	Aquatic or regularly flooded	Cultivated aquatic or regularly flooded	Semiaquatic land without closed forest
			Aquatic area
			Constructed or artificial land
		Natural and seminatural aquatic or regularly flooded	Semiaquatic land without closed forest
			Aquatic area
			Land with closed forest or active forestry
Primarily nonvegetated area	Terrestrial	Artificial surface and associated areas	Constructed or artificial land
		Bare area	Other terrestrial land with human or natural disturbance
	Aquatic or regularly flooded	Artificial waterbodies, snow and ice	Constructed or artificial land
		Natural waterbodies, snow and ice	Aquatic area

Source: Di Gregorio, A, *Land Cover Classification System (LCCS), version 2: Classification Concepts and User Manual.* FAO Environment and Natural Resources Service Series, No. 8—FAO, Rome, 2005.

on the technical limitations of the independent observer, but on the preconditions for the land owner or manager to use the land for a certain purpose, or for other comparable functions to be applicable.

8.5 Implications for Land Use and Land Cover Semantics and for Future Data Collection

We should acknowledge that human activities always interact with the natural processes and other environmental conditions, and that the land surface in itself determines what cover and use can be expected (Cihlar and

Jansen 2001). It is exactly these properties that make such land types suitable for manual interpretation and mapping, making best possible use of the strengths of the human eye and the human brain (Ihse 2007), also taking into account the target object's representation in the source materials used, especially if comparison is done over time from different sources such as maps, aerial photographs, and satellite imagery (Käyhkö and Skånes 2006; Jepsen and Levin 2013). The land types according to our suggestion define the limits for what structural land cover and secondary land use can be expected at a certain site, allowing multipurpose use of a complex dataset with additional structural land cover and secondary land use information as a flexible set of variables, within the frames of a unifying, nominal land type classification. Furthermore, we would like to emphasize that secondary land use and structural land cover manifest themselves in different ways in the landscape and require different methods for data collection.

What can be observed or extracted from remote sensing or other automated measuring devices is structural land cover, and anything other than that must be based on human interpretation or evaluation of natural or human processes. Maybe the common inability to study land use separately from land cover depends partly on the common focus on automated remote sensing as the major data source for mapping (Comber, Wadsworth, and Fisher 2008). In satellite imagery, it is only the apparent physical, structural, or biological conditions at a certain point in time that can be extracted, and the relation to the purpose, activities, or utilization of the area is at best indirect. Thus, the information delivered is too often not determined by the need and the objectives, but relies on technical abilities in data collection, perhaps combined with a certain degree of naivety in planning, interpretation, and use, following arguments from Comber, Fisher, and Wadsworth (2005). The normal procedure is then to improve automatic classification using masks and ancillary data, including polygon information on more land use-related information from other data sources that are in their turn often produced with manual aerial photo interpretation (Jepsen and Levin 2013). The apparent objectivity of automated remote sensing is therefore often quite deceptive.

Examples of land cover categories in different regions show that threshold values based on simple cover values, such as canopy cover, are precarious and potentially confusing (Comber, Fisher, and Wadsworth 2005). Since metadata are typically vague and inconsistent, it is not easy for the user of a map to know the intentions of the producer of the map (Jepsen and Levin 2013, Björk and Skånes 2015, Chapter 3). This problem is discussed by Ahlqvist (2008), highlighting the need for a firmer connection between the conceptual definitions of classes, the group perception of the definitions when creating or using geographical data, and the actual real-world representation of the classified features. This is indicated by the treatment of land cover and land use at the same time as simple categories and as complex multidimensional gradients, as highlighted in the description of *fuzzy categories* (Ahlqvist 2004).

The scientific debate on the distinction between land cover and land use tends to assume that these two concepts are sufficient to describe the state and condition of all land area. However, if land cover is strictly defined as physical structures and land use as human activities, then the natural processes and the physical environmental conditions creating the prerequisites for land use and land cover have no obvious place in the classification system. Such narrow definitions of land cover and land use would justify the distinction of *land type* as a third, complementary concept. The processes forming the shape, content, and function of the earth's surface are an intricate mix of human activities and natural processes. To understand the relationship between these aspects and the physical representation of land cover, it is important to acknowledge the dynamics, not only in the semantics but in the development of the landscape itself (Skånes 1997; Käyhkö and Skånes 2006).

8.6 Concluding Remarks

The identification of which aspects of land cover and land use are truly qualitative and useful for a functional classification of land areas, as summarized by us in the concept *land type*, would, we hope, make the definition of land use and land cover more straightforward and easier to apply and combine in a semantically conscious framework. We believe that the common focus on remote sensing techniques alone partly diverts attention from properties of the landscape that are qualitative and functional, or that the feasibility of automated methods to include such properties is greatly overestimated.

By confining the use of strictly defined hierarchical classes to qualitative and functionally motivated properties, and automated methods to quantitative and structural properties, the best synergies and the best use of both automated remote sensing and human visual interpretation can be achieved, to the benefit of both, and at the same time efficiency and reliability are maximized.

Although our examples in this chapter are focused on Swedish conditions, we argue that this need for coherent and well-defined criteria is valid also in an international context. All countries have their own operative semantics and definitions of common classifications of land use and land cover and need to relate these to international systems such as the FAO Land Cover Classification System, LCCS (Di Gregorio 2005; Björk and Skånes 2015, Chapter 3).

References

Ahlqvist, O. 2004. "A parameterized representation of uncertain conceptual spaces." *Transactions in GIS* 8:493–514.

Ahlqvist, O. 2008. "In search of classification that supports the dynamics of science: The FAO Land Cover Classification System and proposed modifications." *Environment and Planning B: Planning and Design* 35:169–186.

Bakker, M.M. and A. Veldkamp. 2008. "Modelling land change: the issue of use and cover in wide-scale applications." *Journal of Land Use Science* 3:203–213.

Cihlar, J. and L.J.M Jansen. 2001. "From land cover to land use: a methodology for efficient land use mapping over large areas." *Professional Geographer* 53:275–289.

Comber, A., P. Fisher, and R. Wadsworth. 2005. "What is land cover?" *Environment and Planning B: Planning and Design* 32:199–209.

Comber, A., R. Wadsworth, and P. Fisher. 2008. "Using semantics to clarify the conceptual confusion between land cover and land use: the example of 'forest'." *Journal of Land Use Science* 3:185–198.

Davies, C.E., D. Moss, and M.O. Hill. 2004. *EUNIS Habitat Classification Revised 2004*. European Environment Agency and European Topic Centre on Nature Protection and Biodiversity. Paris.

Di Gregorio, A. 2005. *Land Cover Classification System (LCCS), version 2: Classification Concepts and User Manual*. FAO Environment and Natural Resources Service Series, No. 8—FAO, Rome.

Feng, C.-C. and D.M. Flewelling. 2004. "Assessment of semantic similarity and land use/land cover classification systems." *Computers, Environment and Urban Systems* 28:229–246.

Halvorsen, R., T. Andersen, H.H. Blom, A. Elvebakk, R. Elven, L. Erikstad, G. Gaarder, A. Moen, P.B. Mortensen, A. Norderhaug, K. Nygaard, T. Thorsnes, and F. Ødegaard. 2009. *Naturtyper i Norge: Teoretisk grunnlag, prinsipper for inndeling og definisjoner* [Nature types in Norway: Theoretical foundation, principles for classification and definitions]. Naturtyper i Norge versjon 1.0 Artikkel 1:1–210. Oslo. [in Norwegian]

Ihse, M. 2007. "Colour infrared aerial photography as a tool for vegetation mapping and change detection in environmental studies of Nordic ecosystems: A review." *Norwegian Journal of Geography* 61:170–191.

Ihse, M. and H. Skånes. 2008. "The Swedish agropastoral *Hagmark* landscape: An approach to integrated landscape analysis." In *Nordic Landscapes: Region and Belonging on the Northern Edge of Europe*, edited by Jones, M. and Olwig, K., 251–280. Minneapolis, MN: University of Minnesota Press.

Jepsen, M.R. and G. Levin. 2013. "Semantically based reclassification of Danish land-use and land-cover information." *International Journal of Geographical Information Science* 27:2375–2390.

Käyhkö, N. and H. Skånes. 2006. "Change trajectories and key biotopes: Assessing landscape dynamics and sustainability." *Landscape and Urban Planning* 75:300–321.

Normander, B., G. Levin, A.-P. Auvinen, H. Bratli, O. Stabbetorp, M. Hedblom, A. Glimskär, and G.A. Gudmundsson. 2012. "Indicator framework for measuring quantity and quality of biodiversity: Exemplified in the Nordic countries." *Ecological Indicators* 13:104–116.

Rafstedt, T. and L. Andersson. 1981. *Flygbildstolkning av myrvegetation. En metodstudie för översiktlig kartering* [Aerial-photo interpretation of mire vegetation. A study of methods for overview mapping]. Naturvårdsverket [Swedish Environmental Protection Agency], SNV PM 1433. Solna, Sweden. [in Swedish with English summary]

Ramsar Convention 2002. *Guidance for identifying and designating peatlands, wet grasslands, mangroves and coral reefs as Wetlands of International Importance.* Ramsar COP8 Resolution VIII.11, Valencia, Spain.

Rydin, H. and J. Jeglum. 2006. *The Biology of Peatlands.* Oxford, UK: Oxford University Press.

SLU 2011. *Fältinstruktion 2011.* RIS, Riksinventeringen av skog. [Field instruction 2011, National Forest Inventory]. SLU, Institutionen för skoglig resurshushållning [Swedish University of Agricultural Sciences, Department of Forest Resource Management]. Umeå, Sweden. [in Swedish]

Skånes, H., 1997. "Towards an integrated ecological-geographical landscape perspective: A review of principal concepts and methods." *Norsk geografisk Tidsskrift* 51:146–171.

Skånes, H., A. Glimskär, and A. Allard. 2011. *Visual interpretation of vegetation characteristics in laser data.* EMMA T1 initial report. SLU, Department of Forest Resource Management, Report 314. Umeå, Sweden.

Statistics Sweden 1981. *Svensk standard för ägoslagsklassificering av mark för jordbruk och skogsbruk* [Swedish standard classification of land use in agriculture and forestry]. Statistiska Centralbyrån, Meddelanden i samordningsfrågor [Statistics Sweden, Reports on Statistical Co-ordination] 1981:4. Stockholm, Sweden. [in Swedish]

Statistics Sweden 2013. *Markanvändningen i Sverige* [Land use in Sweden], sixth edition. Statistiska Centralbyrån, Sveriges officiella statistik [Statistics Sweden, Official Statistics of Sweden]. Stockholm, Sweden. [in Swedish with English summary]

Ståhl, G., A. Allard, P.-A. Esseen, A. Glimskär, A. Ringvall, J. Svensson, S. Sundquist, P. Christensen, Å. Gallegos Torell, M. Högström, K. Lagerqvist, L. Marklund, B. Nilsson, and O. Inghe. 2011. "National Inventory of Landscapes in Sweden (NILS): Scope, design, and experiences from establishing a multiscale biodiversity monitoring system." *Environmental Monitoring and Assessment* 173:579–595.

Svensson, S.A. 2006. *Ägoslag i skogen. Förslag till indelning, begrepp och definitioner för skogsrelaterade ägoslag* [Ownership types in the forest. A suggestion of classification, terms and definitions for forest-related ownership types]. Skogsstyrelsen [Swedish Forest Agency], Rapport 20:2006. Jönköping, Sweden. [in Swedish]

9

Text Mining Analysis of Land Cover Semantic Overlap

Alexis Comber, Peter Fisher, and Richard Wadsworth

CONTENTS

ABSTRACT This chapter explores the origins and impacts of semantic variation using the example of land cover. It examines the origins of semantic variation in land cover mapping and the philosophical process of categorization, comparing what might be called top-down and bottom-up nomenclatures. In doing so, it illustrates that land cover classifications, as with many geographical concepts, are vague, imprecise, and socially constructed: they represent a coming together of a particular world view *(weltanschauung)*. The context of ubiquitous representation in digital cartographic products and the origins of semantic variation in land cover mapping are examined. This chapter examines the origins of semantic variation in land cover mapping. The chapter considers the influences of specific factors on geographic representation and the need to abstract the infinite complexity of the real world into spatial databases. In so doing, it describes how the need for generalization processes such as abstraction and aggregation are a series of choices. Choices are shown to relate to the commissioning and policy background (who paid for it?), observer variation (what did you see?), institutional variation (why you see it that way?), and variation in measurement variation (how was it recorded?). This problem is tackled using a raw text mining approach that seeks to characterize the semantic overlap between semantically discordant datasets and then to integrate in a land cover change. The results of the text mining are compared with human experts and shown to be more efficient at characterizing the consistency between two land cover maps in the context of land cover change and semantic discordance. A number of critical research areas are identified relating to dynamic metadata and formal ontologies, linguistic issues, and different semantic latency approaches.

KEY WORDS: *Semantics, Probabilistic Latent Semantic Analysis (PLSA), Latent Dirichlet Allocation (LDA).*

9.1 Introduction

The map is an incredibly powerful object, as it describes and formalizes "what is where." Maps represent real-world features and implicitly describe their spatial properties. Their interpretation is supported by the usual cartographic adjuncts: legends, scale bars, north arrows, and so on. Maps are at least superficially understood by most people, regardless of their background or training.* For these reasons, the digital map has become the background noise that has accompanied the shift from a desk-based personal computer–moderated interaction with the Internet, online communities, and applications, to interactions facilitated through GPS-enabled mobile devices, such as smartphones and tablets. The ubiquity of maps, of what those in specialist research communities might call "spatial data," is both a blessing and a curse: a blessing because the outputs of research activities are increasingly accessed by and accessible to a wider public and are increasingly salient to other areas of research. This trend for explicitly spatial analyses will only continue as more and more of the data that are collected in all areas of scientific activity have location attached to them. However, the wider availability of and familiarity with maps, whether in digital or paper format, may also be a curse: those able to easily download and then use the map (or analyze the spatial data) may not fully understand what is recorded by the data they are using, precisely *what* is being represented by the data, the concepts and the world view (*weltanschauung*) that are embedded in the data. This is an important issue as the nature of what is recorded *there* will vary as a result of variations in the semantics and conceptualizations of data creation.

This chapter explores the origins and impacts of semantic variation using the example of land cover. Through a text mining analysis it seeks to determine the degree of overlap concepts associated with classes from divergent land cover classifications and to infer land cover change. It does this by creating a

---

* This is in contrast to Pickles (1995) and others who have long argued that maps are imbued with power and meaning such that map reading (whether poor or skilled) can promote unbalanced power relations, and we note that in this context *not* understanding a map can be in the interest of the mapmaker.

list of *terms*—words used to describe each land cover class—and then applies an explicit measure of overlap inspired by Bouchon-Meunier et al. (1996) for non-ordered qualitative domains. Class-to-class relations are labeled as *expected unexpected* (i.e., land cover change), or *plausible*, and the overlaps generated by text mining are compared with those described by three experts, familiar with both datasets and using the same three-valued logic.

In this way the analysis seeks to explore a generic method that is capable of integrating semantically divergent data about the same phenomenon but without using formal (and useless), standard compliant metadata.

The remainder of this chapter has three sections. The first of these sections briefly examines the philosophical process of categorization and compares what might be called top-down and bottom-up nomenclatures and links these to current approaches to land cover mapping. The second develops a semantic analysis that integrates semantically divergent land cover data to identify land cover change. It does this through a text mining analysis of the class descriptions: lists of terms used in the class descriptions and their frequency of use is created, and then these are analyzed using a measure of overlap for non-ordered qualitative domains. In so doing, this research proposes an approaching for addressing semantically divergent classifications. The last section discusses the results and makes some concluding comments.

9.2 Land Cover Classification

Land cover data describes observable features on the earth's surface, although many land cover datasets also include land use categories for a number of well-documented reasons (Fisher et al. 2005; Comber 2008a,b), and this problem will not be considered here. Common land cover categories include "forest," "water," "artificial surfaces," "bare ground," "tundra," "scrub," and so on, and it is obvious from this short listing that these categories or classes are aggregations of individual features: the forest class will contain different types of trees, the artificial surface class will include buildings, roads, and other infrastructure, and so on.

Land cover data are typically captured either through field survey or by analysis of remotely sensed data. In both cases, the land cover data are constructed by allocating observations and measurements of individual features to specific classes according to a *model* that defines and specifies the characteristics of the land cover classification. In vegetation surveys, every plant growing within a defined quadrat is usually identified (Kent and Coker 1992) and the observations and measurements are eventually used to identify the class of habitat or vegetation community present at a particular location and to produce a map. In remote sensing analyses, pixels of similar spectral properties are grouped or statistically clustered into classes.

Historically, much time and effort was spent preparing written descriptions of the model, the classes (mapping units), and how the classes were or could be derived (keys and schemas). These descriptions were then published in a survey memoir or report. Fisher (2003) has argued that now the map (or spatial data) is frequently seen as the principal information product of the survey when originally it was just one window onto the wealth of information contained in the report.

Thus, the process of classifying land cover is complex, as it depends on the interactions among a number of different factors: the objectives of the analysis or study, the spatial scale or geographic scope of the analysis, the granularity or scale of the processes being observed, and the methods (including data choices) used to map land cover. Simply, land cover classifications do not just "drop out" of observation and measurement (e.g., through remote sensing) but are the result of many processes and decisions made during the creation of land cover data (Comber et al 2005d). Often, these processes are unreported in the data descriptions or metadata,†* but nonetheless have a critical influence over the final data specification.

The process of classification and the allocation of class labels to individual features are fundamental to how people view the world. In land cover classification, two general approaches are used to assign features to categories:

1. A clustering process based on the "closeness" of the feature to other features. This is a bottom-up approach.
2. A characteristics-matching process based on the attributes of the feature to a set of predefined classes. This is a top-down approach.

Research in cognition (Rosch 1978) has shown that in the first case, people compare features to *prototypes* or good examples of a category and the feature is assigned to the category that has the closest prototype. Precisely, what constitutes a good prototype depends on the background of the person. In the second case, a feature belongs to a category when it has all the required characteristics specified in the model as described above. This is the more common situation in land cover classification (Comber et al. 2005c).

The preceding discussion suggests that to generate land cover classifications, there is a need to first identify the kinds of land cover that are of interest based on the objectives of the analysis; second, to decide how to divide what Gardenfors (2000) and Ahlqvist (2004) describe as the *conceptual space* to separate that reality into categories; and, finally, to identify the feature properties that relate to the analysis objectives, conceptual space, and the categories.

Indeed, *categories of objects*, *types of things*, and *classes* are fundamental to how people view the world (Rosch 1975a,b; Rosch and Lloyd 1978). The

---

* And are therefore difficult to formally consider in ontology matching approaches—a point returned to later.

critical issue in relation to land cover categories is that they are deeply connected to the physical space that they occupy but also to the nature of their conceptualization. Consequently the discretization of real world (surface) features into land cover categories can be problematic for a number of well-documented reasons:

- Land cover classes frequently do not have boundaries that correspond to physical discontinuities in the world (Burrough 1986; Burrough and Frank 1996; Smith 1995, 2001; Smith and Mark 2001).
- They exist at particular spatial scales (Fisher 1997; Fisher et al. 2005) relating to the granularity of the class and the features it contains.
- They can depend on the interaction among human perception, spatial arrangement, and properties or characteristics (Smith and Mark 2001).

Thus, the boundaries between classes and what they contain will differ from classification to classification, from culture to culture, and from survey to survey. People from different backgrounds, with different training and different experiences, will conceptualize and categorize landscape features in different ways (Comber et al. 2005b, 2008), creating problems for data integration activities. This has been demonstrated in numerous ethnophysiography studies of land cover* (Derungs et al. 2013; Mark and Turk 2003) and in studies that consider spatial data semantics (Edwardes and Purves 2007; Kuhn 2005), and such variations have been linked to categorization processes (Lakoff 1987; Comber et al. 2005c) and to linguistic and cultural factors (Smith and Mark 1998).

The fundamental problem is that the embedded data semantics and meaning are hidden to potential users of that data (Comber et al. 2008) and not described in metadata. This is what Varzi (2001) refers to the *double-barreled* nature of geographic entities being intimately connected to the space that they occupy and to the manner of their human conceptualization.

The technical and organizational origins of the variations in *what* is recorded *where* in land cover data therefore relate to a series of choices. These are driven by a range of factors: the commissioning and policy background (who paid for it?), observer variation (what did you see?), institutional variation (why you see it that way?), and variation in measurement variation (how was it recorded?). The semantics of any land cover dataset reflect, therefore, a combination of choices, and land cover classifications are vague, imprecise, and socially constructed (as are many geographical concepts): they represent a coming together of a particular world view (*weltanschauung*) with the need to represent and abstract the real world within digital cartographic products.

---

* Although we note that many empirical studies of the concepts held by indigenous people indicate that land use is often more prominently associated with landscape form and properties than to land cover (Mark et al. 2011).

The impacts of such variations can be profound: the ready supply of spatial data, including land cover, encourages the user (sometimes unwittingly) to assume that their conceptual model of the world matches that of the data (Comber et al. 2005d). However, tags, categories, labels, classes, and so on depend on the interactions of human perception and spatial properties (Fisher et al. 2009). Thus, there is the potential for mismatches between the *prototypes* held by the user and similarly labeled land cover classes, as the user may assume that the terms used to describe data classes match his or her conceptions.

The result of this situation is that potential data users are left in the paradoxical situation that on the one hand they have easier access to more data than ever before, but on the other hand they know less about the meaning behind that data in the absence of survey memoirs, monographs, and adequate metadata (Comber et al. 2008). Consider a simple land cover term such as "forest." Almost everyone can provide a prototypical description of a forest; however, some people will consider a forest to be a land use and some a land cover, others will consider it to include specific species and to exclude others, to have different minimum spatial properties and tree densities, and so on. Lund (2002, 2014) lists many national and international definitions of the term "forest," and Figure 9.1 shows that there is huge variation in threshold values of tree height and percentage cover used in the definitions of "forest" in different countries (see also Comber et al. 2005c). Indeed, different parts of the same organization and

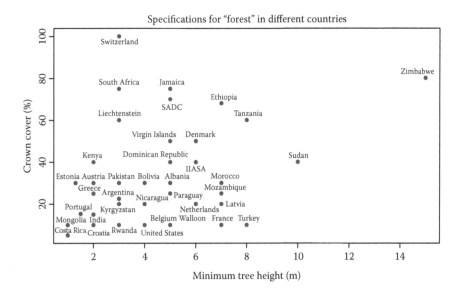

FIGURE 9.1
Variation in the concept of "forest" illustrated through minimum tree height and canopy cover.

multiple organizations in any one country use profoundly different definitions. This variation has many causes but is exacerbated by the fact that "forest" is a term familiar to most people, and without the full definition and model of conceptualization of the category, users will interpret the class from their own preconceptions.

9.3 Text Mining to Integrate Varying Land Cover Semantics

9.3.1 Introduction

Other work considering varying land cover semantics has applied measures of class to class semantic similarity based on expert opinion (Comber et al. 2004a,b, 2005). These have been used to generate measures of semantic overlap and to make inferences about land cover change and relative error when the data are from different time periods, or simply relative error when the data are contemporaneous. There are a number of problems with using experts: they need to spend time thinking about relationships between classes, they have different opinions, they may change their minds, and their reasoning is often opaque, with the result that it is difficult to know why they describe a particular class-to-class relationship in the way that they do. For these reasons, research by the authors has examined automated approaches based on text mining of land cover semantics (Wadsworth et al. 2005, 2006, 2008).

Text mining searches for patterns in unstructured texts; data mining searches for patterns in more organized data (Witten and Frank 2005; Miller and Han 2001), and an extensive summary of the field is presented by Feldman and Sanger (2007). In addition to classifying documents, text mining has been used to identify keywords in documents (Feldman et al. 1998; Nasukawa and Nagano 2001) and for extracting ontologies (Maedche and Staab 2000). Alternative approaches include Natural Language Processing (NLP), by which human-readable text is converted into a form more amenable to computer processing. But, despite a considerable amount of research, the automated application of NLP is a difficult problem because in reality it requires a knowledge of the world outside the text being processed. Central to that problem of NLP is the fact that a single word can have many meanings, depending on context (think of puns, double entrendes, newspaper headlines, etc.), and there are different words that have the same spelling (heteronyms and heterophones); for example, a "figure" might refer to a person, a picture, a number, an idea, and so on. Words can be synonyms in some contexts but not others—for example, "ground" and "land" are often synonymous, but in the two phrases "I spent a few hours on the ground" and "I spent a few hours on land," the first suggests travel by airplane and the second by ship. Because of the complexity of NLP, a number of simpler, more pragmatic approaches have been proposed to categorize documents. By far the most widely used are those employed by Internet search engines to rank millions of documents in terms of their relevance to the phrase or keywords entered by the user. A number of other approaches such as self-organizing feature maps (SOM) (Kohonen 1982, 1995) have been used for categorizing texts as diverse as fairy tales (Honkela et al. 1995) and patent applications (Kohonen et al. 2000). More recently, latent approaches have been explored, but these have their limitations, as discussed below.

9.3.2 Data and Methods

When comparing categories, we should be able to embody the idea that one category is a partial subset or overlaps with another and that the categories can form imperfect hierarchies. Consider two classes: Class A contains the members {α, β, γ, δ, ε} and Class B the members {δ, ε, ζ}; if we use overlap as a measure of relatedness then there is little problem with the idea that Class B has more in common with Class A (2 out of 3) than Class A has with Class B (2 out of 5). The text mining approach described below uses a similar logic.

Text mining was applied to two discordant land cover datasets in an attempt to reconcile their different semantics, meaning, and conceptualizations embodied in the different classifications. Two digital land cover maps of Great Britain were used as a case study. LCM1990 distinguishes 25 "target classes" reported using a 25-m pixel (CEH 2014), and LCM2000 describes 27 "broad habitats" and uses an object-based structure (Fuller et al. 2002). The LCM2000 was issued with a caveat: "NB. The LCM2000 raster dataset is not directly comparable with the LCMGB1990 dataset (i.e., LCM1990), as it has been constructed by different methods. It is 'not' suitable for estimating change over the 10-year period."* The length of the description of the classes in both LCM1990 and LCM2000 is approximately the same (2869 and 2775 words, respectively), representing descriptions that average just over 100 words per category.

Example descriptions from LCM2000 and LCM1990 are shown in Table 9.1 for the classes of Dwarf Shrub Heath (LCM2000) and Shrub Heath (LCM1990). There are clear differences: the 1990 class is much more strongly defined on species, whereas the LCM2007 class is defined in relation to soil and landform, and these differences are exacerbated by the pixel classification in LCM1990 and the object-based classification of LCM2000.

---

* http://www.ceh.ac.uk/documents/lcm2000_product_versions_and_formats.pdf

TABLE 9.1

Examples of the Different Class Descriptions in LCM2007 (Top) and LCM1990 (Bottom) for Shrub Heaths

---

Dwarf Shrub Heath

Dwarf Shrub Heath is characterized by vegetation that has >25% cover of plant species from the heath family (*ericoids*) or dwarf gorse *Ulex minor*. It generally occurs on well-drained, nutrient-poor, acid soils. This habitat type does not include dwarf shrub–dominated vegetation in which species characteristic of peat-forming vegetation such as cotton-grass *Eriophorum spp.* and peat-building *sphagnum* are abundant, or that occurs on deep peat (>0.5 m) as these are included in the "Bog" Broad Habitat type.

Shrub Heath

In the 25 class dataset dense shrub heath and dense shrub moor are kept separate. In the 17 class data they are aggregated into one class.

Dense Shrub Heath: Dense shrub heath refers to communities with high contents of heather (*Calluna*) and ling (*Erica spp.*) but perhaps mixed with broom (*Cytisus scoparius*) and gorse (*Ulex spp.*). It is mostly evergreen, hence different from other scrub communities. Almost invariably, it represents vegetation on sandy soils, in characteristic sites such as the Brecklands, the Dorset and Surrey Heaths, or on extensive coastal dune systems. Fuller key-name: lowland evergreen shrub–dominated heathland. This category carries the label "13" in the 25 "target"-class dataset.

Dense Shrub Moor: The dense shrub moor communities include heather (*Calluna vulgaris*), ling (*Erica spp.*), and bilberry (*Vaccinium spp.*) moorlands. Though dominated by woody shrubs, these may be mixed with herbaceous species, especially those of the moorland grass. The dense shrub moors may be managed by moor burning, in which case they may be bare, for most of the first year after burning; then the grass/shrub heath mixture is found until dense shrub growth again dominates the cover. Fuller key-name: upland evergreen dwarf shrub–dominated moorland.

This category carries the label "11" in the 25 "target"-class dataset.

The methods build on those introduced by Wadsworth et al. (2006, 2008) and the initial stages remain similar to Lin (1997) and Honkela (1997). The stages of the analysis were as follows:

1. Each class description was converted into a list of *terms*. In most cases, the terms are single words but some are phrases. Phrases are concerned with measurements (e.g., "25 m"), geographic locations (e.g., "North York Moors"), and species names (e.g., *Nardus stricta*); these were identified to avoid "double counting;" if two descriptions refer to "North York Moors," then they should count as one term in common, not three. Similarly, if one description referred to "North Africa" and the other to the "North York Moors," it should not be counted as a similarity just because both refer to the "north." The species name of an organism was treated as a single term because related species may have different habits and habitats; for example, *Saxifraga saxifrage* (lowland, moist grassland) and *Saxifraga oppositifolia* (alpine, limestone) occupy different habitats, as does the unrelated *Pimpinella saxifraga* (dry stony grassland). Geographic and species synonyms are

easily identified by inspection, tempting the user to try and find other synonyms, but to do so raises the question of whether the resulting semantic similarity is that between the classes or between the readers' interpretation of what constitutes a synonym. Terms were automatically stemmed to group those with same etymological roots.

2. A matrix of classes versus terms was constructed with the cells in the matrix containing the number of times each term appears in each description. The matrices are sparse because word frequency is very skewed. For example, about 5% of the descriptions consists of the word "the" (minimum 4% for Land Cover and a maximum of 6.6% Soil Taxonomy), more than half (53%) of the terms in the land cover description occur only once, and 40% of the Soil Taxonomy consists of unique terms.

3. Each term in the matrix was weighted so that the unequal length of the class descriptions did not bias the similarity measures generated below. Total frequency times inverse document frequency scheme (*tf. idf*) is the most widely used scheme in information retrieval. It provides a measure of how important a term is to a document. It weights terms such that a term appearing frequently in one short document and nowhere else receives a high weight, and a term appearing in all documents receives a zero weight. Other, less widely used alternatives are discussed by Robertson (2004), but the alternative text weighting schemas that we tried do not generate substantively different results. Formally *tf.idf* is

$$w_{i,a} = \frac{n_i}{L} \ln \frac{D}{n_j} \tag{9.1}$$

where $w_{i,a}$ is the weight of the *i*th word in Class A; n_i is the number of times the word appears in the description of A; L is the length of the description, that is, total number of words describing Class A; D is the total number of classes; and n_j is the number of classes containing the *i*th word.

4. The weighted matrix of terms by classes is used to calculate the similarity between classes. Rodriguez and Egenhofer (2004) used a measure of distance between classes that is symmetric (i.e., Class A is as far from Class B, as Class B is from Class A). However, symmetric measures of similarity fail to characterize partial subsets adequately and so an explicit measure of overlap inspired by Bouchon-Meunier et al. (1996) for non-ordered qualitative domains was applied:

$$O(A, B) = \frac{\sum_i \min(w_{i,a} w_{i,b})}{\sum_i w_{i,a}} \tag{9.2}$$

where $O(A,B)$ is the overlap between Categories A and B and $w_{i,a}$ and $w_{i,b}$ are the weights of the term i in the descriptions of Classes A and B. $O(A,B)$ can vary from 1, when A is a perfect subset of B, to zero, when there is no overlap (no terms occur in both A and B). Consider the two classes introduced earlier, Class A = {α, β, γ, δ, ε} and Class B = {δ, ε, ζ}; for simplicity assume each term (α...ζ) has a weight of 1, then $O(A,B) = 2/5 = 0.4$ and $O(B,A) = 2/3 = 0.67$. That is, B has more in common with A, than A has with B. Note that if the entire "corpus" class descriptions contained just these eight occurrences of the six terms, then from Equation 9.1, δ and ε would both have a weight of zero as they occur in all classes, and so $O(A,B) = 0$ and $O(B,A) = 0$.

5. The measures of overlaps between classes are generated automatically—without expert opinion or without extensive manipulation of the text (some stemming and some manipulation of geographic terms as described above). The overlap calculated in this way is stored in an n by m matrix (where n and m are the number of classes in the two datasets being compared). The overlap weights are then applied to a spatial intersection of the datasets and the weights used to make inferences about change and uncertainty, representing the semantic inconsistency between the two classes.

9.3.3 Results

Comber et al. (2004b, 2005b) developed a statistical–semantic approach for estimating the consistency between the two land cover maps. Three experts—a land cover data producer, a land cover data distributor, and a land cover data user—familiar with both datasets considered the relationships between each pair of LCM1990 and LCM2000 classes. They expressed the relationship as being *expected, uncertain,* or *unexpected* in a lookup table. The tables were used to determine the extent to which the class in 2000 was supported by the class in 1990 without the need to reclassify or thematically aggregate either data set. For each LCM2000 segment, the intersecting pixels from LCM1990 were labeled *expected, unexpected,* or *uncertain,* using the expert tables. The scores were summed over each segment and normalized so that the expected, uncertain, and unexpected scores summed to one. These values were then treated as if they were measures of belief and combined using the Dempster–Shafer theory of the evidence (Dempster 1967; Shafer 1976).

The text mining procedure was to generate a further lookup table. The semantic overlaps between classes in both maps were calculated. These values were then transformed into three discreet categories—*expected, uncertain,* or *unexpected*—for comparison with the scores generated by the tables created by the human experts. Although text overlaps have a theoretical range [0, 1] for the LCM1990 and LCM2000 descriptions, the average overlap was only 0.05 (range 0.001–0.236), and rather skewed (skew = 1.95). The overlaps were low because of the down-weighting of common words and the shortness of the descriptions. Overlap values that were less than the average were labeled as *unexpected*, and values greater than average as *plausible*. The *plausible* results are *uncertain* in the range of the mean to plus-one standard deviation and *expected* above that value (mean and standard deviation calculated after taking the natural logarithm of the values). Thus each LCM2000 segment was labeled as *expected, unexpected,* or *plausible* by evaluating the coincident LCM1990 pixels using the lookup table generated through text mining. The lookup table derived from text mining is shown in Table 9.1.

The inconsistencies (uncertainties) identified by text mining were compared with those generated through the application of human expert lookup tables, generated by a land cover data producer, a data user, and a data distributer. Figure 9.2 shows the different amounts of *uncertainty* in land cover parcels as identified by the three human experts and the text mining. Interestingly the Producer does not express and uncertainty in the overlaps between the classes and the text mining produces similar uncertainties to the User.

The unexpected scores were used to identify LCM2000 parcels that were likely to have changed (or been erroneously classified). Parcels with an *unexpected* score of more than 0.9 are highlighted in Figure 9.3. This shows the spatial distributions of highly inconsistent parcels. From Figures 9.2 and 9.3

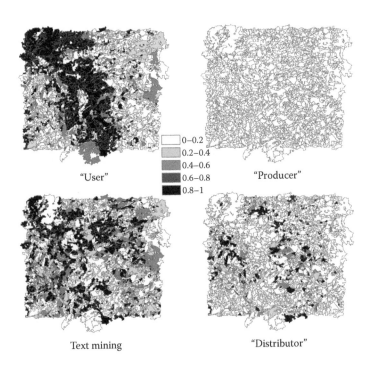

FIGURE 9.2
Different amounts of uncertainty in LCM2000 parcels as calculated through the different expert lookup tables based on the intersecting LCM1990 pixels.

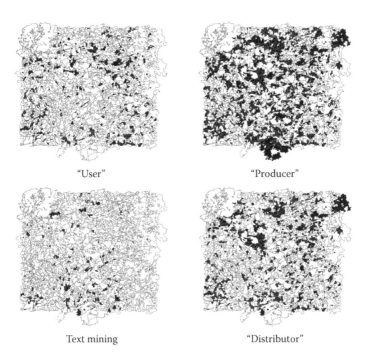

FIGURE 9.3
Highly inconsistent parcels (those with an unexpected score of greater than 0.9) as identified by the different experts and text mining.

it is evident that the different experts have very different ideas about how the datasets relate and how they represent different ideas about land cover mapping and landscape processes; these issues are described in detail by Comber et al. (2005b).

A validation exercise involved visiting 343 parcels in a single 100 × 100-km square in England in 2002. Each parcel was visited and categorized as "changed" or "not changed," and "correctly classified" (in both maps) or "erroneous" (in either or both maps) (Comber et al. 2004b, 2005b). Analysis of the field visits allowed the inconsistencies to be parameterized into proportions of *error* (a misclassification in either map) and *change*. Of the parcels visited in the field, 186 of the *consistent* parcels had not changed, whereas six *inconsistent* parcels had. Table 9.3 shows the proportions of parcels correctly labeled by the analyses using the lookup tables generated by each of the experts and text mining. The Producer and Distributor experts perform less well than the text mining approach in identifying change and no-change. Table 9.4 shows the proportions of the validation parcels correctly identified as having changed and not changed for specific land cover types.

The result suggests that the text mining approach works well in managed landscapes (woodland, arable, improved grass, urban) but less well in upland environments. Overall, the text mining identified many more *uncertain*

TABLE 9.2

The Semantic Overlaps Between LCM1990 Classes (Row) and LCM2000 Classes (Columns) Derived from Text Mining

	Broadleaved Woodland	Coniferous Woodland	Arable	Improved Grassland	Acid Grassland	Neutral Grassland	Calcareous Grassland	Dwarf-shrub-heath	Bog	Built-up
0. Unclass	U	P	P	P	U	U	U	P	P	P
1. Sea	P	P	P	P	P	P	U	U	P	P
2. Water	P	P	U	P	U	P	P	P	U	P
3. Coastal bare	P	P	U	U	U	U	U	U	U	U
4. Saltmarsh	U	U	U	U	U	U	U	P	P	P
5. Grass heath	P	U	U	E	E	E	E	U	P	P
6. Mown	U	P	P	U	U	U	U	U	P	U
7. Meadow	U	E	U	E	E	E	E	U	P	P
8. Rough	P	U	U	U	U	E	U	U	U	U
9. Moorland grass	P	P	P	U	U	U	P	P	P	P
10. OS moor	U	U	P	U	U	U	U	U	U	U
11. Dwarf-shrub moor	U	P	P	U	P	P	P	E	U	P
12. Bracken	P	U	P	U	P	P	P	P	U	P
13. Dwarf-shrub heath	U	P	P	P	P	P	P	E	P	P
14. Scrub	E	E	U	P	U	U	P	P	U	U
15. Deciduous	E	E	U	U	P	P	U	P	P	P
16. Coniferous	E	U	P	P	P	P	P	P	P	P
17. Upland bog	P	P	P	P	P	P	P	E	E	U
18. Arable	U	P	E	U	U	U	P	P	U	E
19. Ruderal	U	E	E	U	U	U	U	P	U	U
20. Suburban	P	U	P	P	P	P	U	P	P	E
21. Urban	P	P	U	P	P	P	P	P	P	E
22. Inland bare	P	P	U	P	P	P	P	P	P	E
23. Felled	U	E	U	U	U	U	U	P	P	U
24. Lowland bog	P	P	P	P	P	P	P	U	E	P
25. OS Heath	U	U	P	U	U	U	E	E	U	U

TABLE 9.3

Comparison of Different LUT on Predicting Consistency

Expert	Correct (%)	Uncertain (%)	Wrong (%)
Text mining	56.0	16.6	27.4
Distributor	45.5	12.8	41.7
Producer	45.2	10.5	44.3
User	54.8	14.3	30.9

TABLE 9.4

Performance of the Different LUT Disaggregated by LCM2000 Broad Habitat Classes

Habitat	Text Mining (%)	Producer (%)	Distributor (%)	User (%)
Broadleaved woodland	80.0	28.6	28.6	65.7
Coniferous woodland	45.0	35.0	40.0	50.0
Arable	75.6	70.0	70.0	50.0
Improved grass	77.1	70.2	70.2	44.7
Neutral, calcareous, and acid grass	34.4	32.8	31.3	60.9
Moorland and bog	12.5	0.0	0.0	66.7
Built environment	47.2	39.6	41.5	49.1

relationships between LCM2000 and LCM1990 classes than the human experts. One of the implications for predicting land cover change is that in more managed, homogenous environments a greater degree of semantic uncertainty between classifications might need to be acknowledged.

9.4 Discussion and Concluding Comments

The results suggest that text mining is of a comparable utility to an expert-based approach to overcoming semantic variation. Although it can be argued that the method of converting the continuous semantic similarity to the three-valued logic of the experts is artificial, there is no easy objective alternative. In a perfect world, the three experts could be re-interrogated to recast their beliefs on a continuous zero-to-one scale, but numeric scales are not a very natural way for humans to think, and more importantly (and typically), the experts were not available after their initial consultation. What this example suggests is that when asked to compare classifications, the experts were looking for direct equivalence between classes wherever possible, and were thereby underestimating the actual uncertainty (fuzziness, vagueness, overlap) in the environment or between the classes.

In other studies of the comparability of land cover classification schemes, Fritz and See (2005) have presented a comparison of two global land cover classification schemes (GLC-2000 and MODIS), and Ahlqvist (2005) has compared the classifications systems of the U.S. National Vegetation Classification Standard and the European CORINE Land Cover System. In neither study would it have been possible to use the text mining approach used here, because in each study at least one of the classification scheme lacks (an accessible) text-rich class descriptions.

In this work we have sought to identify the semantic overlaps and inconsistencies between two datasets purporting to describe the same thing but with very different underlying epistemologies and semantics. We want to be able to reuse data. But are constantly faced with shifting semantic and conceptual sands as our scientific understanding, objectives, and technologies evolve. This makes data reuse very difficult.

We have explicitly tried to avoid delving into and referring to "ontologies"; because this too has specific meanings to specific communities. Ontologies are specified by some to provide slots for formal data descriptors that can then be "matched." However, this is to assume that the slots encapsulate everything that is required by the third-party user. In this work we have chosen to use the language of the class descriptions: we do not seek to subsume *ontology matching* approaches, rather we have sought to examine whether text mining approaches can provide a conceptually and computationally simple method, although we recognize that hierarchical schemas such as the LCCS offer an alternative approach. But they require a lot of manual work! The fully automated approaches presented here offer a robust alternative. Of course, there is always the problem that languages will use different grammatical structures to describe features, and Comber (2011) illustrated this with latent semantic analysis of CORINE classes in Spanish, German, French, and English—each identified different class-to-class overlaps. However, these differences may of course be closely related to local context.

In other work we have explored a number of different approaches, including Probabilistic Latent Semantic Analysis (PLSA) as proposed by Hofmann (1999a,b) and Latent Dirichlet Allocation (LDA) by Blei et al. (2003). We have found the steps in the text mining and use of overlap measure by Bouchon-Meunier et al. (1996) to be more tractable compared to trying to understand the *topics* generated in an LDA. However, we believe that there is still plenty of room for further work in all of these areas: users of geographic data are faced with the paradoxical situation of having more data available more easily than ever before, but of knowing less about the meaning behind the data. Data availability through spatial data infrastructure initiatives has massively increased access to spatial data. However, they have eliminated the need for dialogue between any potential user and the data producer. The intention is that the dialogue is replaced by metadata conforming to spatial data metadata standards: unfortunately, the standards concentrate on exploratory and exploitation metadata, documenting file formats and geometric issues, and very little is said about class descriptions. In contrast, there is a lot of research around the meaning of spatial data classifications within the ontological community, but practical ontologies are only very slowly appearing. The family of text mining approaches above, including PLSA, LDA, and the semantic overlap we applied, are providing solutions to the operational and applications gaps that still pervade the ontology work in this area. In one sense, the text mining could be thought of as an intermediate step between the completely implicit and informal ontology of the class descriptions, toward a slightly more formal "folksonomy." Whether the critical terms and relationships found by text mining could then be assembled into a more formal ontology is a potentially interesting area of research.

In conclusion, for users of spatial data information, the survey report remains the most thorough method of fully describing spatial data semantics. Unfortunately, such detailed reports are rarely produced. At the same time, class descriptions are progressively being reduced to little more than a label or cipher, and there are several global land cover datasets about which nothing other than class labels are available. Some data producers still provide class descriptions that attempt to convey what the data producer knows or means about the categories, and they should be encouraged to continue to do so. However, there is an increasing number of sources of what could be considered informal metadata for spatial data: reports, blogs, scientific papers, user feedback, and so on that are providing an increasing repository of information. So, while text-mining cannot replace the need for some basic understanding of the process being investigated, it can identify the overlap between concepts embedded in elements from different datasets. It provides the uninitiated user with a starting point from which to develop further understanding of the data concepts, and for the more experienced user, the identification of conceptual overlaps through text mining data semantics can enrich existing disciplinary understanding.

References

Ahlqvist, O. 2004. A parameterized representation of uncertain conceptual spaces. *Transactions in GIS*, 8(4): 493–514.

Ahlqvist, O. 2005. Using uncertain conceptual spaces to translate between land cover categories. *International journal of geographical information science*, 19(7): 831–857.

Blei, D.M, Ng, A.Y, Jordan, M.I. 2003. Latent Dirichlet allocation. *Journal of Machine Learning Research*, 3(4–5): 993–1022

Bouchon-Meunier, B., Rifqi, M., and Bothorel, S. 1996. Towards general measures of comparison of fuzzy objects. *Fuzzy Sets and Systems*, 84: 143–153.

Burrough, P.A. 1986. *Principles of Geographical Information Systems for Land Resources Assessment*. Clarendon, Oxford, UK.

Burrough, P.A. and Frank, A.U. (Eds.). 1996. *Geographic Objects with Indeterminate Boundaries*. Taylor and Francis, London, UK.

CEH. 2014. *Land Cover Map of Great Britain (1990): Dataset Information* [http://www.ceh.ac.uk/documents/lcm1990_land_cover_map_of_great_britain.pdf, August 29, 2014].

Comber, A. 2008a. The separation of land cover from land use with data primitives. *Journal of Land Use Science*, 3(4): 215–229.

Comber, A.J. 2011. *Analysing Land Cover and Green Space Consultation Semantics: Concepts, Overlaps and Mappings*, Keynote, 7th International Symposium on Spatial Data Quality (ISSDQ 2011), October 12–14, 2011, Coimbra, Portugal.

Comber, A.J. 2008b. Land Cover or Land Use? Editorial. *Journal of Land Use Science*, 3(4): 199–201.

Comber, A.J., Fisher, P.F., and Wadsworth, R.A. 2005a. A comparison of statistical and expert approaches to data integration. *Journal of Environmental Management*, 77: 47–55.

Comber, A.J., Fisher, P.F., and Wadsworth, R.A. 2004a. Assessment of a semantic statistical approach to detecting land cover change using inconsistent data sets. *Photogrammetric Engineering and Remote Sensing*, 70(8): 931–938.

Comber, A.J., Fisher, P.F., and Wadsworth, R.A. 2005b. Comparing the consistency of expert land cover knowledge. *International Journal of Applied Earth Observation and Geoinformation*, 7(3): 189–201.

Comber, A., Fisher, P., and Wadsworth, R. 2004b. Integrating land cover data with different ontologies: Identifying change from inconsistency. *International Journal of Geographical Information Science*, 18(7): 691–708.

Comber, A.J., Fisher, P.F., and Wadsworth, R.A. 2008. Semantics, metadata, geographical information and users. Editorial. *Transactions in GIS*, 12(3): 287–291.

Comber, A.J., Fisher, P.F., and Wadsworth, R.A. 2005c. What is land cover? *Environment and Planning B: Planning and Design*, 32: 199–209.

Comber, A.J., Fisher, P.F., and Wadsworth, R.A. 2005d. You know what land cover is but does anyone else? An investigation into semantic and ontological confusion. *International Journal of Remote Sensing*, 26(1): 223–228.

Dempster, A.P. 1967. Upper and lower probabilities induced by a multi-valued mapping. *The Annals of Mathematical Statistics*, 38, 325–339.

Derungs, C., Wartmann, F., Purves, R.S., and Mark, D.M. 2013. The meanings of the generic parts of toponyms: Use and limitations of gazetteers in studies of landscape terms. In *Spatial Information Theory* (pp. 261–278), Springer International Publishing.

Edwardes, A.J. and Purves, R.S. 2007. A theoretical grounding for semantic descriptions of place. In *Web and Wireless Geographical Information Systems* (pp. 106–120). Springer, Heidelberg, Germany.

Feldman, R., Dagan, I., and Hirsh, H. 1998. Mining text using keyword distributions. *Journal of Intelligent Information Systems*, 10: 281–300.

Feldman, R., and Sanger, J. (Eds.). 2007. The text mining handbook: Advanced approaches in analyzing unstructured data. Cambridge University Press, Cambridge.

Fisher, P.F. 1997. The pixel: A snare and a delusion. *International Journal of Remote Sensing*, 18(3): 679–685.

Fisher, P.F. 2003. Multimedia reporting of the results of natural resource surveys. *Transactions in GIS*, 7: 309–324.

Fisher, P.F., Comber, A.J., and Wadsworth, R.A. 2005. Land use and land cover: Contradiction or complement. In *Re-Presenting GIS* (pp. 85–98), edited by Peter Fisher and David Unwin. Wiley, Chichester, UK.

Fisher, P.F., Comber, A.J., and Wadsworth, R. 2009. What's in a name? Semantics, standards and data quality. In *Spatial Data Quality* (pp. 3–16), edited by Rodolphe Devillers. CRC Press, London, UK.

Fritz, S. and See, L. 2005. Comparison of land cover maps using fuzzy agreement. *International Journal of Geographical Information Science*, 19(7): 787–807.

Fuller, R.M., Smith, G.M., Sanderson, J.M., Hill, R.A., and Thomson, A.G. 2002. The UK Land Cover Map 2000: construction of a parcel-based vector map from satellite images. *The Cartographic Journal*, 39(1): 15–25.

Gardenfors, P. 2000. *Conceptual Spaces: The Geometry of Thought*. A Bradford Book. MIT Press, Cambridge, MA.

Hofmann, T. 1999a. Probabilistic latent semantic indexing, pp 50–57 in Hearst M, Gey, F., Tong, R. (Eds.) Proceedings of 22nd International Conference on Research and Development in Information Retrieval, Univ Ca, Berkeley, California, Aug, 1999.

Hofmann, T. (1999b). Probabilistic latent semantic analysis, pp 289–296 in Laskey K.B, Prade, H. (Eds.) Proceedings of 15th Conference on Uncertainty in Artificial Intelligence. *Royal Institute of Technolology*, Stockholm, Sweden, Jul 30–Aug 01, 1999.

Honkela, T. 1997. Self-organising maps in natural language processing. PhD thesis, Helsinki University of Technology, Department of Computer Science and Engineering [http://www.cis.hut.fi/~tho/thesis/].

Honkela, T., Pulkki, V., and Kohonen, T. 1995. Contextual relations of words in Grimm tales analyzed by self-organizing map. In *Proceedings of the International Conference on Artificial Neural Networks, ICANN-95* (vol. 2, pp. 3–7), edited by Fogelman-Soulié, F. and Gallinari, P. EC2 et Cie, Paris.

Kent, M. and Coker, P. 1992. *Vegetation Description and Analysis: A Practical Approach*. John Wiley, Chichester, UK.

Kohonen, T. 1982. Self-organized formation of topologically correct feature maps. *Biological Cybernetics*, 43:59–69.

Kohonen, T. 1995. *Self-Organizing Maps*. Springer, Berlin, Germany.

Kohonen, T., Kaski, S., Lagus, K., Salojarvi, J., Honkela, J., Paatero, V., and Saarela, A. 2000. Self organization of a massive document collection. *IEEE Transactions on Neural Networks*, 11(3): 574–585.

Kuhn, W. 2005. Geospatial semantics: Why, of what, and how? In *Journal on Data Semantics III* (pp. 1–24). Springer, Heidelberg, Germany.

Lakoff, G. 1987. *Women, Fire, and Dangerous Things: What Categories Reveal About the Mind* (pp. 39–74). University of Chicago Press, Chicago, IL.

Lin, X. 1997. Map displays for information retrieval. *Journal of the American Society for Information Science*, 48:40–54.

Lund, H.G. 2002. When is a forest not a forest? *Journal of Forestry*, 100: 21–28.

Lund, H.G. 2014. *Definitions of Forest, Deforestation, Afforestation, and Reforestation* [http://home.comcast.net/~gyde/DEFpaper.htm, August 29, 2014].

Maedche, A. and Staab, S. 2000. Discovering conceptual relations from text. In *Proceedings of the 14th European Conference on Artificial Intelligence* (pp. 231–235). ISO Press, Amsterdam, The Netherlands.

Mark, D.M. and Turk, A.G. 2003. Landscape categories in Yindjibarndi: Ontology, environment, and language. In: COSIT 2003. LNCS (vol. 2825, pp. 28–45), edited by Kuhn, W., Worboys, M.F., and Timpf, S. Springer, Heidelberg, Germany.

Mark, D.M., Turk, A.G., Burenhult, N., and Stea, D. (Eds.). (2011). *Landscape in Language: Transdisciplinary Perspectives* (vol. 4). John Benjamins Publishing , Philadelphia, USA.

Miller, H.J. and Han, J. (Eds.). 2001. *Geographic Data Mining and Knowledge Discovery*. Taylor and Francis, London, UK.

Nasukawa, T. and Nagano, T. 2001. Text analysis and knowledge mining system. *IBM Systems Journal*, 40(4): 967–984.

Pickles, J. 1995. *Ground Truth: The Social Implications of Geographic Information Systems*. Guilford Press, New York, USA.

Robertson, S. 2004. Understanding inverse document frequency: On theoretical arguments for IDF. *Journal of Documentation*, 60: 503–520.

Rodriguez, M.A. and Egenhofer, M.J. 2004. Comparing geospatial entity classes: An asymmetric and context-dependent similarity measure. *International Journal of Geographical Information Science*, 18(3): 229–256.

Rosch, E. 1975a. Cognitive representations of semantic categories. *Journal of Experimental Psychology: General*, 104:192–233.

Rosch, E. 1975b. Cognitive reference points. *Cognitive Psychology*, 7:532–547.

Rosch, E.H. 1978. Principles of categorization. In *Cognition and Categorization* (pp. 27–48), edited by Rosch, E. and Lloyd, B. Lawrence Erlbaum Associates, Hillsdale, NJ.

Rosch, E. and Lloyd, B. 1978. *Cognition and Categorization*. Lawrence Earlbaum, Hillsdale, NJ.

Shafer, G. 1976. *A Mathematical Theory of Evidence*. Princeton University Press, Princeton, NJ.

Smith, B. 1995. On drawing lines on a map. *Spatial Information Theory: Lecture Notes in Computer Science*, 988: 475–484.

Smith, B. 2001. Fiat objects. *Topoi*, 20: 131–148.

Smith, B. and Mark, D.M. 1998. Ontology and geographic kinds. In *Proceedings of the 8th International Symposium on Spatial Data Handling (SDH'98)* (pp. 308–320), edited by Poiker, T.K. and Chrisman, N. International Geographical Union, Vancouver, UK.

Smith, B. and Mark, D. 2001. Geographical categories: An ontological investigation. *International Journal of Geographical Information Science*, 15: 591–612.

Varzi, A.C. 2001. Introduction. *Topoi*, 20: 119–130.

Wadsworth, R., Balzter, H., Gerard, F., George, C., Comber, A., and Fisher, P. 2008. An environmental assessment of land cover and land use change in Central Siberia using quantified conceptual overlaps to reconcile inconsistent data sets. *Journal of Land Use Science*, 3(4): 251–264.

Wadsworth, R.A., Comber, A.J., and Fisher, P.F. 2006. Expert knowledge and embedded knowledge: Or why long rambling class descriptions are useful. In *Progress in Spatial Data Handling; Proceedings of the 12th International Symposium on Spatial Data Handling* (pp. 197–213), edited by Riedl, A., Kainz, W., and Elmes, G. Springer, Berlin, Germany.

Wadsworth, R.A., Fisher, P.F., Comber, A., George, C., Gerard, F., and Baltzer, H. 2005. Use of quantified conceptual overlaps to reconcile inconsistent data sets. Session 13 Conceptual and cognitive representation. *Proceedings of GIS Planet 2005* (13 pp.), May 30–June 2, 2005, Estoril, Portugal. ISBN 972-97367-5-8.

Witten, I.H. and Frank, E. 2005. *Data Mining: Practical Machine Learning Tools and Techniques*. Morgan Kaufman, San Francisco, CA.

10

LC3: A Spatiotemporal Data Model to Study Qualified Land Cover Changes

Helbert Arenas, Benjamin Harbelot, and Christophe Cruz

CONTENTS

ABSTRACT Land cover changes caused by humans have reached points
never witnessed in history. Consequences of these changes are environmental
degradation, pollution of water, biodiversity loss, and climate change, among
others. It is in this frame that the interest of the scientific community for this
type of events has increased in the last decades. Nowadays, researchers have
access to sophisticated monitoring tools and techniques. However, the new
threat is that overabundance of information would hide relevant facts and
processes. In our research we use data from the CORINE (Coordination of
Information on the Environment) program. This program has compiled land
cover information about Europe in a standardized form for the years 1990,
2000, and 2006, allowing comparative studies of land cover evolution. In this
chapter we present the LC3 data model to study land cover changes. In our
research we use Semantic Web technologies to create a formal representation
of the CORINE land cover classification. Later, we use the same technologies
to create formal representations of the components involved in land cover
change. In our research, land cover types are organized as a taxonomy. Using
this approach it is easy to aggregate and disaggregate land cover types at different levels, allowing a more flexible analysis. An additional benefit is that it is
possible to create formal descriptions of particular land cover changes, allowing an easy identification of them at different levels of the land cover classification. Our approach has been implemented as a computer program using a
triplestore as its data repository. Our approach enables scientists to easily discover patterns of change that could be hidden due to the large volumes of data.

KEY WORDS: *Spatiotemporal dynamics, spatiotemporal semantics.*

10.1 Introduction

For thousands of years, humans have modified the environment. In many
cases, land cover change is the result of a combination of economic opportunities, national policies, and markets. However, only until recently, scientists
have identified a relation between Land Use/Land Cover Change (LULCC)
and medium-/long-term phenomena like weather pattern modifications
(Lambin et al. 2001). In other cases, the land cover change might pose a more

immediate threat not only to the areas where the change had occurred but also to adjacent ones (Paton and Fantina 2013).

There are several models currently employed to model LULCC. However, there is a continuous need for new approaches to reevaluate current models and improve them (Mahmood et al. 2010).

Researchers studying LULCC need to analyze a vast number of elements and their interactions to identify patterns of change of particular interest. Most of the elements involved have spatial and temporal components, making them dynamic in nature. However, the temporal dimension of elements in LULCC adds a layer of complexity that is often avoided in GIS approaches (Blaschke et al. 2014).

The purpose of our study is to develop a conceptual model for LULCC capable of handling large amounts of dynamic elements and their interactions. A model of this nature needs to be capable to use human expert knowledge encoded as axioms and constraints to identify interesting patterns of change in the LULCC domain. In our research, we propose a model capable to handle both the spatial and the temporal components of dynamic entities identified in land cover datasets.

In this chapter, we present a model based on Semantic Web technologies designed to keep track of dynamic spatial systems. In our approach, we conceptualize the dynamic system as a graph in which the involved elements are linked by relationships. In our model, we conceptualize different types of evolution, in which objects cease to exist or continue their existence after experiencing changes. By keeping track of the evolution of entities, it is possible to determine changes that might increase certain undesirable conditions, for instance, we could detect when certain LULCC increases the fire risk for a certain area.

To show the effectiveness of the model, we implement our ideas using a Java application and a triplestore to manage the data of our model. Our approach is flexible in the sense that it can be used with any spatiotemporal system in which entities have a 2D spatial representation. In this chapter, we present our model, using land cover data for Portugal.

In Section 10.2, we identify previous relevant research in the field of spatiotemporal modeling. Later in Section 10.3, we focus on work carried out in the field of modeling the dynamics of LULCC. In Section 10.4, we present the formalisms of our model followed by the model implementation in Section 10.5. Finally, in Sections 10.6 and 10.7, we discuss our results and present our conclusions.

10.2 Advances on Spatiotemporal Modeling

Several spatial and temporal approaches have been proposed to model environmental processes. These types of models generate complex relationships in space and time. Objects that take part in a dynamic process can move and change shape while maintaining their identity, or evolve into new objects.

Handling the vast number of relations between evolving entities requires the help of software mechanisms capable of analyzing complex networks of relations and inferring new implicit knowledge. Gruber (1995) defined ontologies as conceptualizations of a domain, in a formal, explicit form that can be easily shared among potential users. An ontology modeler attempts to identify concepts and relations that exist within a specific domain. By using formal specifications, it is possible to use inference mechanisms on them capable to identify incoherencies (Gruber 1995).

Ontologies are based on notions of individuals, classes, attributes, relations, and events. In an ontology, entities can be treated as individuals and grouped in defined classes. The definition of a class can be further specialized, creating subclasses. It is possible to define subsumption relations between classes, establishing in this way a hierarchy. The characteristics of entities are modeled as properties. It is also possible to model relations between entities. Any change in the properties or the relations is modeled as an event.

The representation of real-world objects is composed of identity, descriptive, and spatial properties. While an identity property describes a fixed component of the entity, alphanumeric and spatial properties can vary over time and are the entity's dynamic part. When the identity of an entity varies, there is a particular type of evolution in which the spatiotemporal entity is transformed into a new one. In the literature, there are two main types of spatiotemporal entities: (1) moving objects, for example, a boat sailing; and (2) changing objects, for example, a region which boundaries evolve in time (Meskovic, Zdravko, and Baranovic 2011).

There are two main philosophical theories behind models of spatiotemporal dynamic objects: (1) *endurantism* and (2) *perdurantism*. The first approach, *endurantism*, considers that objects (endurants) exist wholly at any given point of time during their lifespan. An alternative modeling approach follows the *perdurantism* paradigm. This approach represents objects as entities comprising several timeslices. Each timeslice is a representation of the object during a finite time. A complete representation of the evolving object is the sum of all its timeslices (Harbelot, Arenas, and Cruz 2013a).

Research using an *endurantism* approach can be found in the work of Bittner, Donnelly, and Smith (2009). In this research, the authors identified three classes of entities: (1) Individual entities, (2) Endurant universals, and (3) Collections of individual endurants. In the ontology developed by Bittner, Donnelly, and Smith (2009), to declare that two entities sustain a relationship, it is necessary to indicate the valid time of the relation.

The alternative *perdurantism* approach can be found in other works such as those of O'Connor and Das (2011), Batsakis and Petrakis (2011), and Harbelot, Arenas, and Cruz (2013a). In the work of Al Debei (2012), the authors compare the *perdurantism* and *endurantism* aproaches. The authors concluded that a *perdurantism* approach when using ontologies offers better expressiveness, handling of time, flexibility, and objectivity.

A *perdurantism* dynamic model requires mechanisms for the representation of dynamic properties. However, the two main Semantic Web languages, OWL and RDF, provide limited support for temporal dynamics, having been designed to define binary relations between individuals (O'Connor and Das 2011). To overcome these limitations, different ideas have been proposed.

Klein and Fensel (2001) present an approach based on *versioning*. In this case, the model constructs multiple variants of the objects to represent their evolution. However, the major drawback of the *versioning* approach is the redundancy generated by the slightest change of an attribute. In addition, any information requests must be performed on multiple versions of the ontology, affecting its performance.

Gutierrez, Hurtado, and Vaisman (2007) present an extension of the standard RDF to name properties and to assign them corresponding time intervals, allowing explicit management of time in RDF. The limitation of this approach is that it uses only RDF triples, lacking the expressiveness of OWL. For example, it is not possible to define qualitative relationships.

Research conducted by Batsakis and Petrakis (2011) uses the so-called 4D-fluent model. Using this methodology, it is possible to express the existence of an entity using multiple representations, each corresponding to a defined time interval. In the literature, 4D-fluent model is the most well-known method to handle dynamic properties in an ontology. It has a simple structure, allowing to easily transform a static ontology into a dynamic one. However, the approach of the 4D-fluent model also has some limitations: (1) it is difficult to maintain a close relationship between geometry and semantics, and (2) it increases the complexity for querying the temporal dynamics and understanding the modeled knowledge. Furthermore, this approach does not define qualitative relations to describe the type of changes that has occurred or to describe the temporal relationships between objects. Thus, it is difficult to identify entities that change and new entities that might emerge as a result of these changes. Batsakis and Petrakis (2011) developed SOWL, which uses 4D-fluent to extend the ontology OWL-time, making it able to handle qualitative relations between intervals, such as *before* or *after*, even with intervals with vague ending points.

O'Connor and Das (2011) developed a lightweight model using *reification*. The proposed model can be deployed using existing OWL ontologies, extending their temporal support. This work also proposes the use of the Semantic Web Rule Language (SWRL) operators. Using *reification*, it is possible to use a triple as the object or the subject of a property. Unfortunately, this method has some limitations: (1) the transformation from a static property into a dynamic one increases the complexity of the ontology substantially, reducing the querying and inference capabilities; and, (2) the approach is prone to redundant objects, which reduces its effectiveness.

The filiation relationship defines the succession link that exists between representations of objects at different instants of time. The analysis of these relations allows us to identify processes such as the division or the merge of entities. Other more complex spatial changes would require the identification of multiple *parents* and *children* entities involved. At this level, filiation relationships are based only on spatial relationships. Therefore, they can be characterized as spatial filiationships in the context of spatial changes. In addition, these spatial changes may reveal an evolution in the nature of the entity. Because of this, the filiation relationship is intimately linked to the notion of identity. This relationship is essential to maintain the identity of an entity that evolves and to follow its evolution along time. In this process, it is also necessary to identify new entities that can emerge from an evolution.

An important concept regarding the evolution of entities is the identity. It can be defined as the uniqueness of an object, regardless of its attributes or values. It is the feature that distinguishes one object from all others. The identity is essential in the conceptualization and modeling of a phenomenon. Its importance while modeling dynamic systems has been identified by previous works by researchers such as Del Mondo et al. (2013), Del Mondo et al. (2010), and Muller (2002). However, this concept is very subjective because it depends on the criteria selected by the user to define the identity of an entity. Usually the criteria for the definition of the identity depends on the domain of study.

Research presented by Del Mondo et al. (2010) describes relationships between objects that exist at different points of time, and how some objects can originate others, creating filiation relationships. Their approach cannot be strictly described as *perdurantistic*, because they do not implement any timeslices. However, it contributes to the formal definition of filiation relationships, and these must respect two constraints: (1) a temporal constraint, that is, the child object must exist after the parent object; and, (2) there must be a spatial relationship between parent and child objects.

Previous works presented by Hornsby and Egenhofer (2000), Stell et al. (2011), Harbelot, Arenas, and Cruz (2013b), and Del Mondo et al. (2013) have identified two general types of filiation relationship: *continuation* and *derivation*. In the first case, *continuation*, the identity remains the same. The entity continues to exist, but undergoes a change. In the second case, *derivation*, a new entity is created from the parent after a certain evolution.

In the study of Del Mondo et al. (2013), the authors further extend the research conducted by Del Mondo et al. (2010) by providing mechanisms to establish filiation relationships at nonconsecutive times, allowing the combination of different graphs. Because of the constraints proposed, the system is not able to deal with geometries defined by multipolygon spatial representations. Del Mondo et al. (2013) implement these ideas in a relational database, and they present an experimental evaluation of their ideas, using cadastral information of the Canton de Neufchatel, in Switzerland, composed by seven snapshots.

A related research is by Stell et al. (2011). Here, the authors use bigraphs to model spatiotemporal dynamics. In this research, the authors apply their ideas to track the evolution of crowds of people, implementing rules to identify splitting and merging of crowds.

There are examples of works in which the authors did not use Semantic Web technologies. For instance, Worboys (1994) presents modeling approaches for spatiotemporal information using relational databases. Similar work is presented by Claramunt, Theriault, and Parent (1997) with the introduction of ideas for the representation of spatiotemporal processes using an object-relationship data model. Hornsby and Egenhofer (2000) present a language designed to follow the identity of objects that represent geographic phenomena, and Egenhofer and Al-taha (1992) use the intersection matrix to identify changes in topological relations between evolving features.

Some models for spatial dynamics are based on discrete approaches such as the snapshot model found in the works of Armstrong (1988) and Chen et al. (2013), the Space-Time Composites model (STC) presented in the work of Langran and Chrisman (1988), and the Spatiotemporal Object model introduced by Worboys (1994). However, there are disadvantages with these approaches as they represent only sudden changes, making it difficult to identify processes such as movement of an entity in a geographical environment.

Another type of model is the so-called *event- and process-based* approach. This approach considers that spatial entities operate under the impetus of an event or a process; the aim of this approach is to analyze the causes and consequences. An example of this type of model is the Event-Based Spatiotemporal Data Model (ESTDM) introduced by Peuquet and Duan (1995). The ESTDM model describes a phenomenon through a list of events; a new event is created at the end of the list whenever a change is detected. However, this model takes into account only raster data, and the causal links between events are hardly highlighted in this model. An alternative to ESTDM is the composite processes as introduced by Claramunt, Theriault, and Parent (1997), dealing with some of the limitations of ESTDM. It is designed to represent the links between events and their consequences; moreover, the authors argue that the data model must differentiate what is spatial, temporal, and thematic. Another example is the model of topological change based on events presented in the work of Jiang and Worboys (2009). This model represents change of a geographic environment as a set of trees. Each tree is connected to the next and the previous through its nodes. The link between two trees is a topological change that reveals the creation of an entity on the geographical environment, the deletion of an entity, division or merger of entity, or no change. The succession of these topological changes enables the representation of complex changes.

The weakness of the last set of models is that they do not use any formal semantics; therefore, the applicability of formal rules or inference mechanisms is limited.

In Section 10.3 we will describe models specifically used in the field of LULCC for spatiotemporal dynamics.

10.3 Spatiotemporal Models for Land Use/Land Cover Change

The field of LULCC, due to data-gathering methods and traditional tools, has developed alternative approaches. A significant part of the information used in LULCC comes from remote sensing (RS) platforms. Currently, the prevailing approach for the analysis of RS data uses pixel-based methods. However, in recent years this approach has been criticized as new tools with an *object-based* approach become more available. Much of the initial work related to object-based image analysis (OBIA) can be traced back to a well-known software called *eCognition*, later renamed as *Definiens* (Blaschke 2010). However, the term OBIA was perceived by some researchers in GeoSciences as too broad, given the fact that similar techniques are used in the medical field. Because of these concerns, a new term, geographic object–based image analysis (GEOBIA), was introduced.

The goal of GEOBIA is to develop automated methods to partition RS imagery into image objects, and analyze their spatial, spectral, and temporal characteristics (Arvor et al. 2013). As a result, it would be possible to generate geographic information from which new spatial knowledge can be obtained. Most of the methods for image classification currently in use were first developed in the early seventies (Blaschke 2010). They are based on the classification of pixels using a multidimensional feature space. In these methods, the spectral values of the pixels are the most relevant characteristics to be considered for any given classification. A pixel is the smallest entity for RS. Image objects are generated by grouping pixels with similar values. Then it is possible to link these groups to real-world objects. When the spectral characteristics of an object are homogeneous, the classification can be straightforward. However, it is more difficult to use this approach when there is heterogeneity among the pixels that compose an object. For instance, an object of the type *urban* could include pixel values of elements that represent *vegetation* (parks, gardens) or *water* (pools, fountains).

Approaches that use only the pixel spectral characteristics do not consider the context and patterns. For instance, considering what exists in the neighborhood of the pixel in spatial and temporal dimensions would provide further information for a given analysis.

The use of classification schemas that are based mostly on pixel spectral values does not take advantage of domain knowledge. For any domain, it is possible to identify the main existing concepts and the relations between them. For instance, relations such as *is part of, is more specific than*, and *is instance of* would provide insights regarding the land use of certain area. These semantics can be formalized, allowing inference mechanisms to operate with the information.

According to Blaschke et al. (2014), the core characteristics of a GEOBIA are (1) data is Earth centric; (2) analytical methods are multisource capable; (3) geo object–based delineation is a prerequisite; and (4) the methods are contextual, allowing for surrounding information, and (5) highly customizable, allowing human semantics and hierarchical networks.

Most of the previous works on GEOBIA focus on the segmentation of images and temporal analysis (for instance, Sheeren et al. 2012 or Herold et al. 2012). However, the use of semantic technologies is scarcer.

One area in which Semantic Web technologies have been used is in the field of interpretation of RS imagery. Such is the case of Morshed, Aryal, and Dutta (2013) and Aryal, Morshed, and Dutta (2014), where the authors capture knowledge from different sources at different scales using RDF.

Another interesting field of use for Semantic Web technologies is for data integration. It is common that GEOBIA practitioners use different classification systems, based on particular conceptualizations of the world. To compare or merge products from different sources, it is necessary to find harmonization tools. Ontologies can fill this gap, having formal description of classes. Then it is possible to link two different classification systems; this process is called mapping or matching. Therefore, there is a need for standardized conceptualizations that would enable the comparison of results from different users and geographic areas (Blaschke et al. 2014).

In the work of Arvor et al. (2013), the authors identify six areas for further research in GEOBIA: (1) the alignment of real-world concepts to image objects, (2) the management of qualitative and quantitative information, (3) the handling of fuzzy geographic entities, (4) the handling of scale, (5) the handling of change and evolution, and, (6) the dichotomy of open-world vs. closed-world assumptions.

Our proposal fits in the fifth research area identified by Arvor et al. (2013). The objective of our model is to provide LULCC practitioners with tools to keep track of evolving land cover entities, allowing researchers to perform queries considering concepts like vicinity in time and space.

10.4 LC3 Model Specification

In this section, we proceed to describe the use of Description Logic for the model, and First-Order Logic for the constraints on the model.

10.4.1 Basic Components

10.4.1.1 Temporal Points

We can think of the temporal domain composed by a set of temporal points. The components of the set follow a strict order $<$, which forces all points between two temporal points *p1* and *p2* to be ordered using the approach presented by Artale and Franconi (1998).

$$\mathcal{P} \tag{10.1}$$

10.4.1.2 Time Intervals

By selecting a pair *[p1,p2]* of temporal points, we can limit a closed set of ordered points (Artale and Franconi 1998). We represent this concept as

$$\mathcal{I} \equiv (=1 \; hasInit.\mathcal{P}) \sqcap (=1 \; hasEnd.\mathcal{P}) \tag{10.2}$$

constraint:

$$\forall i \rightarrow \exists t_o, \exists t_f | (t_o < t_f) \wedge hasInit(i,t_o) \\ \wedge \; hasEnd(i,t_f) \wedge (t_o,t_f \in \mathcal{P}) \wedge (i \in \mathcal{I}) \tag{10.3}$$

10.4.1.3 Time

We can define a generalized class, called Time, which we define as

$$\mathcal{T} \equiv \mathcal{P} \sqcap \mathcal{I} \tag{10.4}$$

10.4.1.4 Geometries

The spatial representation of an object is given by the coordinates representing its geometry. It is represented by G. The spatial topological relations between geometries are defined by the Extended Nine-Intersection model (DE-9IM) (Strobl 2008).

$$G \tag{10.5}$$

10.4.1.5 Object

This component of the model represents the elements that evolve along time.

$$\mathcal{O} \tag{10.6}$$

10.4.1.6 Timeslice

In our model, a TS (timeslice) is a temporal representation of an evolving object. Each TS has four components: (1) an identity that links it to the object it represents, (2) a Geometry that contains its spatial representation, (3) a temporal component that indicates the time point or interval in which this representation is valid, and (4) a set of properties that describes the characteristics of the object during the corresponding temporal component.

$$TS \equiv (= 1 \ hasIdentity.\mathcal{O}) \sqcap (= 1 \ hasGeometry. \ G)$$
$$\sqcap (= 1 \ hasTime.\mathcal{T}) \sqcap \forall \ hasProperties. \ \overline{TS} \tag{10.7}$$

10.4.2 Filiation Relationships and Evolution Processes

When a change occurs a new TS would be generated from a previous one in a parent–child relationship, denominated *filiation* relations. For a filiation relationship it is necessary to compute the existence of a spatial relationship between parent and child. In the case of TSs whose geometries are polygons, the spatial relationship must be an intersection of the type polygon, while the existence of the parent must be previous to the existence of the child.

$$\exists \ Intersection(p.geo, c.geo) \land (p.t < c.t) \rightarrow hasFiliation(p, c) \tag{10.8}$$

where *p* and *c* are instances of Timeslice TS, and *p.geo* and *c.geo* are their geometries and *p.t* and *c.t* are their respective time properties, respectively.

Figure 10.1 depicts the filiation relationships between a set of parent TSs [*p1,p2,p3*] and a set of children TSs [*c1,c2,c3*]. In the example depicted in Figure 10.1, the geometry of *p1* would intersect the geometries of *c1* and *c2*, while the geometry of *p3* would intersect the geometries of *c1, c2,* and *c3*. Each parent has its own identity that might or might not be inherited by one of their children, based on domain-specific rules.

The filiation relationship can be used to describe evolutions of objects along time. An important component of the evolution is the identity inheritance. Previous research, such as that by Hornsby and Egenhofer (2000), Stell et al. (2011), Harbelot, Arenas, and Cruz (2013b), and Del Mondo et al. (2013) has identified two basic types of Evolution based on the inheritance—*Continuation* and *Derivation*. In Figure 10.2, we propose a taxonomy of processes based on the filiation characteristics.

For the definition of the different types of evolution, we use the relations defined in DE-9IM (Equal, Within, Contains) as defined by Strobl (2008).

10.4.2.1 Continuation

In this type of relationship the identity is constant between a parent and a child.

$$hasFiliation(p, c) \land (p.o = c.o) \rightarrow hasContinuation(p, c) \tag{10.9}$$

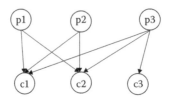

FIGURE 10.1
Filiation relationships between timeslices.

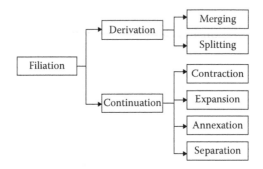

FIGURE 10.2
Evolution types.

where *p.o* and *c.o* are the identities of the parent and child TSs, respectively.

Expansion

$$hasContinuation(p,c) \wedge Within(p.geo,\ c.geo)$$
$$\rightarrow hasExpansion(p,c)$$

(10.10)

Contraction

$$hasContinuation(p,c) \wedge Contains(p.geo,\ c.geo)$$
$$\rightarrow hasContraction(p,c)$$

(10.11)

Separation
This type of process is not limited to a single child. A separation process can result in multiple children TSs. However, only one of them maintains a *Continuation* relationship with the parent.

$$hasFiliation(p,[c_1,c_2...c_n])$$
$$\wedge Equals(p.geo,\ Union(c_1.geo,c_2.geo...c_n.geo))$$
$$\wedge(\exists_{=1}hasContinuation(p,c))\rightarrow hasSeparation(p,[c_1,c_2...c_n])$$

(10.12)

In this case, the process involves multiple parents that result into one single child. However, only one parent has the same identity as the resulting child, thus maintaining a *Continuation* relationship.

$$
hasFiliation\big([p_1, p_2 \ldots p_n], c\big)
$$
$$
\wedge\, Equals\big(Union(p_1.geo, p_2.geo \ldots p_n.geo), c.geo\big) \tag{10.13}
$$
$$
\wedge\big(\exists_{=1} hasContinuation(p, c)\big) \rightarrow hasAnnexation\big([p_1, p_2 \ldots p_n], c\big)
$$

10.4.2.2 Derivation

This kind of relation involves parents and children who do not share the same identity.

$$
hasFiliation(p, c) \wedge (p.o \neq c.o) \rightarrow hasDerivation(p, c) \tag{10.14}
$$

where *p.o* and *c.o* are the identities of the parent and child TSs, respectively.

Derivation processes involving multiple parents or children are Splitting and Merging:

Splitting
This process is similar to *Separation*. However, in this case none of the children shares the same identity with the parent, and the identity of the parent ceases to exist (Del Mondo et al. 2013).

$$
hasDerivation\big(p, [c_1, c_2 \ldots c_n]\big)
$$
$$
\wedge\, Equals\big(p.geo,\, Union(c_1.geo, c_2.geo \ldots c_n.geo)\big) \tag{10.15}
$$
$$
\rightarrow hasSplitting\big(p, [c_1, c_2 \ldots c_n]\big)
$$

Merging
In this case, multiple parents combine into a child. The identity of the child is new, different from the involved parents (Del Mondo et al. 2013).

$$
hasDerivation\big([p_1, p_2 \ldots p_n], c\big)
$$
$$
\wedge\, Equals\big(Union(p_1.geo, p_2.geo \ldots p_n.geo), c.geo\big) \tag{10.16}
$$
$$
\rightarrow hasMerging\big([p_1, p_2 \ldots p_n], c\big)
$$

10.4.3 Identification of the Evolution Process without A Priori Information

In systems in which no *a priori* filiation lineage information exists, it is necessary to identify the relationships between TSs from scratch. Those cases are not unusual if you consider all the systems that rely on RS observations for regular update. In these cases, the new dataset represents a snapshot describing the area of interest at a discrete point of time. However, no link other than geometry is given between the new dataset and previously recorded status of the area of interest. To facilitate the analysis of systems without *a priori* filiation information, we propose the creation of the class *Filiation*. This class links a parent and child TSs and stores information regarding the quantification of their relationships (Figure 10.3).

$$Filiation \equiv \forall hasParentTS.\ TS \sqcap \forall hasChildTS.\ TS$$
$$\sqcap\ \forall has\rho.Double \sqcap \forall has\chi.Double \tag{10.17}$$

constraint:

$$\forall [p,c] \big\| hasFiliation(p,c) \rightarrow \exists Filiation(f) \big\| hasParentTS(f,p)$$
$$\wedge\ hasChildTS(f,c)$$
$$\wedge\ has\rho \left(f, \frac{Area\big(Intersection(p.geo,c.geo)\big)}{Area(p.geo)} \right)$$
$$\wedge\ has\chi \left(f, \frac{Area(Intersection(p.geo,c.geo))}{Area(c.geo)} \right) \tag{10.18}$$

In an evolving system, a parent TS can originate one or many children TSs, while it is also possible that a child TS can be the result of multiple parents. In this case, it would be necessary to identify the most suitable candidates for the identity inheritance. A rule of thumb to solve this problem would be to identify the *parent–child* relationship in which there is the highest spatial similarity between parent and child.

To have a better understanding of this type of relationship, we can analyze the values of the *hasχ* and *hasρ*. In the case of a parent with multiple children, we can identify the *parent–child* relationship in which the child comprises most of the geometry of the parent. This relation would be the one with the highest value for the property *hasρ*.

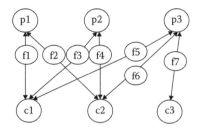

FIGURE 10.3
Instances of the class Filiation.

On the other hand, we can have the case of a child with multiple parents. In this case, we can identify the *parent–child* relationship that corresponds to the one in which the most of geometry of the child corresponds to a certain parent. This relation would be the one with highest value for the property *hasχ*.

The identification of the maximum values for *hasρ* and *hasχ* does not resolve the problems of unknown identity inheritance. In some cases, the identity might evolve regardless of the geometric relationships. However, the comparison of the values of *hasρ* and *hasχ* helps to implement domain-specific rules. Using this information, it is possible to implement rules such as

The identity is only inherited when there is a filiation for which both hasρ and hasχ are maximum values.

This would identify the strongest relation between a parent and a child in an environment where there are multiple parents and children involved.

$$
\begin{aligned}
\forall Filiation(f) \mid\ & hasParentTS(f,p) \\
& \wedge\ hasChildTS(f,c) \wedge isMaxHas\ \rho(f,True) \\
& \wedge\ isMaxHas\chi(f,True) \rightarrow (p.o = c.o) \wedge hasSameIdentity(p,c)
\end{aligned}
\tag{10.19}
$$

More complex rules can be easily defined, for instance, by assigning minimum thresholds for *hasχ* or *hasρ*. For instance

$$
\begin{aligned}
\forall Filiation(f) \mid\ & hasParentTS(f,p) \wedge hasChildTS(f,c) \\
& \wedge\ isMaxHas\rho(f,True) \wedge isMaxHas\chi(f,True) \wedge has\rho(f, \geq 0.9) \\
& \wedge\ has\chi(f, \geq 0.9) \rightarrow (p.o = c.o) \wedge hasSameIdentity(p,c)
\end{aligned}
\tag{10.20}
$$

10.4.4 Land Cover Taxonomy and Identity Inheritance

The identity inheritance process can be defined by more complex rules based on other TS characteristics. There exist classifications for land cover that can be used to qualify a land cover change. For instance, CORINE classification offers a hierarchical classification. Figure 10.4 depicts part of the CORINE land cover taxonomy.

$$
AgriculturalAreas \sqsubseteq TS
\tag{10.21}
$$
$$
ArtificialSurfaces \sqsubseteq TS
$$
$$
ForestSemiNaturalAreas \sqsubseteq TS
$$
$$
WaterBodies \sqsubseteq TS
$$
$$
WetlandAreas \sqsubseteq TS
$$
$$
ArableLand \sqsubseteq AgriculturalAreas
\tag{10.22}
$$
$$
HeterogeneousAgric \sqsubseteq AgriculturalAreas
$$

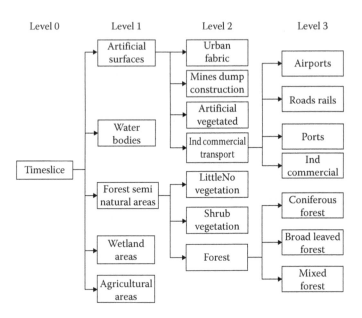

FIGURE 10.4
Partial view of the Corine Land Cover taxonomy.

$$Pastures \sqsubseteq AgriculturalAreas$$

$$PermanentCrops \sqsubseteq AgriculturalAreas$$

$$ArtificialVegetated \sqsubseteq ArtificialSurfaces \qquad (10.23)$$

$$IndustrialCommercialTransportation \sqsubseteq ArtificialSurfaces$$

$$MinesDumpConstruction \sqsubseteq ArtificialSurfaces$$

$$UrbanFabric \sqsubseteq ArtificialSurfaces$$

Using the taxonomy, we can create more complex identity inheritance rules that take into consideration the nature of parent and child TSs. For instance, a change of land cover between a parent and a child would be less severe from *ArableLand* to *Pastures* compared to a change from *ArableLand* to *UrbanFabric*.

Using the model, we can define rules such as

$$\forall Filiation(f) \mid hasParentTS(f,p)$$

$$\wedge \ hasChildTS(f,c) \wedge isMaxHas\rho(f,True)$$

$$\wedge \ isMaxHas\chi(f,True) \wedge ArtificialSurfaces(p) \qquad (10.24)$$

$$\wedge \ ArtificialSurfaces(c) \rightarrow$$

$$(p.o = c.o) \wedge hasSameIdentity(p,c)$$

In this case, we restrict the identity inheritance only to cases in which both parent and child are members of the class *ArtificialSurfaces*. The rule can also be modified to trigger an alert when there is some identity inheritance process that deserves further attention. For instance, in a deforestation scenario, when the parent is of type *MixedForest* while the child is of type *IndustrialCommercial*.

10.4.5 Identification of Process Type

Using the instances of the *Filiation* class, it is possible to identify processes involving one-to-one, one-to-many, and many-to-one TSs. In this section, we rewrite Equations 10.10 through 10.13, 10.15, and 10.16, to be able to identify different spatial processes using the values stored in the related instances of *Filiation*, without using expensive geometric functions, thus reducing spatial processing.

10.4.5.1 *Expansion*

$$TS(p), TS(c) \tag{10.25}$$

$$\forall Filiation(f) \,|\, hasParentTS(f,p)$$
$$\wedge\ hasChildTS(f,c) \wedge hasContinuation(p,c)$$
$$\wedge\ has\chi(f,<1) \wedge has\rho(f,1) \rightarrow hasExpansion(p,c)$$

By using this implementation, we avoid the use of the spatial operation *Within* (Equation 10.10).

10.4.5.2 *Contraction*

$$TS(p), TS(c) \tag{10.26}$$
$$\forall Filiation(f) \,|\, hasParentTS(f,p)$$
$$\wedge\ hasChildTS(f,c) \wedge hasContinuation(p,c)$$
$$\wedge\ has\chi(f,1) \wedge has\rho(f,<1) \rightarrow hasContraction(p,c)$$

In this case, we avoid the use of the operation *Contains* (Equation 10.11).

10.4.5.3 *Separation*

$$TS(p),\ \forall c \in [c_0,c_1...c_n], TS(c) \tag{10.27}$$

$$\forall [p,c] \exists Filiation(f) \,|\, hasParentTS(f,p) \wedge hasChildTS(f,c)$$

$$if \; \left(\forall \; has\chi(f,1)\right) \wedge \left(sum\big(has \; \rho(f)=1\big)\right.$$
$$\wedge \left(\exists_{=1} hasContinuation(p,c)\right) \rightarrow hasSeparation\big(p,[c_0,c_1...c_n]\big)$$

In this case, we identify all the instances of Filiation involved in the process. We assume that children TSs do not overlap between them. By obtaining the summation of the values of the property *hasρ*, we can determine if the combined geometry of the children corresponds to the one of the parent.

10.4.5.4 Annexation

$$\forall p \in [p_0,p_1...p_n], TS(p), TS(p) \tag{10.28}$$

$$\forall [p,c] \exists Filiation(f)\,|\, hasParentTS(f,p) \wedge hasChildTS(f,c)$$

$$if \; \left(\forall \; has\rho(f,1)\right) \wedge \left(sum\big(has\chi(f)=1\big)\right.$$
$$\wedge \left(\exists_{=1} hasContinuation(p,c)\right) \rightarrow hasAnnexation\big([p_0,p_1...p_n],c\big)$$

We assume that parent TSs involved do not overlap between them. To analyze this process, first we identify all the instances of Filiation that link the parents and the child. Then, we calculate the addition of the values of the property *hasχ*. If the result of the addition is equal to one, we can affirm that the combined geometry of all the parents corresponds to the geometry of the child.

10.4.5.5 Splitting

$$TS(p), \; \forall c \in [c_0,c_1 ... c_n], TS(c)$$

$$\forall [p,c] \exists Filiation(f)\,|\, hasParentTS(f,p) \wedge hasChildTS(f,c) \tag{10.29}$$

$$if \; \left(\forall \; has\chi(f,1)\right) \wedge \left(sum\big(has\rho(f)=1\big)\right.$$
$$\wedge \left(\neg \exists hasContinuation(p,c)\right) \rightarrow hasSplitting\big(p,[c_0,c_1...c_n]\big)$$

10.4.5.6 Merging

$$\forall p \in [p_0,p_1...p_n], TS(p), TS(p) \tag{10.30}$$

$$\forall [p,c] \exists Filiation(f)\,|\, hasParentTS(f,p) \wedge hasChildTS(f,c)$$

$$if \left(\forall\ has\rho\left(f,1\right)\right) \wedge \left(sum\left(has\ \chi\left(f\right)=1\right)\right.$$
$$\wedge \left(\neg\exists hasContinuation\left(p,c\right)\right) \rightarrow hasMerging\left(\left[p_0,p_1 \ldots p_n\right],c\right)$$

By using the values stored in instances of *Filiation*, we reduce the processing load. Previous research by Del Mondo et al. (2013) uses a construction similar in nature to the class *Filiation*; however, they use spatial operators such as *Union* or *Equals* to identify the evolution processes. In our approach, we reuse the results of the filiation identification process to identify relevant types of evolution. In our approach, we use basic arithmetic operators, reducing in this way the computing cost.

10.5 Model Implementation

To test our model, we opted for using LULCC information from CORINE. The information was obtained as raster with a pixel resolution of 100 m. The data corresponds to three time points, being the years 1990, 2000, and 2006 (EEA 2014).

The CORINE dataset covers multiple countries. For the purposes of testing our model, we decided to use a portion of the whole dataset. In this research, we use only the information contained within the boundaries of mainland Portugal.

Portugal is an interesting example for the study of LULCC. This country has a high incidence of forest fires compared to other European countries. Research conducted by Varela (2006) identified that by the year 2003, Portugal had 3,200,000 ha. of forests, which represented nearly one-third of the country surface. If we add to this the areas covered by shrub land, this percentage becomes higher than 50% of the country's area. In the same research, the authors identified several conditions that increase fire risk, such as the following: (1) The abandonment of agricultural parcels that become unmanaged shrub and forest lands. This is caused by migration from rural to urban areas, ageing of the rural population, and the loss of value for agricultural products among other things. (2) The existence of large quantities of exotic species such as eucalyptus, which is cultivated for the production of the cellulose pulp, and even the fact that eucalyptus have higher burning rates compared to native species.

In the past, agricultural areas represented a buffer area between urban area and forest. However, due to the abandonment of farms, this buffer area is disappearing, increasing the risk for urban areas (Paton and Fantina 2013).

A forest fire not only represents economic loss but also represents grave disturbances in ecological systems, with losses in fauna and flora, and release of CO_2. After the fire, the affected area loses the coverage that protects it from erosion, leading to further land degradation.

To model the LULCC, we vectorized the original CORINE raster data using ArcGIS. The results, encoded as shapefiles, were then translated into RDF triples using a custom-made JAVA program using the library GeoTools (OSGF 2014). The information in triple format was then uploaded into a Stardog (Clark and Parsia 2014) triplestore.

At the moment, Stardog does not offer support for GeoSPARQL (OGC 2011, Clark and Parsia 2014); therefore, spatial analysis has to be made with tools external to the triplestore. In our case, we have developed a tool based on JAVA/Geotools to perform all the required spatial analysis.

In our research, polygons for each time point were identified and encoded as a TSs. Then our JAVA program queries the triplestore and retrieves the TSs, using a spatial index. Next, it proceeds to identify the filiation relationships, taking into consideration the overlapping between TSs of consecutive periods. Our application also identifies the adjacency relations for TSs that coexist in time. After the relations have been identified, they are translated into triples and uploaded into the triplestore.

After processing the datasets and uploading the information to the triplestore, we have a knowledge base of 3.9 million triples.

10.5.1 Derivation Processes

In this section, we will describe the implementation of *Derivation* processes, assuming that no identity inheritance rule has been defined in the model.

10.5.1.1 Splitting

In this case, we implement Equation 10.29 with the following SPARQL query:

```
select * where
{
{
select ?p ?lc _ Code ?geo (sum(?rho) as ?SumRho)
(count(?c) as ?countC) (sum(?xhi) as ?sumXhi)
where{
?f a checksem:Filiation.
?f checksem:hasParentTS ?p.
?f checksem:hasChildTS ?c.
?f checksem:hasRho ?rho.
?f checksem:hasXhi ?xhi.
?p checksem:hasGeometry ?geo.
?p checksem:hasTime checksem:Time _ 1990.
?p checksem:hasLandCoverCode ?lc _ Code.
FILTER(?xhi = 1)
}
group by ?p ?lc _ Code ?geo
}
FILTER((?SumRho = 1)&&(?countC>1))
}
```

1990   2000

FIGURE 10.5 (See color insert)
Example of a split process. Background map from OSM (2014).

Figure 10.5 depicts a split process between the year 1990 and 2000. In this case, the timeslice ts _ 1990 _ 25040 corresponding to the year 1990, splits into four TSs: (c1) ts _ 2000 _ 25938, (c2) ts _ 2000 _ 26295, (c3) ts _ 2000 _ 26324 and (c4) ts _ 2000 _ 26359, corresponding to the year 2000. In this example, the parent TS has land cover *Mixed Forest*, while two of her children keep the same land cover type; we can see that there are an other two that change land cover to *Shrub Woodland* (Table 10.1).

10.5.1.2 Merging

A similar approach can be used to identify *Merging* processes. The following SPARQL implements Equation 10.30, which detects *Merging* processes that surged in the year 2000 (checksem:Time _ 2000).

```
select * where
{
{
select ?c ?lc _ Code ?geo (sum(?xhi) as ?SumXhi)
(count(?p) as ?countP) (sum(?rho) as ?sumRho)
where{
?f a checksem:Filiation.
?f checksem:hasParentTS ?p.
?f checksem:hasChildTS ?c.
?f checksem:hasRho ?rho.
?f checksem:hasXhi ?xhi.
?c checksem:hasGeometry ?geo.
?c checksem:hasTime checksem:Time _ 2000.
?c checksem:hasLandCoverCode ?lc _ Code.
FILTER(?rho = 1)
}
group by ?c ?lc _ Code ?geo
}
FILTER((?SumXhi = 1)&&(?countP>1))
}
```

Figure 10.6 depicts one of the identified merging processes. In this case, the timeslice ts _ 2000 _ 16622 is the result of the merges of three parent TSs:

TABLE 10.1

Parent and Children Timeslices in a Split Process

	Parent	**LandCover**
p1	ts_1990_25040	Mixed Forest
	Child	**LandCover**
c1	ts_2000_25938	Shrub Woodland
c2	ts_2000_26295	Mixed Forest
c3	ts_2000_26324	Shrub Woodland
c4	ts_2000_26359	Mixed Forest

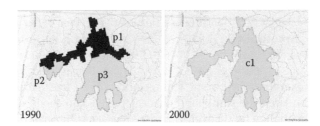

FIGURE 10.6 (See color insert.)
Example of a merge process. Background map from OSM (2014).

TABLE 10.2

Parent and Child Timeslices in a Merge Process

	Parent	**LandCover**
p1	ts_1990_15486	Burned areas
p2	ts_1990_15497	Moorland
p3	ts_1990_15777	Moorland
	Child	**LandCover**
c1	ts_2000_16622	Moorland

(p1)ts _ 1990 _ 15486, (p2)ts _ 1990 _ 15497, and (p3)ts _ 1990 _ 15777. In this example, the geometry of three TSs merge into a new one. In the example, the parent TSs have two types of land cover type, *Burned Areas* and *Moorland*, while the child TS has landcover type Moorland, indicating a land cover change in the areas previously burned (Table 10.2).

10.5.2 Identity Inheritance

In our research, we do not have any *a priori* information regarding the identity inheritance. However, for the sake of the argument and to show the effectiveness of our approach, we implemented some of the rules previously presented in the form of equations. The following SPARQL code implements Equation 10.19, which assigns the identity of a certain TS to a given object based on the maximum values of the properties *hasχ* and *hasρ*.

```
insert
{
?tsC checksem:isTimeSliceOf ?o.
}
where
{
?o a checksem:Feature.
?tsP a checksem:TimeSlice.
?tsC a checksem:TimeSlice.
?tsP checksem:isTimeSliceOf ?o.
?f1 a checksem:Filiation.
?f1 checksem:hasParentTS ?tsP.
?f1 checksem:hasChildTS ?tsC.
?f1 checksem:isMaxHasChi "true"^^xsd:boolean.
?f1 checksem:isMaxHasRho "true"^^xsd:boolean.
}
```

10.5.3 Continuation Processes

These kinds of processes involve the preservation of the identity from a parent to a child.

10.5.3.1 Separation

This process is similar in nature to Splitting. However, in this case one of the resulting children has the same identity as that of the parent.

Let us assume that we implement the rule specified in Equation 10.19 and in Section 10.4.2. Using the same example introduced in Section 10.4.1.1, we have a parent timeslice ts _ 1990 _ 25040, which is a temporal representation of an object identified as feature _ 1990 _ 16174.

After 1990, the parent TS is divided into four children, although only one of them keeps the identity of the parent, representing the same object. The following SPARQL code queries the knowledge base for the children of timeslice ts _ 1990 _ 25040. The results obtained from this query can be seen in Table 10.3.

```
select ?f ?c ?xhi ?rho ?o
where
{
?f a checksem:Filiation.
?f checksem:hasParentTS checksem:ts _ 1990 _ 25040.
?f checksem:hasChildTS ?c.
?f checksem:hasXhi ?xhi.
?f checksem:hasRho ?rho.
?c checksem:isTimeSliceOf ?o
}
```

TABLE 10.3

Identity Inheritance in a Separation Process

Filiation	Child	*hasχ*	*hasρ*	Identity
ts_1990_25040_ ts_2000_25938	ts_2000_25938	1.00	0.031	feature_2000_2619
ts_1990_25040_ ts_2000_26295	ts_2000_26295	1.00	0.870	feature_1990_16174
ts_1990_25040_ ts_2000_26324	ts_2000_26324	1.00	0.062	feature_2000_2726
ts_1990_25040_ ts_2000_26359	ts_2000_26359	1.00	0.037	feature_2000_2734

In Table 10.3, the first column indicates the URL of the filiation that links parent and child, the second column contains the URL of the children, the third and fourth column contain the values of the properties *hasχ* and *hasρ*, and, finally, the fifth column contains the identity of the object that the children TSs represent. By examining the values we can see that the timeslice ts _ 2000 _ 26295 has the same identity as the parent timeslice (feature _ 1990 _ 16174), implementing the rule specified in Equation 10.19.

10.5.3.2 Annexation

This process is similar in nature to Merging. However, in this case the identity of one of the parents involved is preserved in the resulting child.

As in Section 10.5.3.1, we implement the rule specified in Equation 10.19 and in Section 10.4.2. When identity inheritance rules are applied to the example provided in Section 10.4.1.2, we can identify one parent TS that shares the same identity as the child. The following SPARQL code gives us information related to the parents of timeslice ts _ 2000 _ 16622.

```
select ?f ?p ?xhi ?rho ?o
where
{
?f a checksem:Filiation.
?f checksem:hasChildTS checksem:ts _ 2000 _ 16622.
?f checksem:hasParentTS ?p.
?f checksem:hasXhi ?xhi.
?f checksem:hasRho ?rho.
?p checksem:isTimeSliceOf ?o
}
```

Table 10.4 depicts the results from the previous query. In this table we can see in the first column the URL of the filiation. The second column

TABLE 10.4

Identity Inheritance in an Annexation Process

Filiation	Parent	$has\chi$	$has\rho$	Identity
ts_1990_15486_ ts_2000_16622	ts_1990_15486	0.393	1.00	feature_1990_6097
ts_1990_15497_ ts_2000_16622	ts_1990_15497	0.065	1.00	feature_1990_6109
ts_1990_15777_ ts_2000_16622	ts_1990_15777	0.542	1.00	feature_1990_6420

depicts the URL of the parents. The third and fourth columns contain the values for the properties *hasχ* and *hasρ*, respectively. Finally, the fifth column contains the URL of the object that the parent TS represents. By examining the values we can see that the filiation relationship with the highest values for *hasχ* and *hasρ* is ts _ 1990 _ 15777 _ ts _ 2000 _ 16622, then we know that both timeslices ts _ 1990 _ 15777 and ts _ 2000 _ 16622 are temporal representations, at different points of time, of the object feature _ 1990 _ 6420.

10.5.4 Land Cover Change and Identity, Case Example: Increase of Wildfire Risk Areas

Portugal has a high incidence of forest fires compared to other European countries. In 2010, half of the fires in southern Europe were located in Portugal (Paton and Fantina 2013). Combined, forested areas and shrub land represent more than 50% of the surface of the country (Varela 2006).

Research conducted by Paton and Fantina (2013) indicates that in the past farming areas behaved as a buffer between forests and urban areas, protecting in this form towns and cities from forest fires. However, nowadays this buffer is disappearing due to rural depopulation, population aging, and loss of economic value for agricultural activities. Currently, it is possible to see abandoned farms that turn into unmanaged shrub land or forests increasing fire risks and increasing the risk for adjacent urban areas.

A basic three-level classification of fire susceptibility is proposed by Baptista and Carvalho (2002): (1) Null—agriculture riparian vegetation, land burned in the last two years, and urban and irrigated agricultural areas. (2) Medium—Brush land and rock outcrop. (3) High—Forest land and brush land with high density and fuel loads.

Using the model, we can define objects that become fire risks, as objects that in one point have a land cover that represents low or null fire risk, and in a later point evolve, having a land cover that makes them more prone to fires. We can define a SPARQL query that identifies objects with this kind of evolution:

```
select *
where
{
?o a checksem:Feature.
?ts1990 checksem:isTimeSliceOf ?o.
?ts2000 checksem:isTimeSliceOf ?o.
?ts1990 a checksem:AgriculturalAreas.
?ts2000 a checksem:ForestSemiNaturalAreas.
?ts1990 checksem:hasTime checksem:Time_1990.
?ts2000 checksem:hasTime checksem:Time_2000.
}
```

Figure 10.7 depicts an object that evolves, increasing its fire risk. In 1990, object feature _ 1990 _ 9064 is represented by the timeslice ts _ 1990 _ 18156 with land cover *Arable land, non irrigated*. However, by the year 2000, the same object is represented by timeslice ts _ 2000 _ 19078 with land cover *Natural Grass Lands*.

In this case, to answer the query, the ontology navigates through the class taxonomy and infers new statements. Because *ArableNonIrrigated* is a subclass of *ArableLand*, which itself is a subclass of *AgriculturalAreas*, then timeslice ts _ 1990 _ 18156 is also a member of class *AgriculturalAreas*. For the second part of the query, we have timeslice ts _ 2000 _ 19078, which is a member of class *NaturalGrassLands*, which is a subclass of *ShrubVegetation*, which itself is a subclass of *ForestSemiNaturalAreas*; then we can infer that timeslice ts _ 2000 _ 19078 is also a member of class *ForestSemiNaturalAreas* (Equation 10.31).

$$ArableNonIrrigated \sqsubseteq ArableLand \sqsubseteq AgriculturalAreas$$
$$AND \tag{10.31}$$
$$NaturalGrassLands \sqsubseteq ShrubVegetation \sqsubseteq ForestSemiNaturalAreas$$

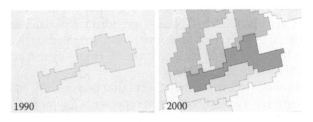

FIGURE 10.7 (See color insert.)
Example of an increase of fire risk process. Background map from OSM (2014).

The fire risk of an object might increase due to changes on their environment. For instance, if the neighbor parcels suffer some land cover change that dramatically increases their own fire risk. Our model is capable of answering queries of this type by using the adjacency information. For instance, the following query would retrieve the objects that are located next to objects that have increased their fire risk by the year 2000.

```
select ?oNeighbour ?tsNeighbour2000
where
{
?o a checksem:Feature.
?ts1990 checksem:isTimeSliceOf ?o.
?ts2000 checksem:isTimeSliceOf ?o.
?ts1990 a checksem:AgriculturalAreas.
?ts2000 a checksem:ForestSemiNaturalAreas.
?ts1990 checksem:hasTime checksem:Time_1990.
?ts2000 checksem:hasTime checksem:Time_2000.
?oNeighbour a checksem:Feature.
?tsNeighbour2000 checksem:isAdjacentTo ?ts2000.
}
```

Figure 10.7 depicts the results of this query.

10.6 Discussion

The starting point of our analysis is data in raster format. We vectorize the information and create objects based on pixel similarity. An important limitation with this approach is that we cannot distinguish between adjacent areas with similar cover but different nature due to factors like management, ownership, and so on. For instance, consider two adjacent areas: the first one is a national park, while the second is a forestry exploitation. In our data, due to their similar cover, both areas would be undistinguishable and would be considered part of the same object. However, due to the different management, the evolution of these adjacent areas would be different. The addition of ancillary information is no challenge from a technical point of view; however, it might be more difficult to obtain. At the time of performing our research, a layer of information of this nature was not available. In future work we propose to include this aspect of LULCC in our analysis.

In our work, we use information from the CORINE program. Therefore, we can safely assume that the class definitions and classification methodology is the same in the three datasets we are using (1990, 2000, and 2006). However, researchers should be aware that this might not be the case when integrating data from different sources. For instance, categories such as deciduous or coniferous or commercial forest can have different meanings in different countries. In addition, it might be the case that land cover datasets corresponding to different time points use different land cover classification schemas. In these cases, it would be necessary to select one classification schema and map all the others into it. Further reading in this topic can be found in the work of Ahlqvist (2005).

The need for the identification of parent–child relations between entities has been mentioned in previous research by Del Mondo et al. (2010) and Del Mondo et al. (2013). To perform this procedure, it is necessary to identify geometric overlaps between entities belonging to different times. Then, in a separate step, the identification of spatiotemporal processes such as splitting, separation, annexation, and merging is performed using topological operators such as Union and Equals. Our approach is similar to previous ones only in the first step. We do identify parent–child relationships using geometric overlaps. However, we reuse the overlapping information to identify spatiotemporal processes, avoiding the use of costly topological operators (Union/Equals). Our approach is computationally low cost, allowing us to work within the operational limits of a triplestore.

In our approach, we use a Stardog triplestore as our data repository. At this point we do not use GeoSPARQL, the spatial extension of SPARQL (OGC, 2011). The reason for this is the limited number of triplestores that support GeoSPARQL. In the case of Stardog, the support for GeoSPARQL is a work in progress (Stardog Support 2014). In our research to supplement the lack of GeoSPARQL, we developed a custom-made Java program that uses the library GeoTools to perform all our spatial operations. The result of any spatial operation is then uploaded into the triplestore. In our implementation, all the information is managed natively as a graph. In future research we would study the best way in terms of performance to include GeoSPARQL in our model.

In our research, after evaluating the LULCC, we decided to loosen restrictions suggested in other studies by Del Mondo et al. (2010). In our model, we allow a child to have multiple parents and a parent to have multiple children, which reflects the reality of spatial entities. However, a child can have only one continuation relationship, while the other relations must be of type derivation. By enforcing this constraint, we ensure that there are no two objects with the same identity existing at the same time.

In Section 10.5.4, we offered basic examples of the use of our model to identify patterns of change related to wildfire risks. More complex examples can be implemented using the model components as building blocks. Using expert knowledge it is possible to identify patterns of change that are associated with events of interest for a given domain. Our model facilitates the encoding of expert knowledge as constraints and rules. Our model is intended to be flexible enough so it can be easily adapted to multiple domains.

10.7 Conclusion

In this chapter, we have presented a model designed to handle dynamic information with spatiotemporal components. The model is implemented using Semantic Web tools. The resulting model represents a dynamic system as a graph, allowing the identification of complex relationships. Using the model capabilities, it is possible to identify spatiotemporal patterns of change. Our approach allows the encoding of domain-specific knowledge as axioms and constraints. Using this knowledge, it is possible to detect complex evolutions, for instance, increase of wildfire risk.

We contribute to the field of GEOBIA by offering an approach to handle the temporal dimension of dynamic spatial information. The temporal dimension is not commonly used in GIS. Blaschke et al. (2014) indicates that in the field of GEOBIA, it adds a layer of complexity that is often avoided. However, using the temporal dimension in the analysis allows the identification of the evolution of entities along time. In our research, we define a flexible methodology that can be used to study spatial objects evolving in time and space. We use our approach to study land cover objects. However, this approach can be used for other applications due to its flexibility.

By using Semantic Web tools, we are able to model spatial dynamics knowledge as axioms and constraints in a format that can be easily shared, thanks to its formal semantics.

Thanks to the use of a Semantic Web approach, we identify each element of the model with a URI. Although at the moment this is not implemented, we could use the URIs to link model elements to external data sources using federated RDF queries. This would significantly increase the model capabilities.

By aggregating pixels into areas (polygons) and performing our analysis at this level, we are able to perform contextual analysis. For each entity of interest, we are capable to identify its neighborhood in space and time, allowing us to better study the evolution of entities. However, this approach can be improved by adding additional information that would help us to differentiate pixels with similar value but of different nature. In future work, we will include additional layers on information such as land ownership and elevation, among others, to increase the sophistication of the analysis.

Acknowledgments

We thank the Direction Generale de l'Armement (DGA) and the Conseil Regional de Bourgogne for the financial support for the development of our work.

References

Al Debei, Mutaz M. 2012. "Conceptual Modelling and the Quality of Ontologies : Endurantism Vs Perdurantism." *International Journal of Database Management Systems* 4(3): 1–19.

Ahlqvist, Ola. 2005. "Using Uncertain Conceptual Spaces to Translate between Land Cover Categories." *International Journal of Geographical Information Science* 19(7): 831–857.

Armstrong, Marc P. 1988. "Temporality in Spatial Databases." In *Proceedings GIS/LIS 88*, 880–889. San Francisco, CA.

Artale, Alessandro and Enrico Franconi. 1998. "A Temporal Description Logic for Reasoning About Actions and Plans." *Journal of Artificial Intelligence Research* 9: 463–506.

Arvor, Damien, Laurent Durieux, Samuel Andrés, and Marie-Angélique Laporte. 2013. "Advances in Geographic Object-Based Image Analysis with Ontologies: A Review of Main Contributions and Limitations from a Remote Sensing Perspective." *ISPRS Journal of Photogrammetry and Remote Sensing* 82: 125–137.

Aryal, Jagannath, Ahsan Morshed, and Ritaban Dutta. 2014. "Land Cover Class Extraction in GEOBIA Using Environmental Spatial Temporal Ontology." *South Eastern European Journal of Earth Observation and Geomatics* 3(2S): 429–434.

Baptista, Manuela, and Josefa Carvalho. 2002. "Fire Situation in Portugal." *International Forest Fire News*.

Batsakis, Sotiris, and Euripides G. M. Petrakis. 2011. "SOWL: A Framework for Handling Spatio-Temporal Information in OWL 2. 0." *Lecture Notes in Computer Science* 6826: 242–249.

Bittner, Thomas, Maureen Donnelly, and Barry Smith. 2009. "A Spatio-Temporal Ontology for Geographic Information Integration." *International Journal of Geographical Information Science* 23(6): 765–798.

Blaschke, Thomas. 2010. "Object Based Image Analysis for Remote Sensing." *ISPRS Journal of Photogrammetry and Remote Sensing* 65(1): 2–16.

Blaschke, Thomas, Geoffrey J. Hay, Maggi Kelly, Stefan Lang, Peter Hofmann, Elisabeth Addink, Raul Queiroz Feitosa, et al. 2014. "Geographic Object-Based Image Analysis: Towards a New Paradigm." *ISPRS Journal of Photogrammetry and Remote Sensing : Official Publication of the International Society for Photogrammetry and Remote Sensing (ISPRS)* 87(100): 180–191.

Chen, Jun, Hao Wu, Songnian Li, Anping Liao, Chaoying He, and Shu Peng. 2013. "Temporal Logic and Operation Relations Based Knowledge Representation for Land Cover Change Web Services." *ISPRS Journal of Photogrammetry and Remote Sensing* 83: 140–150.

Claramunt, Christophe, Marius Theriault, and Christine Parent. 1997. "A Qualitative Representation of Evolving Spatial Entities in Two-Dimensional Topological Spaces." *Innovations in GIS V*, 119–129.

Clark and Parsia. 2014. "Stardog Website." www.stardog.com (accessed August 15, 2014).

Del Mondo, Geraldine, M. Andrea Rodríguez, Christophe Claramunt, Loreto Bravo, and Remy Thibaud. 2013. "Modeling Consistency of Spatio-Temporal Graphs." *Data & Knowledge Engineering* 84: 59–80.

Del Mondo, Geraldine, John G. Stell, Christophe Claramunt, and Remy Thibaud. 2010. "A Graph Model for Spatio Temporal Evolution." *Journal of Universal Computer Science* 16(11): 1452–1477.

EEA. 2014. "CORINE Land Cover." *European Environment Agency Institutional Website*. http://www.eea.europa.eu/publications/COR0-landcover (accessed August 15, 2014).

Egenhofer, Max J., and Khaled K. Al-taha. 1992. "Reasoning about Gradual Changes of Topological Relationships." *Lecture Notes in Computer Science* 639: 196–219.

Gruber, Thomas R. 1995. "Toward Principles for the Design of Ontologies Used for Knowledge Sharing?" *International Journal of Human-Computer Studies* 43(5–6): 907–928.

Gutierrez, Claudio, Carlos Hurtado, and Alejandro Vaisman. 2007. "Introducing Time into RDF." *IEEE Transactions on Knowledge and Data Engineering* 19(2): 207–218.

Harbelot, Benjamin, Helbert Arenas, and Christophe Cruz. 2013a. "A Semantic Model to Query Spatial–Temporal Data." *Lecture Notes in Geoinformation and Cartography, Information Fusion and Geographic Information Systems*: 75–89.

Harbelot, Benjamin, Helbert Arenas, and Christophe Cruz. 2013b. "Continuum: A Spatiotemporal Data Model to Represent and Qualify Filiation Relationships." In *4th. ACM SIGSPATIAL International Workshop on GeoStreaming (IWGS)*. Orlando, FL.

Herold, Hendrik, Gotthard Meinel, Robert Hecht, and Elmar Csaplovics. 2012. "A GEOBIA Approach to Map Interpretation Multitemporal Building Footprint Building Retrieval for High Resolution Monitoring of Spatial Urban Dynamics." *Proceedings of the 4th. GEOBIA*, 252–256. Rio de Janeiro, Brazil.

Hornsby, Kathleen, and Max J. Egenhofer. 2000. "Identity-Based Change: A Foundation for Spatio-Temporal Knowledge Representation." *International Journal of Geographical Information* 14(1): 207–224.

Jiang, Jixiang, and Michael Worboys. 2009. "Event-Based Topology for Dynamic Planar Areal Objects." *International Journal of Geographical Information Science* 23(1): 33–60.

Klein, Michel, and Dieter Fensel. 2001. "Ontology Versioning on the Semantic Web." In *Proceedings of the First International Semantic Web Working Symposium SWWS'01*, 75–91.

Lambin, Eric F, Billie L. Turner, Helmut J. Geist, Samuel B. Agbola, Arild Angelsen, John W. Bruce, Oliver T. Coomes, et al. 2001. "The Causes of Land-Use and Land Cover Change: Moving Beyond the Myths." *Global Environmental Change* 11(4):261–269.

Langran, Gail, and Nicholas R. Chrisman. 1988. "A Framework for Temporal Geographic Information." *Cartographica: The International Journal for Geographic Information and Geovisualization* 25(3): 1–14.

Mahmood, Rezaul, Arturo I. Quintanar, Glen Conner, Ronnie Leeper, Scott Dobler, Roger A. Pielke, Adriana Beltran-Przekurat, et al. 2010. "Impacts of Land Use/ Land Cover Change on Climate and Future Research Priorities." *Bulletin of the American Meteorological Society* 91(1): 37–46.

Meskovic, Emir, Galic Zdravko, and Mirta Baranovic. 2011. "Managing Moving Object in Spatio-Temporal Data Streams." In *Proceedings of the 2011 IEEE 12th. International Conference on Mobile Data Management*, 15–18. Washington, DC.

Morshed, Ahsan, Jagannath Aryal, and Ritaban Dutta. 2013. "Environmental Spatio-Temporal Ontology for the Linked Open Data Cloud." In *12th IEEE International Conference on Trust, Security and Privacy in Computing and Communications*, 1907–1912. IEEE, Melbourne, Australia.

Muller, Philippe. 2002. "Topological Spatio-Temporal Reasoning and Representation." *Computational Intelligence* 18(3): 420–450.

O'Connor, Martin J., and Amar K. Das. 2011. "A Method for Representing and Querying Temporal Information in OWL." *Biomedical Engineering Systems and Technologies* 127: 97–110.

OGC. 2011. "GeoSPARQL: A Geographic Query Language for RDF Data." http://www.opengeospatial.org/standards/geosparql (accessed August 15, 2014).

OSGF. 2014. "GeoTools." Open Source Geospatial Foundation. http://geotools.org (accessed August 15, 2014).

OSM. 2014. "OpenStreetMap" OpenStreetMap Contributors, accessed August 15, 2014, http://www.openstreetmap.org.

Paton, D. and T. Fantina 2013. "Enhancing Forest Fires Preparedness in Portugal: Integrating Community Engagement and Risk Management." *Planet@Risk* 1(1), 44–52.

Peuquet, Donna J., and Niu Duan. 1995. "An Event-Based Spatiotemporal Data Model (ESTDM) for Temporal Aanalysis of Geographical Data." *International Journal of Geographical Information Systems* 9(1): 7–24.

Sheeren, David, Sylvie Ladet, Olivier Ribiere, Bertrand Raynaud, Martin Paegelow, and Thomas Houet. 2012. "Assessing Land Cover Changes in the French Pyrenees Since the 1940s: A Semi-Automatic GEOBIA Approach Using Aerial Photographs." In Proceedings of the AGILE 2012 *International Conference on Geographic Information Science*, 1940–1942. Avignon, France.

Stardog Support. 2014. "Google Groups Stardog Support." accessed August 15, 2014, http://groups.google.com/a/clarkparsia.com/forum/#!forum/stardog

Stell, J., Geraldine Del Mondo, Remy Thibaud, and Christophe Claramunt. 2011. "Spatio-Temporal Evolution as Bigraph Dynamics." *Lecture Notes in Computer Science* 6899: 148–167.

Strobl, Christian. 2008. "Dimensionally Extended Nine Intersection Model (DE-9IM)." In *Encyclopedia of GIS*, edited by Shekhar Shashi and Hui Xiong, 240–245. Springer, US.

Varela, Maria Carolina. 2006. "The Deep Roots of the 2003 Forest Fires in Portugal." 34. Vol. 34. *International Forest Fire News*.

Worboys, Michael. 1994. "A Unified Model for Spatial and Temporal Information." *The Computer Journal* 37(1): 26–34.

11

Applying Tegon, the Elementary Physical Land Cover Feature, for Data Interoperability

Wim Devos and Pavel Milenov

CONTENTS

ABSTRACT Most land cover mapping initiatives have been biased toward optimized data capture and cartographic quality. Interoperability of the resulting data has proven difficult due to the semantic ambiguity embedded in the classification and methodology of each initiative, which often does not correctly reflect and account for the complexity and specificity of the landscape under observation. This chapter describes how the tegon concept can model land cover as a real-world phenomenon. Tegons are instances of the elementary physical components behind any existing mapping unit or legend class, expressed in the Land Cover Meta Language, specified in ISO 19144-2. Two large-scale examples from an agricultural context show how the concept has been used for demarcating the land cover universe of discourse and for harmonization efforts.

The tegon concept was developed by the Monitoring of Agriculture ResourceS (MARS) Unit of the Joint Research Centre (JRC) of the European Commission and first applied during the 2010 quality assessment of the Land

Parcel Identification Systems (LPIS). This exercise required a full description of the European agriculture land cover types of all European Union (EU) Member States (MS), based on the Food and Agriculture Organization of the United Nations (FAO) Land Cover Classification System (LCCS). Later, application of the tegon concept has been expanded toward cross-border land monitoring initiatives, such as the project SPATIAL involving Bulgaria and Romania, where it became the key methodological instrument for the land cover inventory and spatial data harmonization to cover the entire cross-border area representing a major part of Lower Danube Basin.

Both application experiences demonstrate the high potential of the concept, particularly for addressing complex land cover phenomena and for insuring interoperability between existing classifications and their data sets. This potential is most evident at large-scale data. Automation of the tegon modeling will be required to verify the claim of exhaustiveness and universality.

Application of tegon conceptualization during the inception of new classifications or data sets should introduce the correct semantics in those initiatives. However, as tegon modeling in itself does not address the limitations of current data-capture methodologies, impact for ongoing inventories will be limited to improving semantic interoperability.

11.1 Background

The Common Agricultural Policy (CAP) of January 1, 2005, decoupled the farmer aid from the farm output. This made the area of agricultural land the key parameter that controlled the amount of aid. To demonstrate compliance with the eligibility rules for this aid on farmland, the EU MS established a geographic information system (GIS) database called the Land Parcel Identification System (LPIS). The database identifies and quantifies the agricultural areas concerned. By 2009, threatened landscape features such as ponds, stonewalls, hedges, and trees could optionally be considered eligible land and were then added to the LPIS.

As the European institutions monitored and audited the MS' LPIS implementations and aid spending, discussion soon emerged about the nature and hence eligibility for aid on many agricultural lands registered in the LPIS. The discussion revealed the need to uniquely identify agricultural land based on its objective, physical properties. This caused the LPIS to focus on land cover: the observed biophysical cover of the earth's surface (Di Gregorio 2005) or the physical and biological cover of the earth's surface (European Commission 2007).

At the time when this LPIS need for land cover identification became apparent, the land cover domain had been shaped by three drivers:

1. Many land cover initiatives had been launched since the emergence of high-resolution satellite imagery in the early 1970s. Most focused on cartographic products to either extract as much detail from the imagery as possible or to provide maximum correspondence with known maps and data, for example, for change detection. These initiatives left the conceptual framework for the physical aspects largely underdeveloped. Several initiatives to harmonize international mapping initiatives, such as the EU's CORINE (European Commission 1994) and FAO's Land Cover Classification System (LCCS) (FAO 2009), never completely addressed this weak conceptual basis.

2. Advances in the information technology domain changed the expectations on spatial data from cartography toward true data management. This pushed an object-oriented approach for all data, including land cover. Reliance on the available capture methods would thus assume that any spatial information gathered through remote sensing reflects real physical objects on earth. For example, the spatial database features (polygons) would enclose areas with homogeneous properties with respect to land cover, or even slope and altitude. The BULCOVER project (Agency for Sustainable Development and Eurointegration [ASDE] 2010) made an early attempt to move beyond a "visible surface approach," combining remote sensing information with topographic data. Still, the rules of cartography and the coarse spatial resolution of the satellite data very much constrained the semantic information of the resulting data set, since the land cover map, not the dynamic GIS database, remained the basis for most communication and decision making.

3. The launch of reliable mid scale earth observation satellites produced a flood of daily to weekly time series that were processed by semiautomatic classification based on vegetation or phenological cycles. The image processing classes were subsequently correlated to a global legend to produce world land cover maps.

All these developments and successes earned land cover a prominent place in many applications and decisions, but the land cover phenomenon itself was essentially approached as the observed top surface, with a focus mainly on a tessellation of the earth's surface according to the cartographic specification of the particular project. Semantic interoperability between data from different scales, geographical zones, or projects remained a challenge.

Several efforts were therefore made to address this interoperability gap. The FAO LCCS enables a semantic comparison between existing data sets by passing over an exhaustive and mutually exclusive categorization of physical land cover phenomena. This developed into the metalanguage (Land Cover

Meta Language [LCML]), which is a basis for ISO19144-2 (ISO 2011). But LCCS does not challenge the observation as basis for modeling. Neither does a parameterized approach where the land cover observation nor polygon is regarded as a class but as a set of numerical attributes that represent its physical properties (Villa 2008).

The pan-European harmonization of existing land cover data sets, an idea abandoned by CORINE in 1990, was reawakened by the INSPIRE legislation. The INSPIRE Generic Conceptual Model, based on ISO19101, puts an emphasis on spatial objects and their attributes, but, for land cover, the elementary modeling class remained observation (polygon) based. A weak generic harmonization ex post was sought by providing a "pure land cover component nomenclature" (European Commission 2013a, 2013b) to be attached to a given land cover polygon with percentage values. The end result, a parameterization of contributing classes, offers little progress toward the physical conceptualization of land cover.

Most of the above mapping initiatives and harmonization efforts modeled the land cover universe of discourse, not by analyzing the objects that exist in the field, but from analyzing what features can be detected given a mapping unit and the spatial, spectral, and radiometric qualities of an image. The underlying thesis is that image interpretation or classification can offer a correct and complete representation of the physical land cover feature. Nevertheless, the observed land cover features (polygons) determined by the scale and timing of the data sources and their class semantics were biased toward the geographical extent of the particular initiative.

As for the CAP and LPIS, identifying agricultural land appropriately depended on modeling the physical concept of land cover, independent of the mapping initiative or observation method and purely based on what "is" on the ground. The concept of the tegon was introduced in 2010 for this purpose (Devos and Milenov 2013).

11.2 Tegon as the Land Cover Phenomenon

Land cover, the biophysical cover of the earth's surface, varies continuously over space and time, but can still be considered as composed of a series of discrete physical elements (Figure 11.1). This is similar to the variation in soil properties that occurs below the surface. The concept of the tegon (from the Latin words *tegere, tego, tectus,* meaning to cover) is indeed inspired by the notion of the pedon unit from soil science (Johnson 1962).

The tegon is the smallest horizontally homogeneous, physical spatial object with a notable three-dimensional extent and a specific life cycle. It can be visualized as an n-gonal prism that encloses a nongaseous substrate with uniform biophysical characteristics, properties, and life cycle (Figure 11.2).

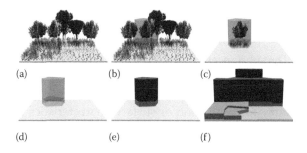

FIGURE 11.1
Modeling land cover with prisms. (a) Illustration of the real-world landscape, including grassland, scrubland, and woodland. (b) Prism reflecting a discrete homogeneous stretch inside the woodland. (c) The woodland prism in isolation, indicating presence of grass, scrub, and tree. (d) The woodland prism above the soil with volumes occupied by a grass layer, a scrub layer, and a tree layer (given in different transparent shades of green). (e) The woodland prism above the soil with volumes occupied by a grass layer, a scrub layer, and a tree layer (given in different solid shades of green). (f) The illustrated landscape of subpart (a) modeled through a contiguous series of such prisms with different strata and dimensions.

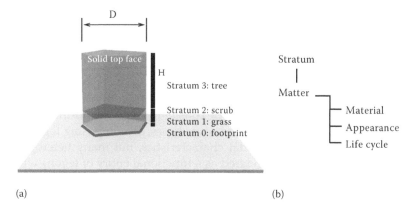

FIGURE 11.2
Schema of tegon characteristics. (a) Illustration of the selected woodland prism of Figure 11.1 with height H and face to face distance D; each stratum has matter with a distinct material, appearance, and life cycle. (b) Schematic representation of the same land cover phenomenon.

The tegon, as a three-dimensional, feature-based element, so represents a distinct and measurable material reality at the earth's surface; it can be considered in itself to be a real-world phenomenon. Its extent typically covers one to several square meters, characterized by the presence of a substrate, in one or more vertical biotic or abiotic strata, from the top of the soil O-horizon (stratum 0) to the solid top face visible from above.

From its base (stratum 0) upward, the tegon exhibits a liquid or solid base stratum, sometimes covered by higher substrate strata holding solid matter. The nadir projection of the n-gonal prism is called the tegon footprint.

Any tegon matter held by a stratum is further characterized by the following:

a. *Material:* The substance(s) that builds up matter. It can be biotic (vegetation) or abiotic (water, artificial construction, mineral deposit) in origin.
b. *Appearance:* The material-specific condition of the matter, such as physiognomy for vegetation or surface characteristics for bare soil, physical state for the water.
c. *Life cycle:* The periodicity or duration of material or appearance.

To model a homogenous land cover object as an individual tegon feature, four modeling rules apply as follows:

1. There must be at least one material in the lowest stratum.
2. Additional materials present on different heights are assembled into sequential strata.
3. The horizontal and vertical extent of the tegon must hold all materials in the strata and their intrinsic relationships, governed by the law of gravity and process of energy exchange (Valeeva 2009).
4. Tegon boundaries cannot be dependent on detection by a single observation approach.

A tegon acts as an undividable, homogenous three-dimensional floor tile. Every land cover phenomenon can be considered as a composition of contiguous tegons. In homogeneous land cover types, all composing tegons share similar biophysical or functional characteristics, but many heterogeneous land cover types can be considered as a functional entity of tegons with different characteristics (Figure 11.3).

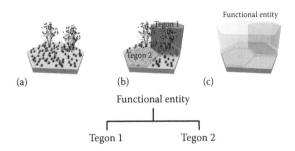

FIGURE 11.3
Schematic illustration of a functional entity with two "types" of tegons. (a) Illustration of the real-world landscape: grassland, with scattered scrubs. (b) Tegons describing discrete scrub and grassland phenomena. (c) The prism reflecting the functional entity of grassland-scattered scrub.

The functional entity identifies an intrinsically new land cover unit or class with distinctive functional characteristics. It derives these from the presence and distribution pattern of all its tegon composites.

Functional entities can originate from a mix with biotic and abiotic characteristics; urban areas are composed of tegons of artificial abiotic material (on buildings and roads), and tegons of biotic material (in parks, gardens, and beds). The prevalence of the composing tegon types can reflect environmental aspects through the proportion of the biotic and abiotic components within the mix. In urban areas, such proportion may provide clues on the quality of life and be indicative of good urban design with sufficient and high-quality public and green spaces (EEA 2010). In the CAP context, the prevalence may quantify the pro rata eligibility of land.

The geometry of a land cover phenomena can be allocated to either an individual tegon or, more frequently, to a series of contiguous tegons, the polytegon. An individual tegon footprint does not usually offer an acceptable graphical depiction of complex land cover phenomena; it can also be too small to exhibit all the characteristics of an individual land cover "object."

The polytegon is a more suitable vehicle for geometry, and it can represent both an occurrence of contiguous similar tegons and a distinctive functional entity of tegons of different types. Either occurrence can be expressed in a single taxonomic land cover class (Figure 11.4). Topological rules on the prisms impose that a polytegon's size cannot be smaller than the smallest individual tegon belonging to that class and that boundaries cannot extend across the boundaries of (poly)tegons of a different taxonomic class.

To ensure that tegon modeling allows for the appropriate feature description of any land cover phenomenon, its descriptors must be exhaustive. This is guaranteed by expressing tegon material and strata components exclusively in LCML terms. LCML aims to be exhaustive and exclusive, fully describing the land cover universe of discourse. Tegons inherit these properties so that any type of land cover phenomenon can be modeled by an appropriate combination of relevant tegons. The complete set of theoretical tegons supports each and every taxonomic land cover class.

As a bonus consequence, this tegon constraint allows it to clarify the relationship between the land cover continuum and any existing land cover taxonomic classes that are expressed through LCML.

The tegon conceptualization above relates to horizontal homogeneity and the structure of vertical strata of substrate matter (with material, appearance, and life cycle). These conceptual elements are independent from any specific mapping, methodology, portrayal requirement, or cartographic scale. These practical considerations merely depend on the user's expectation, available observation method, and product specifications.

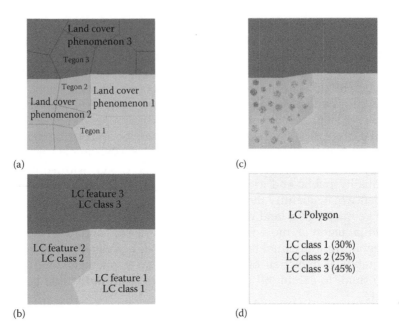

FIGURE 11.4
Application of tegon, land cover feature, and land cover polygon for classical two-dimensional land cover mapping; schematic illustrations based on top view of the landscape of Figure 11.1. (a) Real world: biophysical cover of the earth's surface, composed of three types of tegon (represented by their solid top surfaces). (b) Land cover observation: one observation instance of the land cover surface (top view). (c) A cartographical representation of the real-world 1:1 by three polytegons that represent a grouping of tegons of a similar type. (d) A cartographical representation of the real-world 1:100,000 by a single-mapped polygon that represents a grouping of different polytegons.

11.3 A Practical Tegon Example

The way that the tegon concept is applied in practice can be illustrated with a real LPIS-related use case example. The LPIS reference parcel in Figure 11.5 (contour), surveyed with a Global Navigation Satellite System (GNSS) device, is reported as enclosing normal "grazing" land. However, its northern part is covered by trees, and a traditional land cover observation cannot support the reported grassland delineation.

A field observation (Figure 11.6) conducted by the authors confirmed that the land below the tree cover is indeed different on the east and west sides. Whereas the east side had a grass stratum and no scrubs or low trees, the west sides had a thick layer of low trees but no grass. The observed difference is permanent as confirmed by Google StreetView © 2010 images taken 6 months earlier. On site, the tegon identification rules resulted in the detection in six distinct tegon types, T1 to T6. The set of photos in Figure 11.6b indicates the tegon type for selected ground locations.

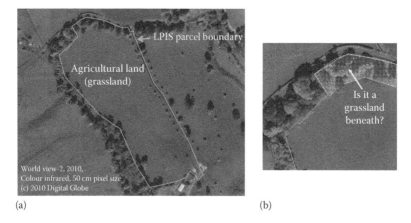

(a)

(b)

FIGURE 11.5 (See color insert.)
Overview image (a): Example of a reference parcel, enclosing "grazed" land; note the extension of the boundary inside the eastern part of the tree-covered area. Inset (b): Enlargement of the northern part. Includes material © DigitalGlobe (2010), all rights reserved.

(a)

(b)

FIGURE 11.6 (See color insert.)
(a) Aerial view of the field observation viewpoint with five arrows pointing the direction of the photographs taken on site (1, 2, and 3) or obtained from © Google Street View (4 and 5). Includes material © BING Maps (2010), all rights reserved. (b) The photographs capturing the substrate of the land cover phenomenon beneath the "solid top face." The abbreviations T1–T6 indicate the footprint location of a particular tegon Ti found on place. Includes material Street View © 2010 Google, all rights reserved.

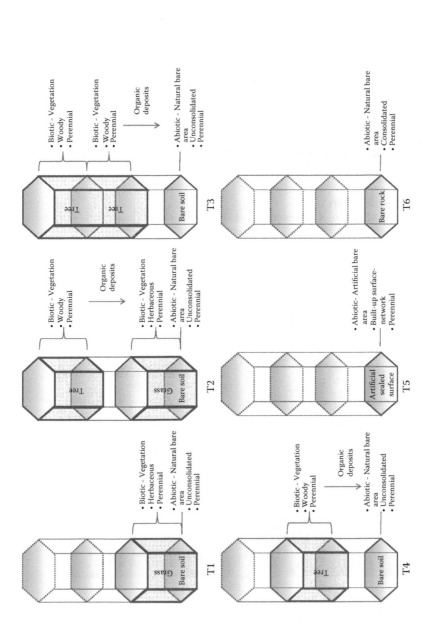

FIGURE 11.7

Schematic representation of the six tegon types positioned in Figure 11.6. The elements and attributes are described with the LCML semantics. The strata with matter are indicated by denser prism edges.

Figure 11.7 schematizes the characteristics of each of the six tegon types. The characteristics are expressed through LCML semantics within the tegon structure introduced in Figure 11.1. The presence of matter in the second and third strata and the nature of the material in those strata determine to which tegon type an individual tegon belongs.

Figure 11.8 shows the footprints of a selection of typical tegons in Figure 11.6b over WorldView-2 satellite and Bing aerial image backgrounds, respectively. Such graphical representations of individual footprints are generally of little practical use; they serve here to visualize the tegon concept.

A delineation methodology should strive to apply the polytegon as the unit of land cover observation, as illustrated by Figure 11.9a. This way, the polytegon corresponds to the spatial element (polygon) captured and portrayed by most land cover products (Figure 11.9b).

In general, the tegon–polytegon combination allows any land cover inventory to define the proper mapping units or legend classes in terms of the profusion of tegon types that can be identified inside the observed polytegons. The appropriate mapping rules then ensure that each polytegon occurrence results in a single polygon. Table 11.1 illustrates how four of the six tegon types of Figure 11.7 were combined to form the three polytegons relevant in this particular use case.

Although a polytegon can be associated with only one land cover class, two of the classes imply a continuous herbaceous layer, and their joined polygons would correspond to the polygon of the initial reference parcel, as shown in Figure 11.5a.

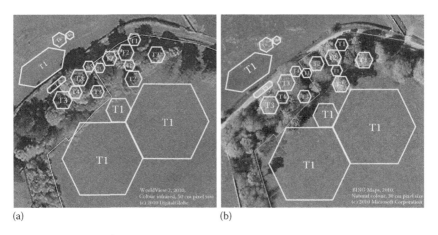

(a)   (b)

FIGURE 11.8 (See color insert.)
(a) Footprints of the tegons from Figures 11.6 and 11.7 presented on the area of interest over WorldView-2 satellite image. Includes material © DigitalGlobe (2010), all rights reserved. (b) The same footprints over an aerial image. The difference of visual appearance of the tegons is due to the different fields of view of the two acquisitions, Includes material © BING Maps (2010), all rights reserved. For simplicity and compliance with Figure 11.7, tegon footprints are drawn as hexagons.

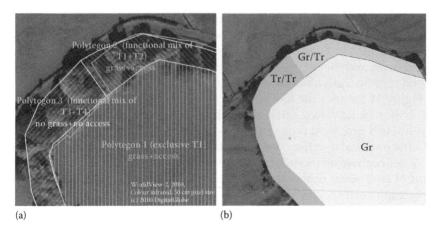

(a) (b)

FIGURE 11.9 (See color insert.)
(a) The resulted polytegons on the area of interest. (b) The resulted spatial features (polygons) on the area of interest stored in the GIS with their correspondent map codes. Includes material © DigitalGlobe (2010), all rights reserved.

TABLE 11.1

LC Class—Polytegon Relationship

Land Cover Class (LCCS)	Map Code	Polytegon
Trees	Tr/Tr	T3(50%) + T4(50%)
Grassland with Sparse Trees	Gr/Tr	T1(50%) + T2(50%)
Grassland	Gr	T1(100%)

11.4 Operational Tegon Cases

11.4.1 Describing Land Cover Types for the Common Agricultural Policy, Accommodating Grassland Complexity

When the tegon concept was first introduced in 2010 (Devos and Milenov 2013), it was tested through the 2010–2011 LPIS quality assessment (QA). The latter required the mapping of a sample of any agricultural land considered eligible for farmer aid and was technically coordinated by the authors in the MARS unit. The exercise was rather limited in area totals, but all MS (or where appropriate, their regions) had to participate; the EU territory is effectively covered by 43 + 1 systems. On the basis of the CAP regulations' land cover concepts (arable land, natural grassland, permanent crops, specific arable crops but also some generic land use expressions), the authors had compiled a full set of 27 predefined agriculture land cover classes. This was based on the tegon philosophy with the semantics of the FAO Land Cover Classification System (v.2). To offer a simpler alternative, the authors produced a minimum set with 10 generic agricultural polytegon seeds by aggregating the 27 detailed classes (Milenov and Devos 2012).

The first 2 years (2010 and 2011) of implementation of this pan-European mapping exercise revealed that neither the 27-class nor the 10-class set were considered adequate by many MS but that the tegon concept had allowed either classification to be complemented with any MS-defined agricultural land cover class (Figure 11.10). Such additions never affected the original seed classes. Many of the additions related to land cover phenomena associated with complex cultivation patterns and seminatural vegetation. By the second year, 11 MS had thus identified 62 country-specific classes that could not be allocated to any of the 10 JRC predefined seed polytegons. These classes of heterogeneous lands were well modeled through a combination of agriculture and non-agriculture-related tegons and often related to either multistrata lands or to a functionally uniform mixture of life forms. This definition of new classes was accompanied by a decline in the reporting of the arable crop and generic land use classes.

The following 2 years (2012–2013) of this LPIS QA showed a slight overall decrease in the number of the newly introduced classes but an increased use of specific classes related to complex grassland and seminatural vegetation (Figure 11.11). Two out of forty-four systems chose to revert to fewer but more aggregated land cover definitions. One might indeed have defined too many (16) grassland types with very specific prevalence of herbaceous and nonherbaceous vegetation. Still, neither system returned to any of the 27 predefined agriculture land cover classes nor 10 generic polytegon seeds, but both resorted to a country-tailored class with specific definitions. In parallel, four systems introduced new classes, all related to complex grassland, representing functional entity between grassland and trees and grassland and unconsolidated surfaces.

The 4 years (2010–2013) of operational experience evidences that tegons are capable of modeling each and every type of the highly specific regional land cover. Tegons enabled the compilation of a complete and standardized list of all agricultural land cover types identified for all MS.

11.4.2 Cross-Border Harmonization of Land Cover Nomenclatures: Experience from the Danube Region

After this initial implementation in the scope of LPIS QA exercise, a second opportunity to explore the operational applicability and usability of the tegon concept emerged in the scope of the Cross-Border Cooperation (CBC) Project "Common Strategy for Sustainable Territorial Development of the Cross-Border Area Romania–Bulgaria—Project SPATIAL." Funded by the European Regional Development Fund, the project, MIS-ETC 171, started in 2012 and will end in 2015. An important prerequisite for the successful elaboration of this strategy was the establishment of common geospatial data and Information and Communication Technology (ICT)

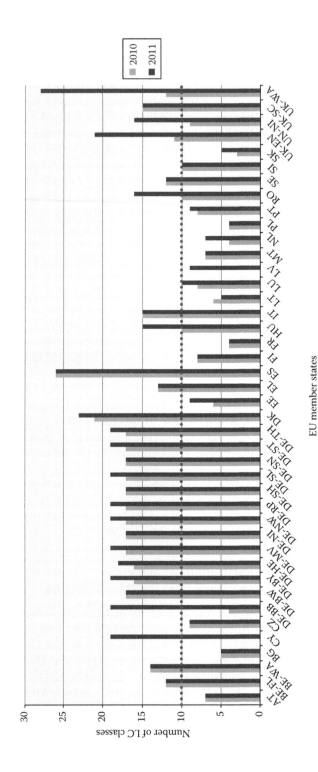

FIGURE 11.10

Abundance of the agriculture land cover types applied in the LPIS QA per EU MS for 2010 and 2011. The dotted line marks the 10 "seeds" limit.

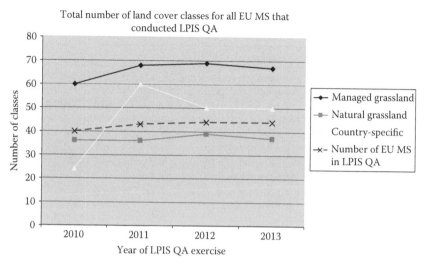

FIGURE 11.11

Evolution of the number of agriculture land cover types related to managed grassland, natural grassland, and specific seminatural land cover types applied by the EU MS in the LPIS QA in the period 2010–2013. The number of EU MS participating is given with dashed line.

resources for territorial planning and analysis over an area of more than 70,000 km², shared by both countries. A land cover data set was considered essential, as it should not only serve as the reference for various sector thematic layers, the backbone of the spatial data infrastructure (SDI) needed for territorial planning, but also support a common, integrated, and holistic approach for the management of the cross-border territory.

The requirements for this reference land cover were more stringent and demanding than many other specifications of classical land cover mapping. What was needed was a management system to effectively monitor the land changes and to support impact assessment of sectoral policy interventions and EU funds expenditures at regional level. The reference land cover data set had to deliver a tessellation of the earth's surface on homogeneous, continuous, and nonoverlapping segments that follow visible and stable boundaries. No controlled gaps were allowed. During the process, the project team had to consider and integrate all relevant, available, and reliable data, and map sources, and it had to take into account a multitude of land cover data already captured by the many pan-European and national initiatives (ADSE-BSDI 2013). More information on the project and product specification is given in Table 11.2.

The three-dimensional concept of tegon was an instrumental element in almost every step of the elaboration of the reference land cover product. As a first step, the tegon model was applied to generate the set of basic land cover types (polytegons). This set was not derived from interpreting a particular

TABLE 11.2

Project SPATIAL. Reference Land Cover Product Specifications

General SPATIAL Information	
Project objective(s) as part of cross-border program	Provision of a comprehensive and accurate overview of the social, economic, and territorial condition and the evolution of the Romania–Bulgaria cross-border area between 2012 and 2015
Priority axis	Priority axis 1: Accessibility—Improved mobility and access to transport, information, and communication infrastructure in the cross-border area
Key area of intervention	Development of information and communications networks and services within the cross-border area
Partners	Lead partner: Ministry of Regional Development and Tourism (MRDT), Romania 4 other Romanian partners: public and local administrations, and Nongovernmental organizations 7 Bulgarian partners: Ministry of the Regional Development, regional and local administrations, and Nongovernmental organizations
Territorial– administrative scope	7 Romanian counties; nine Bulgarian regions—all adjacent to the Danube (one exception only).
Project duration	40 months (February 2012–May 2015)
Work package	6 Work packages (WPs) in total; WP3—Development of common resources for territorial planning analysis and strategy (Database and Information System—Project partner P9—ASDE). The period of implementation of WP3 was from February 2012 to February 2014. Methodology was developed by ASDE in collaboration with the MARS Unit of JRC.
Reference land cover product specifications	
Area of interest	72,000 km^2
Spatial resolution	Minimum cartographic scale of 1:25,000. Minimum mapping unit of 0.25 ha. Minimum width of linear features is 12 m
Thematic resolution	Common cartographic legend of 31 classes (compliant with Annex G of INSPIRE Data Specification on Land Cover)
Rules for creation of a land cover polygon	A land cover polygon should enclose one of the following: • a single/individual land cover phenomenon • a multitude of individual land cover phenomena of the same type • a multitude of individual, independent, and nonrelated land cover phenomena of different type that are small enough to be represented separately at the given cartographic scale (cartographic mix) • a multitude of individual land cover phenomena of different type that jointly form a "functional entity" (see Figure 11.2)
Alphanumeric data associated to each spatial segment	Type of land cover detected. If more than one type of land cover is present, the first three most dominant land cover types are reported together, with the ratio of each of them given as percentage from the total area of the polygon. In addition, the land use information is given in a separate attribute field.

(Continued)

TABLE 11.2 (*Continued*)

Project SPATIAL. Reference Land Cover Product Specifications.

General SPATIAL Information	
Data sources	LPIS, national orthophoto, COPERICUS Core Data sets, GMES Initial Operations (GIO) High-Resolution Layers, field data, national and pan-EU land cover data (CORINE Land cover [CLC], FAO TCP/BUL/8922 and FAO TCP/ROM/2801), soil data, and so forth.
Resources and implementation	The land cover elaboration and database harmonization required a multidisciplinary staff of 15 people working 18 man-months each.
Data encoding and format	INSPIRE Compliant

Source: Modified from Work Package 3 (WP3) Common Methodology, http://www.cbc171 .asde-bg.org.

observation, but from analyzing the occurring land phenomena themselves (Figure 11.12). This allowed for a standardized and unambiguous three-dimensional description of the land cover types based on their biophysical aspect. The generation of the common set of land cover classes was reduced to an iterative but straightforward process of tegon recombination until the necessary level of thematic detail was achieved.

For the resulting set of 31 classes, 20 different tegon types were defined and used; 18 classes were associated with land cover types that can be represented by polytegons with one single type of tegon, whereas the other 13 referred to land cover classes that are functional entities of two or more different types of tegons. The fact that land cover classes were intended to represent the feature on the ground rather its cartographic abstraction greatly simplified and standardized the photointerpretation and classification process. These land cover types and their class representations were indeed modeled through objective and material-based properties and not biased by subjective descriptors. Such descriptors, when originating from a land use perspective, may bias a land inventory toward particular user needs, for example, wetland mapping.

More information on the characteristics of the classification set is given in Table 11.3. The principles of the design of the common set of land cover classes are in accordance with the best practices of the LPIS QA Framework and INSPIRE data specifications on land cover (European Commission 2013c).

The conceptual design of the land cover classes, based on the tegon model, supported comprehensive interpretation keys that went beyond the traditionally applied "top-view" photointerpretation or image classification approaches. This introduced greater flexibility in using various data sources and data capturing methods. It also allowed a clearer distinction between cartographic mixtures and functional entity as well as an unambiguous differentiation between land cover and land use components. The latter was

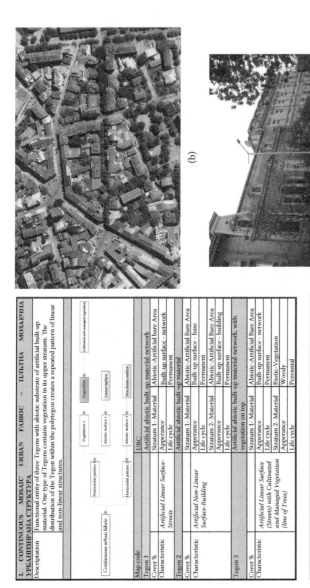

FIGURE 11.12 (See color insert.)

Illustration of the land cover class "Continuous urban fabric," used in the project SPATIAL. (a) Land cover profile: application of the tegon following the classification concepts of ISO 19144-2 (LCML). (b) As seen from above (National orthophoto map, 2011, provided by the Bulgarian Ministry of Agriculture and Food). (c) As seen from the ground (includes material © ASDE [2014] all rights reserved).

TABLE 11.3

Characteristics of SPATIAL's Common Set of Land Cover Classes.

Properties	Comments
Flatness	No hierarchical multilevel classification is available. All classes are on the same level of semantic description.
Independency from scale and product type (map)	All classes in the legend represent either single or "pure" land cover type, or a functional entity. Legend doesn't include cartography-related mixes between different land cover types, as classes.
Ground-based reference	The class definitions don't involve classifiers or characteristics that are related to the abstract representation/observation/polygon of the given feature in a mapping product. It involves the feature on the ground as the basis for the reference.
Exhaustiveness	The set of class entries represents any type of land cover that can be found in the Area of Interest (AOI) of the project.
Mutual exclusiveness of classes	There is no overlap in the semantic concepts. A given land cover feature or functional entity could be associated to only one class.
Pure land cover semantics	Involves class components that purely based on land cover semantics. Any relevant land use information is included as a separate characteristic to the land cover element.
LCML—compliancy	Based on ISO 19114-2 and LCCS ver.3. Listing only class entries that are conceptually modeled in the semantic apparatus of LCML

Source: Modified from WP3 Common Methodology, http://www.cbc171.asde-bg.org.
Note: This set of class definitions is not specific to the project area, but can be applied elsewhere.

essential for the semantic conversion of classes from the national land cover data sets into the common set of land cover classes of the project.

The tegon-derived interpretation keys could be categorized in three groups:

1. Tegon related—considering strata, material, and appearance of tegons. For example, the presence of woody life form in the tegon's upper strata can be an indication for a forest.

2. Polytegon related—considering shape, pattern and density of tegons. For example, a rectangular patch of densely located natural trees can be an indication for a tree plantation.

3. Context related—considering location, spatial connection, and adjacency of tegons/polytegons. For example, a patch of natural trees located inside an urban settlement is an indication of urban vegetated area.

The resulting data set provided a meaningful tessellation of the whole territory. Tessellation segments were homogeneous from the perspective of the interpretation rules (a unique set for each land cover type) and relatively stable as for profusion of the land cover features inside, rooted in the tegon recombination. In consequence, all spatial features had clear physical meaning, providing a harmonized interpretation of the territory by the local, regional, national, and European policy makers. The data set satisfactorily provided the robust common spatial framework for systematic monitoring of the land and impact assessment of policy interventions—offering much more than statistical data and cartographic illustrations.

The land cover data set was subjected to comprehensive validation. By applying stratified random sampling of all class types, 2085 spatial units were individually checked against reference data (field visits, national orthophoto, and independent cartographic material). The inspection comprised three quality measures as follows:

1. Trueness of the dominant land cover class.
2. Correctness of the cartographic mix, if present in the unit.
3. Validity of the boundary of the spatial unit (unit boundaries follow visible and stable features, present on the land).

The first results of the truthfulness (thematic accuracy) reported between 86% and 100% accuracy for 25 out 31 classes. For the remaining six classes, the thematic accuracy was between 78% and 84%, with the class related to scrubland being the least accurate. For 80% of the verified spatial units, the nature and ratio of the cartographic mix was correct. For the other 20%, the class codes were correct, but the ratio between them inside the segment was questionable. Less than 2.5% of the spatial units were found not to follow the physical land boundaries exactly.

Table 11.4 presents the thematic accuracy of the tegon-derived land cover product from project SPATIAL and some pan-European and national land cover products. Although a direct comparison of the results is not straightforward, the table suggests that the reference land cover from CBC project performs equal or better than these other land cover products.

At the time of writing, the project SPATIAL is still ongoing, but its reference land cover data set is accepted by the project partners and stakeholders. Their feedback was highly positive. Both Bulgarian national authorities (Ministry of Agriculture and Food) and Commission Services (Joint Research Centre) considered the tegon-based method for generating this reference land cover data set to be very powerful for ensuring harmonized and interoperable data for the whole Danube Region. The data set will become part of the Danube Reference Data Service Infrastructure (DRDSI) (Figure 11.13).

TABLE 11.4

Thematic Accuracies of Various Land Cover Products

LC Project	Overall Thematic Accuracy (%) from Confusion Matrices	Class Nomenclature	Cartographic Scale	Source
SPATIAL (overall)	84.4	31 LC classes	1:25,000	(ASDE 2013)
SPATIAL (urban only)	95.8	8 Urban-related LC classes only	1:25,000–1:10,000	(ASDE 2013)
CLC 2000	80.1	42 LCLU classes	1:100,000	(EEA 2000)
BULCOVER	85	10 generic LC classes	1:50,000	(ASDE 2010)
Urban Atlas (overall)	88.2	14 LCLU classes	1:50,000–1:25,000	(Urban Atlas 2011)
Urban Atlas (urban only)	89.7	12 Urban-related classes only	1:50,000–1:25,000	(Urban Atlas 2011)

FIGURE 11.13 (See color insert.)
Extract and legend of the reference transborder land cover data set between Bulgaria and Romania, derived from the Geoportal of ASDE. Note that the agricultural lands are subdivided into LPIS production blocks, whose stable borders are consistent with polytegon edges. Contains material © ASDE (2014), all rights reserved.

11.4.3 Practical Challenges and Opportunities from the Common Agricultural Policy Reform

The LPIS and SPATIAL experiences gave the tegon concept its "baptism of fire" as it moved away from a theoretical environment to real-world conditions. The LPIS QA showed that it ensures exhaustiveness for describing and modeling land cover types from opposite corners of the European continent. The cross-border project demonstrates applicability in an operational mapping project requiring the integration of earlier data. Both cases required a very high level of interoperability between essentially local land cover implementations.

The 2004–2013 CAP (European Commission 2013d) dealt with decoupled income support for agricultural land; the 2014–2020 policy places land and environmental services in an even more central position. However, much more priorities and implementation choices are now left to the 28 MS. The 2014 legislation also introduces a series of new land cover–related terms and rules on "agricultural land" that, inevitably and unfortunately, are defined in a legal domain vocabulary, mixing land use with land cover aspects.

Some of these agricultural land definitions that at one point will require spatial identification, although excluding any "area not taken up by agricultural activities," are as follows:

- Permanent grassland, including species such as shrubs and/or trees, which can be grazed, provided that grasses and other herbaceous forage remain predominant.
- Land that can be grazed and that forms part of the established local practices where grasses and other herbaceous forage are traditionally not predominant in grazing areas.
- Parcels with scattered trees.
- Permanent grassland with scattered ineligible features.
- Features part of the good agricultural and environmental condition obligations or the statutory manage requirements.
- Ecologic Focus Area elements that persist for 3 years. (Pending on the choice of the MS, these EFA can, among others, include terraces, all landscape features [hedges, tree lines, ponds, etc.], agroforestry lands, strips along forest edges, short rotation coppice fields, and afforested lands.)

For many of the above land types, it will be up to MS to decide whether or not they include them in their integrated administration and support system and LPIS. Any expansion of the agricultural land framework will follow a local political decision to include lands not previously supported. If it does so, there will be a strict requirement to document and report the nature, control activities, and levels of support (in terms of precise area) on these lands.

(a)   (b)

FIGURE 11.14 (See color insert.)
(a) Example of class "Artificial non-built up surface" as seen from the ground. Includes material © ASDE (2014), all rights reserved. (b) Example of class "Artificial non-built-up surface" as seen from above. (COPERNICUS CORE 03 Image Dataset 2011). © European Union, 2014, all rights reserved. Includes material provided under http://gmesdata.esa.int/web/gsc/terms_and_conditions, SPOT 5 © CNES (2010–2013); distribution Astrium Services/Spot Image S.A., all rights reserved.

It is difficult to envisage how these obligations can be achieved in a traditional land cover/land use approach to area mapping. In fact, the authors are convinced that the need for the common identification of each type of agricultural land, as supported by the tegon approach, becomes much stronger.

One of the most challenging issues in the CAP management and control remains the proper detection of the "area not taken up by agricultural activities," a pure land use term that masks a range of various land cover types. If approached from the tegon model, the associated set of land cover phenomena that prevent agricultural activity would be easily defined. For example, a gravel pit which is permanently bare land subject to extraction of building material can be modeled through a tegon with one stratum holding abiotic material with unconsolidated aspect, where the current state of the material and the appearance of the substrate are permanent. This clearly differs from arable land, which has a similar state and appearance only at the time after plowing. The artificial nature of the gravel pit will be further manifested through the polytegon and context-related properties of the tegon-specific shape and connection of road network (Figure 11.14).

11.5 Conclusions

Adopting the three-dimensional, feature-based tegon concept to replace the traditional top-view observation concept opens a wider range of possibilities for the handling of the earth's surface data. The three-dimensional nature of the land cover substrate is often not detectable by remote sensing techniques with passive optical sensors, but it remains an important property of the observed material (e.g., anisotropic reflectance from volume scattering) (Schaepman-Strub et al. 2006). By contrast, it is an essential factor for active sensors such as Laser Imaging Detection and Ranging (Lidar) and multidimensional Synthetic Aperture Radar (SAR).

Our practical applications demonstrate that the three-dimensional tegon concept is very workable. In these applications, the use of a top-view constrained, legend-based methodology could never be seriously considered. We identified several advantages to our tegon-based approach to land cover, as follows:

1. It deals with complexity better. As the biotic/abiotic nature of the material no longer applies on the class as a whole but on the stratum, the resulting classes much better reflect the true complexity of the earth. Indeed, many "biotic" classes can have an abiotic footprint— tree lines on city lanes are an example that has tree canopies over impermeable surface. Tegon encoding applies the location context to identify the presence of both strata but no longer imposes a separate "land use" concept for defining such individual class (as, e.g., "urban green area" would).

2. It is a pure biophysical and material approach. As a result, most "land use" inspired descriptors could become obsolete. For instance, "wetlands" have such high ecological value that this class is a "must" for any land cover system, and such a class is often defined to match areas designated for nature protection or special management. But from a purely biophysical or material perspective, wetlands represent little more than a typical mix of rock, water, trees, shrub, and grass forming a diverse habitat, often with seasonal variation of water levels. Tegons can describe and quantify that mix precisely by the presence and life cycle of every individual material in its strata, without any need for invoking the much more contextual land use perspective.

3. The tegon approach is similar to the soil classification methodology. The soil reduction phenomena in these wetlands illustrate how closely soil and land cover were linked before human interaction irreversibly changed the surface of the earth. The land cover substrate was one key factor in the formation of soil horizons, and the analogy between pedon (Johnson 1962) and tegon supports this interrelationship very well.

4. It adds structure to LCML applications. The tegon's demonstrated interoperability for land cover concepts came from the exclusive use of exhaustive LCML qualifiers, solely defined on physiognomic and globally valid properties. This makes the tegon semantics nearly universal. Compared to stand-alone LCML, tegon offers additional structure and also simplicity through the introduction of the matter with strata, their life cycle, and the footprint. This structure is

equally universal and enables a much more generic approach to some of the issues that always troubled the top-view classes.

5. Compatibility with any mapping initiative is guaranteed. Defining land cover in tegon and thus in physical terms will support any top-view approach and will greatly facilitate the elaboration of proper interpretation keys; our need for understanding keys from each MS in the scope of LPIS was a trigger of tegon development. By contrast, vertical and temporal dimensions can rarely be appended to snapshot(s) such as top-view observations. For mapping initiatives, it should pay off to ensure the correct description the relevant land cover types and correspondent substrate in tegon terms before geometries are delineated and spatial data sets are produced.

6. It provides a universal foundation for land cover change. Approaching the land cover from its stable physical characteristics, beyond the limitations of the rather volatile observation methods, will improve the framework for detecting and describing land cover change. In general, monitoring land cover change provides the most direct indicator of so many spatial and environmental policies. The top-view cartographic approach to land cover mapping that currently remains predominant should be underpinned by tegon conceptualization.

However, to reach its full potential, the tegon approach could benefit from additional developments. As a novel method, it requires more investigations and feasibility tests in different operational cases. The bottlenecks and suggestions for further research can be identified as follows:

1. The tegon concept needs further formalization with support of a tegon design and classification tool. Tegon applies the LCML semantics through a strata structure. The LCML design software (LCCS version 3) applies these LCML semantics for a given land cover class to model its definition within a mapping product. A tool that applies these semantics through the tegon strata structure would be better able to tackle the physical feature itself. Such a tegon design and classification tool should introduce additional rules to implement elements for stratum and functional entity in a consistent manner, allowing automation of the semantic classification. New, universal tegon codes and tegon type codes, analogue to the LCCS land cover code, could ensue.

2. The biophysical extent of the tegon prism needs further investigation. The tegon rules that drive the design and classification tool can also be used to predict all possible classifier permutations or tegon type instances. This set should be able to prove the claims of exhaustiveness and inclusiveness. When a known land cover class rejects

either of these claims, the model or even tegon concept itself should be investigated. Obviously, regions outside the EU should thereby be considered.

3. The assumptions of the authors that the number of theoretical tegon types is finite, and that these combine into any type of land cover, need to be verified. Such verification implies tackling some technical issues such as the maximum number of the strata and the exact limitation to the base of the tegon prism (When does the tegon base include the upper soil horizons?).

4. The application scope of the tegon concept is still not well defined. In which cases does the proposed approach apply? Where does it not add any value? The introduction of tegon had neither the objective nor the ambition to address the restrictions of the current methods of land cover capture and delineation. Tegon experience indicated that the traditional top-surface approach used in remote sensing cannot reveal the true three-dimensional nature of the land cover phenomena. But this is a well-known challenge for the remote sensing community. Investments are being made in object-oriented image classification techniques and holistic approaches, relying on synergies of different sensors and data sources. Tegon will not dramatically change these methods, but should greatly help land cover data producers and users to understand what information a land cover data set provides, given the data capture description.

5. The impact at smaller scale data sets will most likely be limited. The tegon concept was derived to tackle problems of the LPIS implementation and land definition at large and very large cartographic scale (better than 1:5000). The tegon conceptual model allows specification of land use semantics for very high-detail data; its primary application can obviously be expected to be at regional to local planning and mapping rather than at pan-European or worldwide land monitoring products such as Corine Land Cover or GlobeLand30. Pan-European land cover and land use harmonization efforts oriented toward supporting EU environmental reporting through CLC data, such as EAGLE (Arnold 2013), should be little affected. For practical reasons, such smaller-scale initiatives will in the shorter term remain focused on providing products with predefined and user-tailored cartographic classes. And although the spatial units of these data sets (polygons, points) can occasionally be populated with series of numerical attributes to provide further information on the land components found within, the collected information would remain for the most part "top-view" derived.

In the future, the demand for better and more comprehensive land information can be expected to push the creation of more and more large-scale data sets with high levels of detail and accuracy. These will represent opportunities where the tegon concept will be able to deploy its full potential. Obviously, capturing this three-dimensional information by remote sensing will be dependent on the development of new techniques, on the better usage of synergies between different sensors, and on an integration of Earth Observation with in situ data. Last but not the least, it will require three-dimensional GIS functionalities for effective handling and manipulation of the collected information. The choice between the classical top-view two-dimensional method and the comprehensive three-dimensional tegon approach will ultimately depend on pragmatism, cost-efficiency, and data availability.

Acknowledgments

The authors would like to thank the Agency for Sustainable Development and Eurointegration for the kind technical support and information exchange with respect to the project SPATIAL during the preparation of this manuscript.

References

Arnold, S., Kosztra, B., Banko, G., Smith, G., Hazeu, G., Bock, M., Valcarcel Sanz, N. 2013. The Eagle concept—A vision of a future European Land Monitoring Framework. *Towards Horizon 2020*. Matera, Italy.

ASDE. 2010. ADSE-BSDI Portal, Bulcover Project. http://bsdi.asde-bg.org/lccs .php.

ASDE. 2013. ASDE-BSDI Portal, PP9-[WP3]—Development of Common Resources for Territorial Planning Analysis and Strategy—General Methodology. http:// cbc171.asde-bg.org/docs/.

Devos, W., Milenov, P. 2013. Introducing the Tegon as the elementary physical land cover feature. *Agro-Geoinformatics*, 562–567, doi: 10.1109/ Argo-Geoinformatics.2013.6621939.

Di Gregorio, A. 2005. Land Cover Classification System—Classification Concepts and User manual, Food and Agriculture Organization of the United Nations, Rome, Italy.

European Commission. 1994. CORINE Land Cover Technical Guide, EU, CEC, Luxembourg. ISBN 92-826-2578-8.

European Commission. 2007. Directive No. 2007/2/EC, OJ L 108, 25.4.2007, pp. 1–14 (INSPIRE).

European Commission. 2013a. Commission Regulation (EU) No. 1089/2010 of 23 November 2010 implementing Directive 2007/2/EC of the European Parliament and of the Council as regards interoperability of spatial data sets and services 08.12.2010.

European Commission Joint Research Centre. 2013b. INSPIRE Data Specification on Land Cover—Technical Guidelines 10.12.2013.

European Commission. 2013c. Inspire: Interoperability of Spatial Data Sets and Services. Regulation (EU) No. 1253/2013, OJ L 331, 10.12.2013, p. 43.

European Commission. 2013d. Common Agricultural Policy: Horizontal Regulation Regulation (EU) No. 1306/2013, OJ L 347, 20.12.2013, p. 549.

European Environmental Agency (EEA). 2000. The Thematic Accuracy of Corine Land Cover 2000. Assessment using LUCAS. EEA Technical Report No. 7/2006, 22, ISSN 1725-2237, Copenhagen.

EEA. 2010. *Urban Environment—SOER 2010 Thematic Assessment*. ISBN 978-92-9213-151-7. Copenhagen, Denmark.

Food and Agriculture Organization (FAO). 2009. Land Cover Classification System, vol. 3, FAO Global Land Cover Network.

International Organization for Standardization (ISO). 2011. ISO/FDIS 19144-2 Geographic Information—Classification systems—Part 2: Land Cover Meta Language (LCML).

Johnson, W. 1962. The Pedon and the Polypedon. Presented before Div. 5. Soil Science Society of America, at St. Louis, MO. November 27.

Milenov, P., Devos, W. 2012. *LPIS Quality Assurance Framework—ANNEX III*. The Concept of Land Cover and "Eligible Hectares." ISBN-13 978-92-79-22804-9.

Schaepman-Strub, G., Schaepman, M. E., Painter, T. H., Dangel, S., Martonchik, J. V. 2006. Reflectance quantities in optical remote sensing—definitions and case studies. *Remote Sensing of Environment* (103): 27–42.

Urban Atlas. 2011. Delivery of Land Use/Cover Maps of Major European Urban Agglomerations. Final Report, vol. 2.0, 50, November. Villeneuve, France.

Valeeva, E., Moskovchenko, D. 2009. Zonal characteristics of the vegetation cover of Tazov Peninsula and its technogennic transformation. UDK 581.5, 504.73.06.

Villa, G., Valcarcel, N., Arozarena, A., Garcia-Asensio, L., Caballero, M.E., Porcuna, A., Domenech, E. and Peces, J.J. 2008. Land cover classifications: An obsolete paradigm. *Remote Sensing and Spatial Information Sciences*, vol. XXXVII, Part B4. Beijing, China.

12

Resolving Semantic Heterogeneities in Land Use and Land Cover

Nancy Wiegand, Gary Berg-Cross, and Naijun Zhou

CONTENTS

ABSTRACT This chapter introduces our prior and current work in resolving semantic differences in land use/land cover codes, along with a discussion and description of our current activities from a National Science Foundation INTEROP grant awarded to the Spatial Ontology Community of Practice group (SOCoP). The INTEROP grant allowed continued work on semantics in general and on furthering its application to land use and land cover codes.

KEY WORDS: *Land use, land cover, semantic technology, ontology, query, ontology design pattern, ontology repository.*

12.1 Background

Land use data have importance in many applications, including land use planning. However, land information data developed by local, regional, state, and federal governments and the private sector are not homogeneous across jurisdictional boundaries. Spatial integration of land use datasets represented in Geographical Information System (GIS) format can be accomplished by translating the different coordinate systems. But, attribute data, such as a land use code, are typically not compatible, limiting integrated planning use. As one moves across jurisdictional boundaries, the format of data tables and attributes varies, and the definitions and values change significantly. In Wisconsin, for example, many jurisdictions develop their own land use coding system. But, to be able to plan across jurisdictional boundaries, data distributed over a wide spatial area and diverse in organization and composition need to be integrated.

12.1.1 History of Land Use Coding Systems

Historically, land use classification systems evolved out of the need to describe certain observable conditions in the landscape. They offer a systematic way to codify and categorize these conditions. Two common examples of land use classification schemes are exhaustive lists and hierarchical models (American Planning Association 1994). For example, the hierarchical example in Table 12.1 is from North Carolina land use and land cover classification (1994). Structurally these two systems are quite different, with the hierarchical model being more highly organized than the exhaustive list, which facilitates aggregation of land uses.

In the United States, the Standard Industrial Classification (SIC) system could be considered the cornerstone from which subsequent land use classification systems were derived. The SIC came from the need to standardize and

TABLE 12.1

Classification Schemes

Exhaustive List	Hierarchical
009 Shopping center	1 Urban and developed land
010 Open water	1.01 Residential
111 Single family	1.01.01 Single family detached or duplex
113 Two family	1.01.02 Mobile homes (not in parks)
115 Multiple family	1.01.03 Multi-family dwellings
116 Farm unit	1.01.03.01 Low-density multi-family
129 Group quarters	1.01.03.02 Med.-density multi-family
140 Mobile home	1.01.03.03 High-density multi-family
142 Mobile home park	1.01.04 Mobile home parks

Source: Wiegand, N. et al., *Proc. National Conf. Digital Gov. Res., dg.o2002*, 115–121, 2002.

codify the industrial sector (Pearce 1967). By 1940, a Technical Committee, established under the direction of the Central Statistical Board, had published two volumes of code lists for manufacturing industries and nonmanufacturing industries. Domain experts were used to decide where questionable activities lay within the classification system.

Because the SIC was designed primarily to classify industry, it is an imperfect surrogate for land use classification. In response to the SIC shortcomings in this regard, the Federal Highway Administration and Department of Housing published the Standard Land Use Coding Manual (SLUCM) in 1965. While based on the SIC framework, "the SLUCM coding was to provide an exhaustive set of land uses … and a limited set of attribute data to further define some of the land-use categories" (Everett and Ngo 1999). The manual, which was not a mandated standard, proved popular throughout the nation for over a decade before it was abandoned for failing to account for the rapidly evolving nature of land use planning (Everett and Ngo 1999).

In 1999, the SIC codes were deemed outmoded and were subsequently replaced by the North American Industrial Classification System (NAICS). The NAICS codes are far more detailed than the SIC codes and account for new types of industry that did not exist 60 years prior. The codes are also used beyond the borders of the United States, in both Mexico and Canada (Jeer 1997). The NAICS codes do not, however, correct the shortcomings associated with the SIC regarding suitability for land use planning; the codes are oriented toward industry and are similarly an imperfect system for land use planning purposes. As a result, despite these efforts to advance new classification systems, jurisdictions continue to rely heavily on site-specific systems and the use of domain experts to help classify land uses.

Currently, there are myriad types of classification systems in use across the nation. Some of these may be based around the SIC or SLUCM framework, and others may be entirely new and unique systems. While these many systems allow for greater customization, they also create difficulties in the integration of land use information across geographic areas. Critics of this individualism, and the host of problems this creates, felt another standard for land use classification was needed. In response to this, in 1994 the American Planning Association (APA), with the support of a variety of federal sponsors, spearheaded an effort to develop a new land use classification system that would not only update the SLUCM codes but also go beyond them to address the changed landscape of land use planning. As a result, the APA created a multidimensional system, the Land Based Classification System (LBCS, https://www.planning.org/lbcs/), which codifies land uses in a hierarchical system for each dimension. The five dimensions are activity, function, ownership, site, and structure. Although it was not expected that complete information on all five dimensions would be available or even needed in all jurisdictions, the standards provide a database structure and coding system to accommodate each dimension. The multidimensional aspect of the LBCS allows for a greater precision of land use information to be captured, such as for natural resources, the existing built environment, and ownership/development rights (Jeer 1999). Adoption of the LBCS was slow, however, because while most planners recognize the need for a unified system, the new standard is complex and conversion from existing systems is time-consuming. Recently, however, Montenegro et al. (2012) created an OWL representation of the LBCS classification system along with its dimensions to develop the land use ontology LBCS-OWL2.

12.1.2 Example Land Use Coding Systems

To study issues in integrating land use coding systems, we collected data from seven jurisdictions in Wisconsin, including counties, regional planning commissions (RPCs), and a city. In these coding systems, there does not seem to be a common ancestor (e.g., SIC or SLUCM) from which systems arose, probably because Wisconsin does not have a mandated land use standard to which codes must adhere. This results in classification systems and their associated codes varying from place to place and attuned to local conditions. For example, one jurisdiction could have five separate codes to describe agricultural land uses while another could have only two. It is likely that the first jurisdiction is a more rural and farm-oriented community.

Differences in the level of detail in codes affect the specificity of queries. For example, for some jurisdictions, only the code agriculture can be returned, compared to other jurisdictions that have separate codes for types of crops (e.g., Table 12.2). Another example is differences in codes relating to commercial lands for Dane and Racine counties. If a user were to query the Dane County RPC data for all the commercial lands, there would be 22 possible codes that would satisfy the query. The user could either take all codes returned or possibly refine the query to hone in on a specific type of commercial use, such as financial institutions (which has its own unique land use code). However, the Southeastern Wisconsin Regional Planning Commission (SEWRPC), which covers Racine County, has only three codes for commercial lands. These codes are very general and not split into the same basic categories as that of Dane County. The user would not be able to refine the query to find financial institutions because there is no unique code to describe that use; it is lumped together with many other types of commercial uses. However, there is a descriptive document that accompanies the SEWRPC code set. A search capability over that document could allow the user to scan for words resembling financial institutions (such as "bank") to determine the broad category that would contain financial institutions (here, 210 Retail Sales & Service—Intensive). However, all the returned parcels containing the code 210 would be a large superset of what the user wants. And, in general, many code sets do not have written descriptions. In fact, refinements of categories may not be recorded anywhere and may only be known to a few specialists in the jurisdiction.

TABLE 12.2

Semantic Heterogeneity

Jurisdiction	Land Use Code	Description of Code
Dane County RPC	8120–8139	Separate code for corn, soybean, etc.
Racine County	811	Cropland
(SEWRPC)	815	Pasture and Other Agriculture
Eau Claire County	AA	General Agriculture (including crop farming, dairying, etc.)
City of Madison	81	Agriculture
	811 and 8110	Farms

Source: Wiegand, N. et al., *Proc. National Conf. Digital Gov. Res., dg.o2002*, 115–121, 2002.

A typical query in the effort to promote regional and statewide land use planning consists of a predicate applied over multiple jurisdictions. An example is, Where are all the row crop fields in Dane, Racine, and Eau Claire counties? A query of this kind is relatively straightforward when using one dataset but more difficult when posed over a larger geographic area. Table 12.2 again illustrates the semantic heterogeneity of codes that would satisfy the query's criteria over these areas.

In our data sets, each jurisdiction uses different attribute identifiers and coding representations. But these are easily resolved compared to determining whether the codes share common definitions. Unfortunately, often these definitions are not exact matches; each description slightly varies from one another. For example, the 8110 code from the city of Madison covers almost all types of agriculture, whereas the Dane County RPC has almost 60 codes covering types of agriculture. The description of Eau Claire County's AA code states crop farming and dairying and then says "etc." without specification. Also, Madison makes no distinction for farm buildings, whereas Dane County has six codes for types of farm buildings. Eau Claire County has three codes for farm buildings: AD is farm duplex, AH is farm housing, and AR is farm residence.

Another example query could be, Where are all the lands in conservation uses for Racine County? Potential source datasets have different definitions for conservation. Racine County has more than one jurisdictional authority with land use classification systems: the RPC and a federal agency operating in the area, such as the Natural Resources Conservation Service (NRCS) or the Federal Farm Service Agency (FSA). Conservation, as defined by one of these authorities, may not mean the same thing to the others operating in the same political boundary. For the RPC, there is not a specific code that describes "conservation." In lieu of an explicit code, they assign separate codes for wetlands, woodlands, unused pastures, and fallow agricultural lands—all of which embody their concept of "conservation." In contrast to this, the NRCS's National Resources Inventory (NRI) has specific land use codes for Conservation Reserve Program (CRP) lands, which is a program available to farmers, run under the FSA, for the purpose of conserving marginal agricultural lands. It is likely that if both datasets were queried for conservation lands, there would be discrepancies between areas satisfying the query criteria. For instance, wetlands were excluded from the NRCS codes. Ideally, a query system would inform the user of the potential for multiple datasets with dissimilar definitions for a given land use query. Then, the user would either choose the appropriate dataset for a specific area of interest or accept the results along with system-supplied information regarding its accuracy.

(Section 12.1 was mostly taken from Wiegand et al. 2002.)

12.2 Ontology-Based Query System to Resolve Semantic Differences in Land Use Codes

As an initial project to resolve heterogeneity in land use codes, Wiegand and Zhou (2005) added ontologies and mappings to a database query system. At that time, query processors in Database Management Systems (DBMSs) did not accommodate anything other than an exact match for strings. For example, if a user wanted to find all "agriculture" parcels but a local data set had subclasses of agriculture listed and not the string "agriculture," no results would be returned. The purpose of this project was to show how ontology querying could be added to a large DBMS code base.

The ontology-based query system was built on top of the Niagara XML Internet Query System (Naughton et al. 2001), which had been developed with the emergence of XML data being put on the Web. Niagara crawled remote XML data and stored it in inverted indexes. The ontology extension that we added to the Niagara code base included a custom interface, lookups using predefined mappings, and query rewriting. The data used were geospatial parcel datasets that contained land use codes. The data were originally in GIS shapefile format that we first converted to XML format. We then put the parcel data in XML format on the Web, such that it would be crawled and indexed by the Niagara crawler. Our custom ontology interface allowed the user to pose queries in either local terms or ontology terms (for which lookups would be done, Figure 12.1a). The user also specified jurisdictions over which the query would range using a map or drop-down lists, for example, to choose certain counties (Figure 12.1b). Our ontology extension code rewrote queries, adjusted for each jurisdiction using local terms, that were then sent to the Niagara query engine for processing. Then, using the spatial GIS information for each parcel, we postprocessed the results using

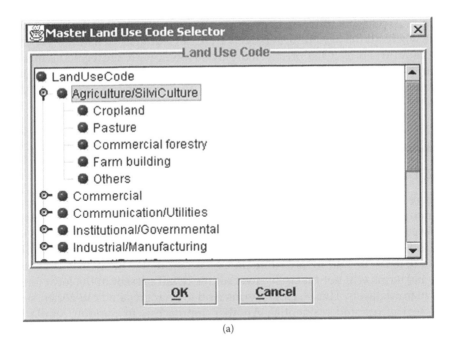

(a)

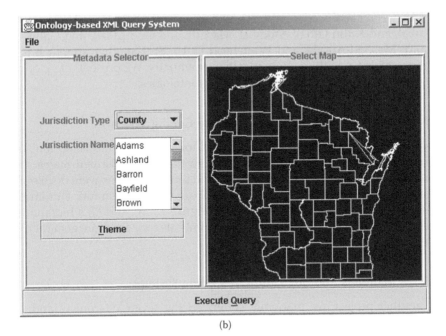

(b)

FIGURE 12.1
(a) Selecting ontology terms. (b) Choosing jurisdictions. (From Wiegand, N. and Zhou, N., *Next Generation Geospatial Information: From Digital Image Analysis to Spatio-Temporal Databases*, ISPRS Book series, Balkema, 2005. With permission.)

ESRI's ArcObjects software to generate a map display so that the user could see where all the agricultural parcels were, for example. In addition, results were returned as a list of selected parcels along with aggregated sums of number of acres for the selected land use. As stated, this project illustrated ontology-based querying by implementing an extension to a research DBMS.

12.3 Mapping and Merging Land Use Code Ontologies

There are different mapping approaches to resolve semantic heterogeneity among multiple datasets. One is to create a global set of terms or an ontology that can then be mapped to local terms in each dataset, as was done in Section 12.2. But, this method requires developing global terms, which can be difficult to agree on and involve domain experts. And, it is likely that one set of global terms will not precisely cover all the detail present in the local terms of many datasets. That is, global terms tend to be aggregated categories, such as agriculture and residential. Another approach is to compute similarity measures between local codes. Examples of this approach include Cruz et al. (2010), who developed an automated mapping tool to compute similarities between classification systems. Other work has been done by Janowicz, Raubal, and Kuhn (2011); Ahlqvist, Macgill, and Guo (2004); and Gahegan, Smart, Masoud-Ansari, and Whitehead (2011). Still another approach is to merge ontologies either automatically or manually. We present a few of our experimental methods to map and merge ontologies.

12.3.1 Automatically Mapping Land Use Coding Systems

Experimental algorithms developed by Zhou and Wei (2008) map four different land use classification systems from the Eastern United States. Two automated ontology mapping methods and representations were developed. These are (1) a semantic network in which quantitative semantic similarities between land categories are calculated (Figure 12.2a), and (2) a hierarchical structure in which characteristics of a parent category are more general than the children's (Figure 12.2b).

12.3.2 Manually Merging Land Use Coding Systems

In another project, we manually combined the seven Wisconsin land use coding sets mentioned at the beginning of Section 12.1.2 into one large set of codes (Wiegand 2012). The reason for creating a merged ontology, rather than a global set of terms and mappings to local terms, was to not have to develop global terms and to not lose any detail for the user's choice of query terms. Using a merged ontology, the most detail can be returned for a query

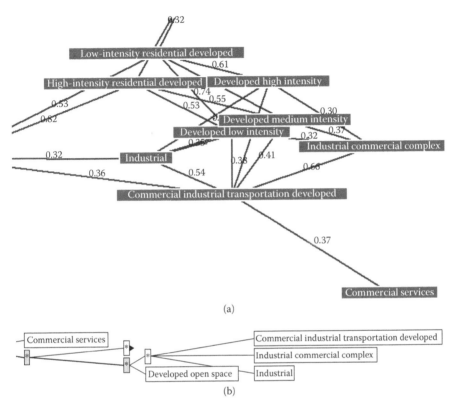

(a)

(b)

FIGURE 12.2

(a) Visualization of mapped ontologies with similarities shown as a semantic network.
(b) Visualization of mapped ontologies as a tree: Commercial services.

covering multiple jurisdictions. The reason we worked manually was to use human expertise in understanding the codes and make the best choices as to how a code related to other codes. This level of accuracy seems appropriate for making critical land use decisions. Now that the merged ontology exists, however, if additional code sets are added, automated methods could be used to merge them because enough variety in codes probably now exists to enable automatic matching. That is, the chance of a newly added term matching an existing term is now high.

The merged ontology has multiple hierarchical levels for each of the main land use code groupings, for example, residential, agriculture, and commercial, and keeps all codes from each jurisdiction. We put the ontology into a formal ontology web language representation called OWL (Smith, Welty, and McGuinness 2004). Each term is annotated with a list of the jurisdictions that have that term along with the jurisdiction's particular land use code value, for example, Group Quarters ([Bay Lakes, 170], [ECWRPC, 942], [Dane County, 129], [Madison, 12]). This allows querying the combined code list using custom software and the OWL API. For example, a search for Group Quarters over any of the above jurisdictions would return the actual land use code values needed to generate a subquery for each jurisdiction's dataset. For example, in Madison, Group Quarters is coded as 12. An issue is how to handle a jurisdiction in the query that does not have the query land use code. To avoid returning a null result, in the effort to provide as much land use information as possible for decision making, our software first searches down the hierarchy in the merged ontology from the specified land use code to see if any subset terms have that jurisdiction listed. This is the case, for example, with SEWRPC that has Local Group Quarters and Regional Group Quarters as subclasses. In this case, the subclass terms are returned along with a message stating that these are subclasses of the query term. However, if no subset term for the jurisdiction is found, then the software returns the first superset term that lists the jurisdiction, along with a statement that this is a superset term. In some cases, just the root term can be returned, for example, residential. We created a Web demo to illustrate semantic resolution for land use codes: http://www.ssec.wisc.edu/landuse/.

This ontology resolution software is intended to sit between the user and additional code that has access to actual parcel data. An example vision for the merged land use code ontology and software is for it to be used as part of a larger information system, say for a statewide GIS parcel layer. With the resolution of heterogeneous land use codes along with spatial parcel boundaries being available, statewide questions could then be asked and visualized, such as to show on a map all the parcels in the state growing corn, or any land use. Using our software, the most detailed level of codes that are available would be returned, not just all agricultural parcels, for example.

12.4 Spatial Ontology Community of Practice

Further work on resolving land use codes along with advancing other semantic technology for the geospatial domain was enabled by a National Science Foundation INTEROP grant awarded to the Spatial Ontology Community of Practice (SOCoP). SOCoP was officially created in October 2006 and chartered by the Federal Chief Information Officer Council as a best practice and volunteer organization specifically focused on semantics for the geospatial domain. The SOCoP mission has been to

- Foster collaboration among researchers, technologists, and users of spatial data to develop spatial ontologies
- Support open collaboration and open standards for increased interoperability of spatial data across various communities

Activities of SOCoP through the INTEROP grant are described in Sections 12.5 through 12.8. These activities include software demonstrations and tool building, such as creating an Open Ontology Repository, remotely accessing the repository to resolve land use code heterogeneity, and designing and implementing a GeoSPARQL query interface. Further activities include conducting workshops, stand-alone or at conferences. Some workshops were GeoVoCamps, which have the purpose of creating small ontologies and modular pieces called Ontology Design Patterns (ODPs).

Ontologies form the core of work in semantic technology. Ontologies specify concepts, terms, meanings, and relationships for a specified domain. Ontologies can be formalized using an ontology language, such as OWL, mentioned in Section 12.3.2. OWL is a World Wide Web Consortium standard for a computational logic-based language allowing automation. That is, reasoners can automatically find subclasses, transitive relationships, and so forth in the ontology. Uses of ontologies include resolving differences in terms within and across domains, as discussed in Sections 12.2 and 12.3. Also, it is anticipated that existing ontologies will provide terms for others to use in new datasets, creating more homogeneity.

12.5 Open Ontology Repositories

Once developed, ontologies, such as the merged Wisconsin land use code ontology described in Section 12.3.2, need to be available so that others can use or extend them. For this purpose, the Open Ontology Repository (OOR) initiative (http://oor.net) was created (Baclawski and Schneider 2009). Example OOR repositories include the BioPortal, COLORE, Ontohub, and others (http://oor.net/index1.html).

The OOR effort has largely been a volunteer effort to promote the vision of a common, easily used infrastructure for shared "ontologized metadata." An OOR instance is a persistent registry and storage repository that reduces some of the barriers for collaborative research communities to provide standard ontology/metadata annotations for their data.

All OOR work is in compliance with open standards and uses, including

- Open source technology featuring light rules
- Open knowledge content
- Open collaboration, that is, use of a transparent community process
- Open to integration with "non-open" repositories via an open repository interface
- Simple gatekeeping mechanisms for user login, editing, download, upload, etc.

Specific ontology capabilities include an ability to

- Retrieve ontologies for use in domain applications
- Retrieve ontologies for integration with other ontologies
- Retrieve ontologies that can be extended to create new ontologies
- Determine whether or not an ontology is a good candidate for integration with given ontologies

SOCoP developed its own OOR instance specifically to store ontologies for the geospatial domain. The BioPortal code was used as a base for SOCoP's OOR. The SOCoP OOR was used to collect several dozen geospatial ontologies to promote their use. Initial population of ontologies was taken from the study by Ressler, Dean, and Kolas (2010). Populating the portal provided simple and familiar examples that could be understood by domain matter experts and potentially be used for new work sessions, presentations, and as educational baseline material. A relevant example of one such ontology is ISO 19107 (*Spatial schema*), which defines a large, useful set of standard spatial/geometric data types and operations for geometric and topologic spaces. Other examples are SWEET (https://sweet.jpl.nasa.gov/), a comprehensive ontology for the earth and environmental domains, and our Wisconsin land use code ontology. Figures 12.3 and 12.4 illustrate some of the user interfaces of the OOR for publishing and mapping ontologies. These were used in the project described in Section 12.6. Note, the SOCoP ontologies have now moved to Ontohub (https://ontohub.org/repositories/socop).

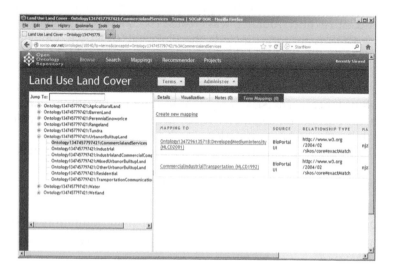

FIGURE 12.3
Open Ontology Repository interface to create mappings.

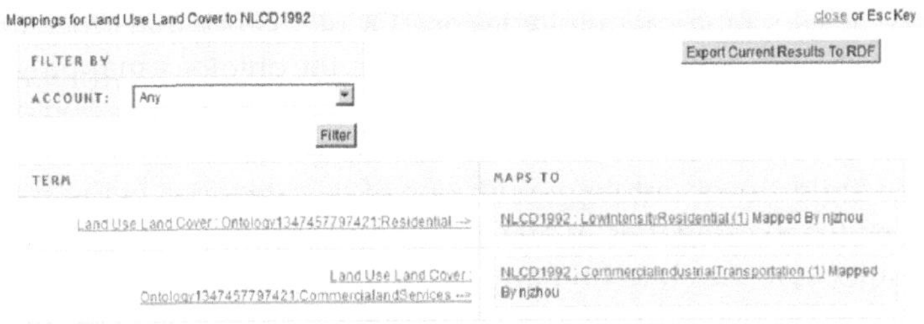

Mappings for Land Use Land Cover to NLCD1992   close or EscKey

FILTER BY   Export Current Results To RDF

ACCOUNT: Any   ▼

Filter

TERM   MAPS TO

Land Use Land Cover : Ontology1347457797421:Residential →   NLCD1992 : LowIntensitvResidential (1) Mapped By njzhou

Land Use Land Cover :   NLCD1992 : CommercialIndustrialTransportation (1) Mapped
Ontology1347457797421,CommercialandServices →   By njzhou

FIGURE 12.4
Open Ontology Repository interface to find mappings.

12.6 A Prototype of Querying Heterogeneous Datasets Using OORs

In addition to other features listed, the Open Ontology Repositories provide an API such that software can remotely access ontologies and mappings. We developed a demo to illustrate this feature to resolve land use codes on the East Coast (United States). The value of remotely accessing ontology information is that each software application does not have to include specific ontologies and various mappings itself. Instead, needed information can be found using web services.

This prototype demo uses U.S. Geological Survey (USGS) land cover classification systems and the National Land Cover Dataset (NLCD) to illustrate the semantic problem of land categories. As mentioned earlier, in different land use/cover classification systems, the same or similar land cover categories can be represented in different text, and searching the data for a land code should be based on its meaning (semantics) instead of the text string representation. In this prototype, we simulate to identify the areas that will be flooded within a certain distance of the coastline of Chesapeake Bay. NLCD data were collected for Maryland (1992), Virginia (2001), and Washington, DC, (1997) together with their land categories. When users wish to search for "high-density residential" lands within 1 mile of the Chesapeake Bay coastal line, due to land categories' heterogeneity, there are "high-intensity residential" lands in the Maryland dataset, "residential" lands in the Washington, DC, dataset, and "developed, high intensity" lands in the Virginia dataset. These categories have the same or similar semantics and should be retrieved in the result. On the other hand, instead of searching each individual dataset, an integrated view of the categories and their semantic relations is desired, particularly when a large amount of data is involved in the application.

The prototype is designed as a service-oriented architecture consisting of the services of data, ontologies, ontology mappings, visualization, and a web interface that can be distributed over the Internet. The three semantically heterogeneous land cover classification systems mentioned above are published as geographical ontologies on SOCoP's geospatial Open Ontology Repository (OOR). An ontology expert creates the ontology mappings among the land category ontologies. The OOR allows for the publication of ontologies that can be accessed as web services. Figure 12.3 shows the East Coast Land Use Land Cover ontology with one of its categories being mapped as an exact match to categories in NLCD1992 and NLCD2001. Figure 12.4 illustrates results of a search to find any existing mappings, showing, for example, that the Residential category is mapped to Low-Intensity Residential land in the NLCD1992 classification system.

Figure 12.5 shows the land use code DevelopedHighIntensity represented in the OWL ontology, and Figure 12.6 shows ontology mappings encoded in OWL. The mappings are published as XML REST services. An online semantic visualization tool can retrieve the REST services and graphically present the ontological mappings of the semantics of the land categories. A web GIS user interface integrates the services to provide an integrated view of the multiple land category ontologies, and allows users to form queries and analyses against the ontologies and data. Both the semantic network and

```xml
<?xml version = "1.0" encoding = "UTF-8"?>
<success>
  <accessedResource>/bioportal/concepts/10033</accessedResource>
  <accessDate>2012-09-17 16:48:40.213 PDT</accessDate>
  <data>
    <classBean>
      <id>Ontology1347296135718:DevelopedHighIntensity</id>
      <label>Ontology1347296135718:DevelopedHighIntensity</label>
      <type>class</type>
      <isObsolete>0</isObsolete>
      <relations>
        <entry>
          <string>rdfs:subClassOf</string>
          <list>
            <classBean>
              <id>Ontology1347296135718:Developed</id>
              <label>Ontology1347296135718:Developed</label>
              <type>class</type>
            </classBean>
          </list>
        </entry>
        .....
      </relations>
    </classBean>
  </data>
</success>
```

FIGURE 12.5
Ontology of the land use type "Developed: High Intensity" hosted on the Open Ontology Repository.

```
<mappings:One_To_One_Mapping
rdf:about = "http://purl.bioontology.org/mapping/
d6a690b2-6e7a-492f-905c-7179a963da74">
    <mappings:source
rdf:resource = 'http://www.semanticweb.org/ontologies/2012/8/
Ontology1347457797421.owl#Residential'/>
    <mappings:target
rdf:resource = 'http://www.semanticweb.org/ontologies/2012/8/
Ontology1347456113812.owl#LowIntensityResidential'>
...
    <mappings:source_ontology_id rdf:datatype =
"xsd:int">1032</mappings:source_ontology_id>
    <mappings:target_ontology_id rdf:datatype =
"xsd:int">1031</mappings:target_ontology_id>
...
    <mappings:comment rdf:datatype = "xsd:string">0.32</
mappings:comment>
...
</mappings:One_To_One_Mapping>
```

FIGURE 12.6
Representation in an Open Ontology Repository for the ontology mapping of "Residential" to "Low-Intensity Residential" (similarity degree: 32%).

the hierarchy in Figure 12.2, are delivered via REST services and visualized in an interactive semantic visualization tool.

12.7 GeoSPARQL and the GeoQuery Tool

As another aspect of our SOCoP work, we created a graphical web interface to query geospatial data. The W3C format to publish data on the web is a linked data format, RDF (http://www.w3.org/RDF/). It is a vision of the Semantic Web to have geospatial data also on the web in a linked data format. Data in RDF can be queried using the SPARQL query language. Because SPARQL does not include spatial operators, GeoSPARQL was developed (Battle and Kolas 2012). This effort enhances SPARQL and helps facilitate a Geospatial Semantic Web. GeoSPARQL allows users to pose spatial queries over RDF data on the web using spatial operators, such as "within." However, although SPARQL and GeoSPARQL are reasonable to use for simple queries, the syntax can quickly become complex. To enable anyone to pose GeoSPARQL queries, we developed the GeoQuery tool (Grove, Wilson, Kolas, and Wiegand 2014). GeoQuery, http://geoquery.cs.jmu.edu/geoquery/, provides drop-down lists for choosing feature types to query, optional filtering specifications, and the spatial operator to be applied. Queries are sent into Parliament, an RDF store, for processing. Results are text-based as well as shown on a map. This tool is web based and even further enables a Geospatial Semantic Web.

12.8 Ontology Design Patterns

12.8.1 SOCoP GeoVoCamps

Contrary to developing large domain ontologies, small, modular ontologies called Ontology Design Patterns (ODPs) have been found to be useful. In addition to organizing other kinds of workshops, SOCoP hosted vocabulary camps (VoCamps) focused on geospatial and related topics. SOCoP's GeoVoCamps are part of a series, http://vocamp.org/wiki/. This section describes the process of conducting a GeoVoCamp and discusses an example concept pattern.

For these workshops, preliminary topics have generally been cooperatively worked on by interested parties ahead of time, which allows development of background material and determination of issues for particular topics. Background material may include domain vocabularies and even related extant ontologies. The list below, for example, was developed before sessions started by organizers of a topic on useful geopatterns for a path/route idea:

- An ontology for grounding vague geographic terms (Bennett et al. 2008)
- Ontology for spatiotemporal databases (Frank 2003)
- Understanding geographic feature types (Janowicz 2011)
- Microtheories for spatial data infrastructures—accounting for diversity of local conceptualizations at a global level (Duce and Janowicz 2010)
- Constructing geo-ontologies by reification of observation data (Adams and Janowicz 2011)

When sessions start, proposed topics are further defined by group members by identifying critical vocabulary at the center of the work, including terms and models from Linked Data. Conceptual components are also identified through discussions of variable names associated with available data needed to describe the target topic, where to find vocabularies and domain data relevant to the topic, which elements to use, and how much information can be reused (Freitas et al. 2012).

Small (usually three to eight people) workgroup teams detail the goals of work for the remaining sessions, which are focused on creating small ontologies. Preliminary session work employs lightweight methods, grounded by scenarios and domain interests to reduce the entry barrier. Activities reflect both a practical limitation of what can be accomplished in a few days of work and also the observation that comprehensive (domain or foundational) quality ontologies, such as DOLCE (Masolo et al. 2003), have proven difficult to reuse. For large ontologies, it is hard to follow their complexity, find useful pieces in a wealth of axioms expressed in logical languages, find only the "useful pieces" with local semantics, and extract these properly for use or insertion (Gangemi and Presutti 2009). Further, many large ontologies lack a clear modular structure where a module is considered some subset of a "whole" that makes sense by itself to formally address some questions (Doran et al. 2007). For all these reasons, the cost of reusing large ontologies may be higher than opportunistically developing a scoped-purpose modular ontology from scratch. This is especially important recognizing time limitations of efforts within small time frames such as 2- to 3-day workshops. A well-founded basis for using top-down approaches on large ontologies to make them easily comprehended and modular remains a research issue (Rector 2003; Stuckenschmidt and Klein 2003; Grau et al. 2007).

An alternate approach is to start with the idea of a modular, coherent piece around a domain or modeling issue and build around it. The essential idea is to develop a small module that captures an intuitive but critical aspect of some domain of interest. Such modules may reuse pieces of other ontologies if this is easy to do. Adopting this stance, GeoVoCamps have built modular pieces from the bottom up, using the idea of an ODP and opportunistically leveraging existing ontologies and vocabularies along the way. ODPs (aka Conceptual/Content patterns [CPs]) are based on the working hypothesis that, just as in other domains, such as architecture and software, small problem and solution patterns reoccur and can be reused for formalization of useful knowledge.

One example of an extremely simple CP is the idea of "Part Of," which occurs often to express part–whole relations. This can be applied in the geo-spatial domain to places, such as Chicago, which may have neighborhoods or districts. In addition to being transitive, the idea of Part Of may or may not imply a timeless property of transitivity for parts. Typically, for endurants, if an object is a part at time $t1$, it is assumed to also be a part at time $t2$. If time is used in the concept pattern definition, however, a part could be temporary.

We see this part relation in another simple, but slightly more complex CP—a participation pattern (Gangemi 2005). A participation pattern is made up of some objects existing in some region of space, such as Chicago, that take part in events at some regions of time. This general idea of participation is a small constellation of related concepts that can describe something as simple as a sensor used in a research study or more complex sets of objects participating in a flood event. A flood involves objects such as roads, building structures, fields, and cars. In such a pattern, we may indicate that a parcel of land or a road is "Part Of" a whole entity, such as a flood, for a particular period and not permanently. A road or parcel may participate in, that is, be Part Of, a flood for some period but not the entire flood period.

In practice, it seems the case that a community of expertise can develop and encode its own CP(s) in small, relatively homogeneous and modular/autonomous ontologies (Gangemi and Presutti 2009). Once captured, the structured pattern can be used as a template for modularly growing new and larger ontologies, as with the basic motion ODP shown in Section 12.8.2.

12.8.2 Examples of Ontology Design Patterns

Various aspects of the ODP proposition have been employed as part of SOCoP workshops to develop vocabularies organized as ODPs. The first VoCamps, such as GeoVoCampSB2012 (2012), started with schema patterns for very basic processes such as Motion and related objects such as Path. For this effort, motion was conceptualized generally as a process or as some type of change event (a perdurant). A richer conceptualization denotes the "From-To" aspect of a motion along a path by some object participating in the motion. An example of the visualized conceptual model with observable elements is shown in Figure 12.7 along with the beginnings of some linkages to OWL classes.

To make the model of a generic motion pattern, all slots such as motion parts are optional and can have multiple instances. Motion description is also general, although particular attributes such as velocity could be added with numerical values. Likewise, some modal characterizations of the motion, such as gyrating or wavering, could be added. Event is used here as a class but is a placeholder for a more general event pattern such as that of Van Hage et al. (2011), which would bracket movement in time as well as space. Thus, a movement may be part of some larger event such as a transportation event. The model was formalized in OWL, and a portion of this formalization is shown below with Motion identified as a subclass of Event (Motion Pattern 2012):

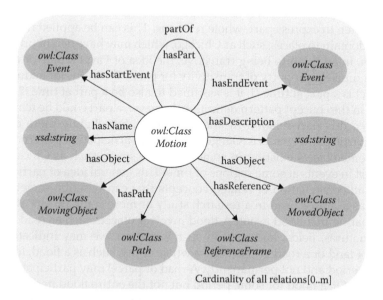

FIGURE: 12.7

Motions: most general pattern, developed by Gary Berg-Cross, based on work at GeoVoCampSB2012.

```
:Motion a owl:Class;
     rdfs:subClassOf event:Event;
     rdfs:label "Motion";
     rdfs:comment "A motion is an event in which some entity
     moves through space" ;
     rdfs:subClassOf [
          a owl:Restriction;
          owl:onProperty :startEvent;
          owl:allValuesFrom event:Event
     ];
```

Later VoCamps, GeoVoCampDayton2012 (2012) and GeoVoCamp SOCoP2012 (2012), built on these early patterns of Motion, Path, and Points of Interest (POIs) to develop additional ODPs. An example of one such ODP is a 3D semantic trajectory defined by a sparse set of temporally indexed positions or fixes (Hu et al. 2013). The basic pattern deals with annotated locations (along a path or segment) of moving objects over time, including start and end points. Such fixes are readily available from sensing devices, including navigational devices in cars, and can therefore be used to model different modes of movement where fixes are captured. To the idea of motion along path-segments was added the idea of a semantic annotation about POIs. These provide useful information, such as a label to the movement and its fixed points, the sensing devices, and the nature of the moving object. Annotations might include some information about POIs along a trajectory of the moving object (e.g., rest stops) and/or segment attributes such as toll road. Different attributes may be of interest, for example, depending on the type of travel (e.g., commute or migration) or on the nature of the traversing vehicle (e.g., whale or ship).

12.8.3 Land Use Model

Some of the above approaches to ODP work have been applied to land use codes and semantic mapping. A start on a pattern was made at a recent SOCoP Geospatial Semantics Workshop and GeoVoCamp (2014) to model parcel data, its legal boundary and geometry, ownership, and land use code (Figure 12.8). Adding boundary geometry as a formal concept makes data for this pattern able to be queried using GeoSPARQL. The land use code used for each parcel is based on the jurisdiction's land use coding system. Because each jurisdiction may have its own land use classification, mappings need to be done between coding systems. As with the motion pattern, formal axioms can be crafted to logically specify the constraints that are understood and agreed to by the workgroup and that are visualized in the conceptual model.

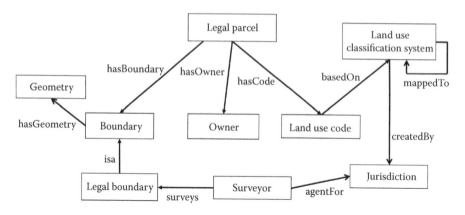

FIGURE 12.8
Example ODP for a parcel developed at the Geospatial Semantics Workshop and GeoVoCamp.
(From http://www.ssec.wisc.edu/meetings/geosp_sem/.)

12.9 Conclusions

Land use and land cover represent a challenging example of the need to resolve semantic heterogeneity. Each country, or smaller unit such as a local jurisdiction, uses its own classification system, which may even change over time. Being able to do regional or worldwide studies of land use, or changes in land use, requires resolving semantic differences. Although research of ontology-based semantic integration has been conducted for more than a decade (Bishr 1998), automated and accurate semantic similarity measurements, semantic integration, and ontology mapping still remain challenges for semantic interoperability. Formal ontologies and ODPs will continue to be developed to describe basic concepts in the land use or land cover domain, capturing information on parcels or other types of regions to which classification codes are assigned. The ontology repository work provides a valuable platform for ontology sharing via Web services. Other semantic technologies are still also needed, such as ontology aware architectures, ontology-based query processors, federated ontology repositories, and more comprehensive semantic representations and languages.

Acknowledgments

This work was partially supported by the National Science Foundation's Office of Cyberinfrastructure (OCI), INTEROP Grant No. 0955816. In addition to Dave Kolas and James Wilson, already referenced, we also acknowledge the work of Mike Dean (deceased November 2014) and Peter Yim on our INTEROP grant, along with consulting provided by John Moeller.

References

Adams, B., and K. Janowicz. 2011. Constructing geo-ontologies by reification of observation data. *In Proceedings of the 19th ACM SIGSPATIAL International Conference on Advances in Geographic Information Systems.* ACM, New York, NY.

Ahlqvist, O., J. Macgill, and D. Guo. 2004. Exploration of semantic uncertainty in parameterized land use/land cover classifications. http://www.geovista.psu .edu/publications/2004/AhlqvistGuoMacgillExploration_Proof.pdf.

American Planning Association. 1994. Toward a standardized land-use coding standard. Working Paper, Research Department for the Federal Highway Administration, U.S. Department of Transportation, 30 March 1994. https:// www.planning.org/lbcs/background/scopingpaper.htm.

Baclawski, K., and T. Schneider. 2009. The Open Ontology Repository initiative: Requirements and research challenges. In *Proceedings of Workshop on Collaborative Construction, Management and Linking of Structured Knowledge at the ISWC 2009.*

Battle, R., and D. Kolas. 2012. Enabling the Geospatial Semantic Web with Parliament and GeoSPARQL. *Semantic Web Journal* 3(4): 355–370.

Bennett, B., D. Mallenby, and A. Third. 2008. An ontology for grounding vague geographic terms. *FOIS 183.* In Formal Ontology in Information Systems— Proceedings of the Fifth International Conference (FOIS 2008), pp 280–293.

Bishr, Y. 1998. Overcoming the semantic and other barriers to GIS interoperability. *International Journal of Geographical Information Science* 12(4):299–314.

Cruz, I. F., C. Stroe, and M. Caci, et al. 2010. Using AgreementMaker to align ontologies for OAEI 2010. Fifth International Workshop on Ontology Matching, co-located with the International Semantic Web Conference, Shanghai, China, November (OAEI track paper).

Doran, P., V. Tamma, and L. Iannone. 2007. Ontology module extraction for ontology reuse: An ontology engineering perspective. *16th ACM International Conference on Information and Knowledge Management,* 61–70. ACM Press, New York, NY.

Duce, S., and K. Janowicz. 2010. Microtheories for spatial data infrastructure— accounting for diversity of local conceptualizations at a global level. In *Geographic Information Science,* 27–41, edited by S. I. Fabrikant, T. Reichenbacher, M. J. van Kreveld, and C. Schlieder. Springer Berlin-Heidelberg, Berlin, Germany.

Everett, J., and C. Ngo. 1999. Land-based classification standards: Federal role. In *Proceedings of the 1999 American Planning Association National Conference.* Seattle, Washington.

Frank, A. U. 2003. *Ontology for Spatio-Temporal Databases.* Springer, Heidelberg, Germany.

Freitas, A., E. Curry, and S. O'Riain. 2012. A distributional approach for terminological semantic search on the linked data web. In *Proceedings of the 27th Annual ACM Symposium on Applied Computing.* ACM, New York, NY.

Gahegan, M., W. Smart, S. Massoud-Ansari, and B. Whitehead. 2011. A semantic web map mediation service: Interactive redesign and sharing of map legends. In *Proceedings of the 1st ACM SIGSPATIAL International Workshop on Spatial Semantics and Ontologies (SSO'11),* 1–8. ACM, New York, NY.

Gangemi, A. 2005. Ontology design patterns for semantic web content. In *The Semantic Web–ISWC 2005*, 262–276. Springer, Heidelberg, Germany.

Gangemi, A. and V. Presutti. 2009. Ontology design patterns. In *Handbook of Ontologies* (Second edition), edited by S. Staab, and R. Studer. Springer, Berlin, Germany.

Geospatial Semantics Workshop and GeoVoCamp, 2014. University of Wisconsin-Madison, Madison, Wisconsin. June 2–3. http://www.ssec.wisc.edu/meetings /geosp_sem/.

GeoVoCampDayton2012. 2012. Dayton, Ohio. http://vocamp.org/wiki /GeoVoCampDayton2012#Formed_breakout_groups.

GeoVoCampSB2012. 2012. Santa Barbara, CA. http://vocamp.org/wiki /GeoVoCampSB2012.

GeoVoCampSOCoP2012. 2012. http://vocamp.org/wiki/GeoVoCampSOCoP2012 and http://vocamp.org/wiki/GeoVoCampSB2012#A_possible_approach_ to_a_Motion_and_Path_design_pattern_based_on_the_PATH_image_schema.

Grau B. C., I. Horrocks, Y. Kazakov, and U. Sattler. 2007. Just the right amount: Extracting modules from ontologies. *16th International Conference on World Wide Web*, 717–726. ACM, New York, NY.

Grove, R., J. Wilson, D. Kolas, and N. Wiegand. 2014. GeoSPARQL query tool: A Geospatial Semantic Web visual query tool. *International Conference on Web Information Systems and Technologies (WebIST)*, April. Barcelona, Spain.

Hu, Y., K. Janowicz, D. Carral et al. 2013. A geo-ontology design pattern for semantic trajectories. In *Spatial Information Theory*, 438–456, edited by T. Tenbrink, J. Stell, A. Galton, and Z. Wood. Springer International Publishing, Cham, Switzerland. http://geog.ucsb.edu/~jano/semantic_trajectories.pdf.

Janowicz, K. 2011. Understanding geographic feature types. Presentation at Harvard Center for Geographical Analysis, April 14. https://cga-download.hmdc .harvard.edu/publish.../Janowicz.pdf.

Janowicz, K., M. Raubal, and W. Kuhn. 2011. The semantics of similarity in geographic information retrieval. *Journal of Spatial Information Science* 2: 29–57. http://geog. ucsb.edu/~jano/semofsim2011.pdf.

Jeer, S. 1997. Land-based classification standards project. LBCS Discussion Issues, American Planning Association Research Department. Revised February 20, 1997.

Jeer, S. 1999. Land-based classification standards: Session outline. In *Proceedings APA National Planning Conference*.

Masolo, C., S. Borgo, A. Gangemi, N. Guarino, A. Oltramari, and L. Schneider. 2003. Dolce: A descriptive ontology for linguistic and cognitive engineering. WonderWeb Project, Deliverable D17 v2 1.

Montenegro, N., J. Gomes, P. Urbano, and J. Duarte. 2012. A land use planning ontology: LBCS. *Future Internet* 4(1): 65–82. doi: 10.3390/fi4010065.

Motion Pattern. 2012. http://ontolog.cim3.net/file/work/SOCoP/Workshops /GeoVoCampUCSB2012/movement.ttl.

Naughton, J., D. DeWitt, D. Maier, et al. 2001. The Niagara Internet query system. *IEEE Data Engineering Bulletin* 24(2): 27–33.

North Carolina land use and land cover classification. 1994. http://gis.idaho.gov /portal/pdf/Framework/LULC/StdClassSysLCinNC.pdf.

Pearce, E. 1967. History of the Standard Industrial Classification.

Rector, A. 2003. Modularisation of domain ontologies implemented in description logics and related formalisms, including OWL. In Proceedings of the 2nd international conference on Knowledge capture, *Knowledge Capture* 121–128. doi:10.1145/945645.945664.

Ressler, J., M. Dean, and D. Kolas. 2010. Geospatial ontology trade study. *Ontologies and Semantic Technologies for Intelligence* 213: 179–211.

Smith, M., C. Welty, and D. McGuinness. 2004. OWL Web Ontology Language Guide, W3C Recommendation 10 February 2004, http://www.w3.org/TR/2004/REC-owl-guide-20040210/.

Stuckenschmidt, H., and M. C. A. Klein. 2003. Integrity and change in modular ontologies. In *Proceedings of the International Joint Conference on Artificial Intelligence: IJCAI '03*. Morgan Kaufmann, Acapulco, Mexico.

Van Hage, W. R., V. Malaise, R. Segers, et al. 2011. Design and use of the Simple Event Model (SEM). *Web Semantics: Science, Services and Agents on the World Wide Web* 9.2: 128–136.

Wiegand, N. 2012. Preserving detail in a combined land use ontology. *GIScience LNCS* 7478: 284–297.

Wiegand N., and N. Zhou. 2005. Ontology-Based Geospatial Web Query System. In *Next Generation Geospatial Information: From Digital Image Analysis to Spatio-Temporal Databases*, 157–168, edited by P. Agouris, and A. Croitoru. ISPRS Book series, Balkema.

Wiegand, N., E. D. Patterson, N. Zhou, S. Ventura, and I. Cruz. 2002. Querying heterogeneous land use data: Problems and potential. In *Proceedings National Conference on Digital Government Research*, dg.o2002. May. 115–121.

Zhou, N., and H. Wei. 2008. Semantic-based data integration and its application to geospatial portals. In *Proceedings of GIScience 2008* (Extended Abstracts), 331–334, edited by T. J. Cova, J. M. Harvey, K. Beard, A. U. Frank, and M. F. Goodchild, Park City, UT.

13

Crowdsourcing Landscape Perceptions to Validate Land Cover Classifications

Kevin Sparks, Alexander Klippel, Jan Oliver Wallgrün, and David Mark

CONTENTS

> If every object and event in the world were taken as distinct and unique—a thing in itself unrelated to anything else—our perception of the world would disintegrate into complete meaninglessness. The purpose of classification is to give order to the thing we experience.
>
> **—Abler, R., Adams, J. S., and Gould, P. 1971**

Of all the countless possible ways of dividing entities of the world into categories, why do members of a culture use some groupings and not use others? What is it about the nature of the human mind and the way that it interacts with the nature of the world that gives rise to the categories that are used?

—Malt 1995

ABSTRACT. This chapter analyzes the correspondence between human conceptualizations of landscapes and spectrally derived land cover classifications. Although widely used, global land cover data have known discontinuities in accuracy across different datasets. With the emergence of crowdsourcing platforms, large-scale contributions from the crowd to validate land cover classifications are now possible and practical. If crowd science is to be incorporated into environmental monitoring, there needs to be some understanding of how humans perceive and conceptualize environmental features. We are reporting on experiments that compare crowd classification of land cover against an authoritative dataset (National Land Cover Dataset), and crowd agreement between participants using novices, educated novices, and experts. Results indicate misclassifications are not random but rather systematic to unique landscape stimuli and unique land cover classes.

KEYWORDS. *crowd science, classification, land cover*

13.1 Introduction

The way humans understand their natural environments—landscapes—either as an individual or as a collective, frames prominent research topics in several disciplines. Landscape perception for instance has been analyzed for land management and planning purposes to characterize landscape aesthetics and objective scenic beauty. Understanding earth surface processes through terrain analysis has been relevant to military and civil engineering. From a geographic perspective, it can be argued that the *man–land tradition* or in more recent terms *human–environment relations* is nothing less than one of the four intellectual cores of geography (Pattison 1964; see also Mark et al. 2011).

Land cover data has been used for much more than looking up land cover at a given location, including uses for climate modeling, food security, and biodiversity monitoring. The focus of this chapter is on how the abundance of freely available high-resolution imagery of the earth's surface and the maturing of crowd science offers new opportunities for an unprecedented access to environmental information. We will examine how crowd science and human perceptions can be used for the purposes of improving overall quality of land cover datasets. First, we will discuss variation between land cover classifications and explore why they exist. Second, we will discuss issues and debates surrounding the accuracy of land cover datasets. Last, we will discuss shortcomings with current assessment processes and opportunities for new methods to assist in the overall quality of land cover datasets.

Although widely used, many global land cover datasets have unique characteristics that lead to differences between classifications. Variety is represented through differing land cover classes, changed meanings of shared terminology, and differences in interpretation and perception of the land cover classes. Ahlqvist (2008) discussed the importance of standardizing terminologies in science, using the topic of variation in land cover classifications as an example. He stresses the need for more interpretability, reflecting on the subjectivity not only in the creation process of land cover classifications but also in the interpretation of the classification from the user. This point builds off of research by Comber et al. (2005) discussing the varying conceptualizations of the world that geographic data are mapped into. They list examples of how terms such as *Forest* and *Beach* have varying meanings based on the purpose that the land cover dataset was created for, and how these terms are interpreted differently across cultures and among users. There is a growing demand for harmonization of data, especially in class descriptions (Jepsen and Levin 2013). Recognizing this becomes more important as local environmental knowledge is increasingly being incorporated into land use and land cover analysis. Robbins (2003) exemplifies this through differences shown in land cover and land use classification choices made by foresters and herders, enforced by their respective cultural and political role in their community. This creates variation between the classifications generated by the producers, and variation between how a unique land cover class is perceived by the users.

Gaining a deeper understanding of the perception of land cover classes addresses significant challenges in the effectiveness of classification interpretability (Ahlqvist 2012). Comber et al. (2004) use the example of the Great Britain Datasets LCM1990 and LCM2000 to illustrate how changes in methodology and semantics cause unknown variation between datasets, creating uncertainty between either observing land cover change, or simply observing a change in how the land cover is represented.

To reduce variability in interpretation of classification, land cover classifications must be concerned with users' natural concepts and perceptions of the land cover, and be aware of formal cognitive models about the common-sense geographic world. Coeterier (1996) concludes that even when comparing landscapes of great differences between inhabitants of those landscapes, there is agreement among the importance of higher level attributes of the landscape. Some of these attributes include the unity of the landscape, its use, maintenance, naturalness, and spaciousness. More so, these attributes are not necessarily independent from each other. Habron (1998) analyzes the perceptual differences of Wild Land across varying demographics in Scotland. He concludes that human presence/influence has a large effect on the perception of Wild Land. Furthermore, what is considered Wild Land varies between sections of the population, with a consensus on a core definition and variation at the periphery.

Along with classification and interpretation variation, land cover datasets have accuracy-related issues. Foody (2002, 2008) has discussed the state of land cover dataset quality along with their corresponding accuracy assessments and has noted the debate surrounding accuracy expectations. Once accuracy assessments are performed, of which there are multiple different methods of assessment, the vast majority of land cover datasets do not meet the commonly recommended target of 85% accuracy. He further discusses that 85% is perhaps unrealistically high. This number was historically specified by Anderson et al. (1976) for mapping general land cover classes (Anderson, Level 1). In addition, that accuracy rate was inspired by work associated with United States Department of Agriculture's Census of Agriculture in which 85% "would be comparable to the accuracy of land-cover maps derived from aerial photograph interpretation." (Foody 2008, p. 3140). Yet, in the face of potentially harsh critiques of accuracy, land cover nonetheless can benefit from new approaches of classification and assessment. Wilkinson's (2005) 15-year survey of published papers on satellite image classification revealed no upward trend in classification accuracy. This creates an opportunity for unconventional classification approaches to assist in a severe lack of advancing accuracy rates.

As previously mentioned, once land cover datasets are created, accuracy assessments are often performed to measure the quality of the dataset. Comber et al. (2012) note that a common approach of measuring land cover accuracy is to compare the dataset with corresponding data that is considered to be of a higher accuracy. This could mean a collection of relatively sparse ground data acquired in the field used as control points. Comber discusses that these approaches of assessment overlook the spatial distribution of errors, leading to possible localized subregions of high inaccuracy that distort the global accuracy. Furthermore, Foody (2002) reiterates that land cover is dynamic. The earth's surface will inevitably change in the time it takes to update datasets. This creates an opportunity for new methods of assessment and classification that are flexible, have the capability of covering large areas, and are quick relative to classic approaches.

New (unconventional) methods need evaluation. Although there is a lot of interest surrounding the opportunities of the crowd, examples of which will be explained in detail in Section 13.2, there is a high demand for systematic evaluations of how much improvement in land cover classification can be achieved using crowd-based assessments. In response to these challenges, this chapter analyzes the correspondence between human conceptualizations of land cover and spectrally derived land cover datasets. If crowdsourced human participants are to be incorporated into the evaluation of land cover data, there needs to be a more rigorous understanding of how humans perceive and conceptualize land cover types and a more detailed assessment of how well humans perform in recognizing predefined land cover classes. We analyze crowdsourced human participants' ability to recognize existing land cover classes given on the ground photographs. We are reporting on three experiments that provide insights on the relationships between human conceptualizations of land cover and land cover classifications using novices, educated novices, and experts. Our findings suggest misclassifications are not random but rather systematic to unique landscape stimuli and unique land cover classes. By comparing novices and experts, we are able to evaluate the potential for using crowdsourcing in aiding the advancement of land cover classifications.

13.2 Background

Recent growth and improvement to online crowdsourcing platforms now allow for large-scale contributions to scientific research. These crowd contributions have shown to be successful across many disciplines, including the discovery of protein structures (Khatib et al. 2011) and identification of new galaxies (Clery 2011). In the context of land cover, using crowdsourced human participants to validate global land cover datasets has been recognized by previous research. This use of crowdsourcing shows promise, especially as the use of more than one dataset often provides more accurate land cover mapping (Aitkenhead and Aalders 2011). The Geo-Wiki project (Fritz et al. 2009) asks online participants to use aerial imagery via Google Earth as well as any local knowledge they may have to make classification choices on which land cover type they are observing given a predefined classification scheme. This volunteer geographer approach complements the classification and accuracy assessments in use but, at the moment, fails to guarantee a level of quality in the volunteered data.

The Land Use/Cover Area Frame Survey (LUCAS) (http://www.lucas -europa.info/) is an example of a more authoritative, non-crowd-based attempt at capturing land cover data. LUCAS, commissioned by Eurostat, deploys land surveyors to many locations across the European Union to determine land cover/land use, record transects, and take photographs of the landscape. By virtue of LUCAS's means of data collection, creating a comprehensive dataset using this method would be highly improbable. Using these data as a means of validation, however, is more likely.

For the purposes of measuring land cover, due to the complexity of the earth's surface, all measurements contain error to some unknown extent. It is thus very difficult to precisely describe and categorize features of land cover. This error is true for both remote sensing classification and classification via human interpretation of aerial imagery. Foody (2002) discusses this from the perspective of remote sensing to the degree that ground truth measurements are still a classification and thus contain some degree of error. Kinley (2013) and Hoffman and Pike (1995) discuss this from the perspective of volunteered data and terrain analysis, stating that these data and terrain descriptions are often critiqued harshly for not meeting an impossible ideal. Yet, in the face of inescapable error in land cover data and volunteered data, steps must be taken to ensure the methods for collecting data allow for the opportunity of the highest quality products. This means understanding humans' concepts and perceptions of land cover to assist in the classification process.

The majority of experiments measuring quality of crowdsourced volunteered land cover classifications come from experiments run through the Geo-Wiki project (Perger et al. 2012; See et al. 2013; Foody et al. 2013a; Comber et al. 2013, 2014). See et al. (2013) most notably report on an experiment which expert and nonexpert participants during a Geo-Wiki campaign were asked to classify land cover given aerial imagery for the purposes of measuring participant accuracy rates, and comparing expert and non-expert results. Control points generated by three experts visually classifying land cover from aerial imagery were used to measure how accurate the crowdsourced participants' classifications were. Averaged accuracy rates range from 66% to 76% for the full set of participants, with experts reaching a maximum of 84%, and nonexperts reaching a maximum of 65%. Comber et al. (2013) also uses crowdsourced classification data gathered from Geo-Wiki, but focuses on the level of agreement between expert and nonexpert classification of land cover type, rather than reporting accuracy rates measured against control points. They conclude by illustrating map outputs that show obvious visual differences between expert and nonexpert classification choices, and call for "further investigation into formal structures to allow such differences to be modeled and reasoned with" (Comber et al. 2013, p. 257). Comber et al. (2014) further states that expertise in classification has a general influence but is varied across land cover classes.

Similarly to Geo-Wiki, the OpenStreetMap (http://www.openstreetmap.org) dataset comprises crowdsourced geographic information that research has identified as potential data to assist, support, and validate other land use mapping projects. Arsanjani et al. (2013) have analyzed OpenStreetMap contributions to analyze the accuracy of participants' land use (opposed to land cover) classifications in an urban setting compared to other non-crowdsourced land use datasets. He concludes that OpenStreetMap, and in general other forms of crowdsourced geographic data, can be valid data sources for mapping land use.

Perger et al. (2012) notes how land cover can be difficult to classify when given only aerial imagery. Deviating from classification via aerial imagery, others have attempted to measure the effectiveness of using on-the-ground photographs for the purposes of land cover classification (Iwao et al. 2006; Foody et al. 2013b). The data source of these ground-based photographs comes from the Degree Confluence Project (DCP) which will be explained in detail in section 13.3.1.1. While the works of Iwao and Foody both report land cover classification accuracy rates, the main intention of their research was to test the validity of using DCP data to classify land cover. Both conclude DCP data is a valid data source when attempting to classify land cover.

To summarize, land cover classifications have not experienced significant accuracy improvements in the past 20 years. Research has recognized an opportunity to benefit from advancing technologies and improvements in crowd science to assist in the evaluation of land cover. This has largely been experimented through providing crowdsourced participants aerial imagery of the earth surface and asking for their classifications of the land cover. Aerial imagery, however, can sometimes provide a lack of information when distinguishing between similar land cover classes. Other research has proven the validity of using on-the-ground photographs for land cover classification, but has failed to test it with crowdsourced participants and a wide range of land cover classes.

13.3 Experiments

We conducted three experiments to shed light on humans' understanding of and ability to classify land cover according to official National Land Cover Dataset (NLCD) 2006 classes (Fry et al. 2011). The first two experiments involve laypeople (without and with intervention), whereas the third uses experts.

13.3.1 Experiment 1: Laypeople, No Intervention

The first experiment addresses the question whether laypeople can classify images of land cover according to existing land cover classes. Although the ground truth itself, that is, the NLCD 2006, only has a level II accuracy of 78% (Wickham et al. 2013), it serves as a starting point for improving the understanding of how humans perceive land cover classes.

13.3.1.1 Materials

Two datasets were used for this experiment: on-the-ground-photographs of landscapes provided by the DCP (confluence.org), and the NLCD 2006 provided by the Multi-Resolution Land Characteristics Consortium (http://www.mrlc.gov).

The DCP is a site that provides a platform for collecting crowdsourced photographs of landscapes at confluence points across the world in a systematic way. The word *confluence* as defined for the purposes of the DCP is the location where two integer latitude and longitude coordinate lines meet. An example of this would be "latitude 42 N, Longitude 100 W" as opposed to "latitude 42.65 N, longitude 100.23 W." Users are encouraged to visit these locations, take photographs of the landscape, and upload the images with metadata such as date visited and travel information.

For the scope of these experiments, we constrained our data collection to the contiguous United States. A total of 799 photographs were collected out of a possible 856. In an attempt to be consistent in data collection, north-facing photographs were collected when at all possible. Two sampling criteria restricted the data collection process: First, scenes that included snow in the photograph were excluded as this is not reflective of the land cover but rather temporal weather conditions. Second, images that included human presence were excluded. Outside of these sampling restrictions, few confluences do not have photographs uploaded to the website, and as such, could not be collected.

Latitude and longitude coordinates from the DCP dataset were extracted and converted into a point shapefile to be used in the Environmental Systems Research Institute's (ESRI) ArcGIS software (Figure 13.1). This allowed for the extraction of the corresponding land cover class from NLCD level II (16 land cover classes) for each confluence point and its corresponding image. Out of the 16 possible classes from NLCD level II, 11 were used in the experiments: we aggregated *deciduous forest, evergreen forest,* and *mixed forest* into one *forest* class. In addition, the following three classes did not provide sufficient sampling points to ensure balanced class representation, and a suitable number of total images: *developed medium intensity, developed high intensity,* and *perennial ice/ snow.* From the remaining 11 classes, 7 locations and associated images were randomly selected (stratified random sampling), resulting in 77 images shown partly in Figure 13.2. To ensure that confluences were not on the boundary of two land cover classes, confluences were selected when located in a homogenous land cover region of at least 90 m (3 NLCD pixels) in the direction the photo was taken. Although land cover change has the possibility of influencing incorrect land cover extraction, each of the 77 images were analyzed together with their corresponding land cover class to ensure consistency between land cover features in the images, and assigned land cover classes. It is important to note that the accuracy assessment of NLCD 2006 by Wickham et al. (2013) for the contiguous United States concludes that level I (8 aggregated land cover classes) accuracy is equal to 84% and level II accuracy is equal to 78%.

13.3.1.2 Participants

Twenty lay participants (nonexperts, five female) were recruited through the crowdsourcing platform Amazon Mechanical Turk (AMT); average age 32.2 years; reimbursement: $1.25.

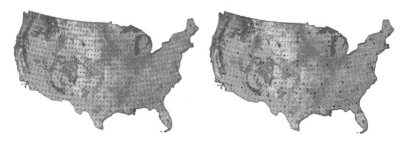

FIGURE 13.1 (See color insert.)
The NLCD 2006 overlaid by confluence points (left). Stratified random sampled confluence points, 77 total sampled, 7 in each land cover class (right).

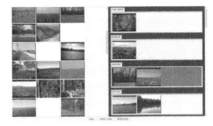

FIGURE 13.2 (See color insert.)
Screenshot of the CatScan interface of an ongoing mock-up experiment.

13.3.1.3 Procedure

The experimental software CatScan (Klippel et al. 2008) used for the experiment has been designed to be serviceable in combination with AMT (Figure 13.2). In the experiment, each participant performed a non-free classification task. During the non-free classification, all images were initially displayed on the left panel of the screen. On the right side of the screen, the 11 land cover classes were displayed into which participants were able to drag icons from the left panel into the classes on the right panel. It was possible to leave classes empty.

13.3.1.4 Results

The classification results should be interpreted with consideration to Wickham's (Wickham et al. 2013) accuracy assessment in mind. To reiterate, our sample from the NLCD was taken from the level II classification, which Wickham concludes is 78% accurate. There exists, however, accuracy variation among classes in the NLCD, and Wickham stresses the need for improved distinction among grass-dominated classes (*develop, open space* [dO], *grassland* [GS], *pasture/hay* [PH], *cultivated crops* [CC], and *emergent herbaceous wetland* [EW]), as they account for higher classification error relative to the other classes.

Participants used an average of 10.25 classes (out of the possible 11) with a standard deviation of 1.07. The average grouping time was 665.86 seconds (11 minutes 5 seconds) with a standard deviation of 263.73 seconds (4 minutes 23 seconds).

To analyze the classification results, we created a confusion matrix (Figure 13.3) that not only shows the number of correctly classified land cover images but additionally reveals how images were misclassified; the confusion matrix shows in which class an image was placed and whether or not this was the correct class. We performed chi square tests to corroborate the interpretation statistically. Several main observations can be summarized as follows.

Overall classification accuracy for Experiment 1 is approximately 40.19%. Against the relatively low overall accuracy of the classification task, the following land cover classes were significantly classified correctly more frequently than expected by having a standardized residual value greater than 1.96 (Table 13.1): *developed, low intensity* (dL), *forest* (FO), and *open water* (OW). In contrast, the following land cover classes were significantly classified less correctly than expected by having a standardized residual value less than −1.96: *emergent herbaceous wetlands* (EW), *pasture/hay* (PH), and *woody wetlands* (WW).

As participants proceed through the experiment, CatScan records the land cover class that an image is placed in. Correct classification is assumed based on the land cover class the image is sampled from (see Figure 13.1). Organizing this data in the form of confusion matrices allows for reviewing the classification behavior of all participants and assessing both correct and incorrect classifications. The confusion matrix below (Figure 13.3) shows the classification behavior in percentages; results can be summarized as follows: The *Woody wetlands* (WW) class is almost exclusively confused with *forest* (FO). Participants are generally successful in recognizing *developed* land cover but confuse *developed, open space* (dO) and *developed, low intensity* (dL), having more success classifying *developed, low intensity* (dL). Participants almost exclusively confuse *barren* (BA) and *shrub/scrub* (SS) with each other. The *emergent herbaceous wetlands* (EW), *grassland* (GS), and *pasture/hay* (PH) classes are confused across many classes.

	BA	CC	dL	dO	EW	FO	GS	OW	PH	SS	WW
BA	46.43	2.86	0	0	7.14	0.71	2.86	0	2.86	36.43	0.71
CC	9.29	37.14	0	2.14	2.86	0	34.29	0.71	11.43	2.14	0
dL	0	0	57.86	30	0	0	7.86	0	2.86	1.43	0
dO	0.71	0	46.43	35.71	0.71	0	2.86	0	7.86	5.71	0
EW	6.43	5	0	0	2.14	33.57	12.14	0.71	17.86	16.43	5.71
FO	0.71	0.71	0	0	2.86	72.14	0	0	1.43	20	2.14
GS	23.57	13.57	0.71	0.71	0.71	0	35	0	15	10	0.71
OW	0	0	0	0	0.71	0.71	0.71	92.14	0	0	5.71
PH	14.29	2.14	2.86	3.57	3.57	12.14	34.29	0	9.29	14.29	3.57
SS	45	0.71	0.71	0	0.71	0.71	10	0	3.57	37.14	1.43
WW	0	1.43	1.43	0	1.43	71.43	0	0	0	7.14	17.14
Total	13.31	5.78	10	6.56	2.08	17.4	12.73	8.51	6.56	13.7	3.38

FIGURE 13.3 (See color insert.)
Confusion matrix for Experiment 1 (lay participants with no intervention) showing percentages of correct (diagonal) and misclassified landscape images (rows). Misclassified classes between 5% and 25% are indicated by light pink, misclassifications between 25% and 50% are light orange, and misclassifications above 50% are red. The "Total" row indicates the percentage of classification choices made in each class.

TABLE 13.1

Standardized Residuals for Experiment 1

	BA	CC	dL	dO	EW	FO	GS	OW	PH	SS	WW
correct	1.57	−0.77	4.47	−1.13	−9.63	8.08	−1.31	13.14	−7.82	−0.77	−5.83

13.3.1.5 Discussion

Comparing Experiment 1 to Wickham's (Wickham et al. 2013) analysis of the NLCD 2006 accuracy, both human classification and NLCD classification have relative difficulty in classifying *emergent herbaceous wetlands* (EW) and *pasture/ hay* (PH). From this we can speculate that both visual stimuli and spectral characteristics of the land cover feature in *emergent herbaceous wetlands* (EW) and *pasture/hay* (PH) are not well defined and cause confusion between classes.

Using the *woody wetlands* (WW) class as an example, visually classifying certain land cover features cannot be done with relatively high levels of confidence, whereas spectrally it can. This could be a case of remote sensors' ability to collect data outside the visual spectrum, leading to clear distinctions between, say, forest and woody wetland via soil and vegetation moisture. Visually recognizing this distinction from DCP data proves to be very difficult for human classification (17% accuracy for WW).

Although human classification and NLCD classification may start to have similar relative inaccuracies for *developed, open space* (dO), the confusion matrix shows humans being very successful in generally identifying *developed*. While remote sensors may have difficulty distinguishing spectral characteristics between developed features and natural features, this may not be as difficult a task for human classification. Developed features, although potentially spectrally similar to certain natural land cover features surrounding them, become easily identifiable for humans to visually interpret and distinguish from surrounding natural features.

13.3.2 Experiment 2: Laypeople, Intervention

Intrigued by the findings of Experiment 1, especially by the overall low number of correctly classified images, we designed an intervention described in Section 13.3.2.3. The goal of this intervention was to reduce confusion between land cover classes and increase classification accuracy.

13.3.2.1 Materials

Same as Experiment 1.

13.3.2.2 Participants

Twenty new lay participants (nonexperts, 11 female) were recruited through AMT; average age 34.2 years; reimbursement: $1.25.

13.3.2.3 Procedure

The main procedural difference between Experiment 1 and Experiment 2 was the inclusion of the NLCD land cover class definitions as defined on the Multi-Resolution Land Characteristics Consortium website (http://www .mrlc.gov) and associating prototypical images for each land cover class with the definition (Figure 13.4). The images were sourced from the DCP and were assigned to each definition based on their associating extracted NLCD class. These definitions and prototypical images were shown to the participants before they began the experiment and were available to revisit throughout the entire experiment.

13.3.2.4 Results

Participants used an average of 10.65 classes (out of the possible 11) with a standard deviation of 0.59. The average grouping time was 822.01 seconds (13 minutes 42 seconds) with a standard deviation of 326 seconds (5 minutes 26 seconds).

The overall accuracy is 44.35%. The improvement in classification by lay participants after the intervention is statistically significant ($X^2 = 5.2807$, df = 1, $p = .02$), with *developed, open space* (dO) specifically benefiting from the intervention, increasing its accuracy 22.15% from Experiment 1. This relatively high accuracy of *developed, open space* (dO) contrasts with the confusion between grass-dominated classes in NLCD that Wickham et al. (2013) note is relatively inaccurate.

Against the relatively low overall accuracy of the classification task, the following land cover classes were significantly classified correctly more frequently than expected by having a standardized residual value greater than 1.96 (Table 13.2): *developed, open space* (dO), *forest* (FO), and *open water* (OW). In contrast, the following land cover classes were significantly classified less correctly than expected by having a standardized residual value less than −1.96: *emergent herbaceous wetlands* (EW), *pasture/hay* (PH), and *woody wetlands* (WW).

When examining the confusion matrix below (Figure 13.5) in comparison to the confusion matrix for Experiment 1 (Figure 13.3), general relationships between classes persist but changes occur in magnitudes of accuracy. As

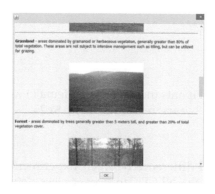

FIGURE 13.4 (See color insert.)
An example of what the laypeople see before and during the experiment.

TABLE 13.2

Standardized Residuals for Experiment 2

	BA	CC	dL	dO	EW	FO	GS	OW	PH	SS	WW
correct	−0.55	−0.01	0.87	3.37	−8.75	8.72	−0.72	12.29	−8.22	0.34	−7.33

	BA	CC	dL	dO	EW	FO	GS	OW	PH	SS	WW
BA	42.14	0.71	0.71	0	7.14	0	0.71	2.14	0.71	45.71	0
CC	8.57	44.29	0	0.71	3.57	0	30	1.43	10	0.71	0.71
dL	0	0	47.86	47.14	0	0	3.57	0	1.43	0	0
dO	0	0	37.86	57.86	0.71	0	2.14	0	1.43	0	0
EW	3.57	1.43	0	0	9.29	34.29	6.43	0	20.71	17.14	7.14
FO	1.43	0	2.86	0.71	4.29	79.29	0	0	1.43	5.71	4.29
GS	15.71	14.29	0	0	0.71	0	41.43	0	13.57	13.57	0.71
OW	0	0	0	0	0	0	0	93.57	0	0	6.43
PH	12.14	4.29	2.86	1.43	6.43	10.71	32.86	0	11.43	17.86	0
SS	35.71	0.71	0.71	0	5	2.14	6.43	0	3.57	45.71	0
WW	0	0	0.71	0.71	3.57	77.86	0	0	0	2.14	15
Total	10.84	5.97	8.5	9.87	3.7	18.57	11.23	8.83	5.84	13.5	3.11

FIGURE 13.5 (See color insert.)
Confusion matrix for Experiment 2 (lay participants with intervention).

in Experiment 1, Experiment 2 also results in almost exclusive confusion of *woody wetland* (WW) being misclassified as *forest* (FO), *barren* (BA) and *shrub/scrub* (SS) being confused with each other, the *developed* classes being confused with each other, and the *emergent herbaceous wetlands* (EW), *grassland* (GS), and *pasture/hay* (PH) confused across many classes. Differences between the experiments were as follows: Participants classified *developed, open space* (dO) more accurately than *developed, low intensity* (dL) in Experiment 2, compared to participants classifying *developed, low intensity* (dL) more accurately than *developed, open space* (dO) in Experiment 1. Participants confused *barren* (BA) with *shrub/scrub* (SS) more often, and confused *shrub/scrub* (SS) with *barren* (BA) less often in Experiment 2, compared to Experiment 1.

13.3.2.5 Discussion

Referring to the grass-dominated classes that Wickham et al. (2013) note are the cause for most confusion in NLCD 2006, while all grass-dominated classes increase in varying degrees of accuracy from Experiment 1 to Experiment 2, *developed, open space* (dO) by far benefits the most from the intervention, increasing 22.15%. The inclusion of the intervention changes *developed, open space* (dO) from a cause of confusion in Experiment 1, similar to NLCD relative confusion, to a class that is relatively accurate. We can speculate then that even though *developed, open space* (dO) may need more distinction to decrease confusion for NLCD, human classification of this class is relatively accurate when provided land cover class definitions. This further would indicate that the land cover class definition for *developed, open space* (dO) creates more clarity, whereas the land cover class definition for *developed, low intensity* (dL) introduces more confusion.

Participants confused *shrub/scrub* (SS) with *barren* (BA) less, but confused *barren* (BA) as *shrub/scrub* (SS) more. This indicates that the intervention convinced participants that *shrub/scrub* (SS) includes more land cover possibilities than perhaps initially thought, while the intervention narrowed the possibilities of what might be considered *barren* (BA).

13.3.3 Experiment 3: Experts

Given the potential for errors based on the accuracy of the level II NLCD data (78%), we also investigated how experts would classify the images we sampled.

13.3.3.1 Materials

Experts were provided the class definitions and visual prototypes of each land cover class, just like in Experiment 2, but on printed-out sheets of paper.

13.3.3.2 Participants

Four experts were solicited that have ecological and geographic information science backgrounds with experience in working with land cover data.

13.3.3.3 Procedure

Each expert viewed the original DCP images on a computer screen, one at a time. As previously mentioned in Section 13.3.3.1, they were each given a printed-out copy of the class definitions and visual prototypes of each land cover class. Each expert viewed the original DCP images on a computer screen one at a time, and recorded their classification choice on a sheet of paper.

13.3.3.4 Results

The classifications by each expert were compared against those from each other expert to establish levels of agreement between experts. We represent agreement as Cohen's kappa coefficient (Figure 13.6) and percent agreement (Figure 13.7). Cohen's kappa coefficient is a measure of inter-rater agreement for categorical objects. It expands on general percentage agreement and takes into account the likelihood of random agreement. The coefficient is defined by the following equation:

$$\kappa = \frac{p_o - p_c}{1 - p_c} \tag{13.1}$$

where p_o is the observed proportion of agreement and p_c is the proportion of agreement expected by chance. If the raters are in perfect agreement, then $\kappa = 1$. If the raters' agreement is what would be expected by chance, then $\kappa = 0$. Foody (2013b) uses this coefficient as an index of the level of inter-rater agreement in an experiment of classifying presence of forest (forest, or nonforest) given DCP images.

The overall accuracy is 48.37%. There is no statistically significant difference between educated lay participants (Experiment 2) and experts ($\chi^2 = 1.52$, df = 1, $p = .22$). The most notable change from Experiment 2 to Experiment 3 is the increase of *cultivated crop* (CC) accuracy (23.57%).

Against the relatively low overall accuracy of the classification task, the following land cover classes were significantly classified correctly more frequently than expected by having a standardized residual value greater than 1.96 (Table 13.3): *cultivated crops* (CC), *developed, low intensity* (dL), *forest* (FO), and *open water* (OW). In contrast, the following land cover classes were significantly classified less correctly than expected by having a standardized residual value less than –1.96: *barren* (BA), *emergent herbaceous wetlands* (EW), *grassland* (GS), *woody wetlands* (WW).

	A	B	C	D
A		0.552	0.595	0.594
B			0.617	0.588
C				0.586
D				

FIGURE 13.6
Cohen's Kappa coefficient values between the experts, A–D.

	A	B	C	D
A		61%	63%	65%
B			66%	63%
C				63%
D				
Full agreement				43%

FIGURE 13.7
Percent agreement between the experts, A–D. Full agreement indicates the percentage that all four experts agreed on the same classification given a DCP image.

TABLE 13.3

Standardized Residuals for Experiment 3

	BA	CC	dL	dO	EW	FO	GS	OW	PH	SS	WW
correct	–3.78	2.16	3.35	–0.21	–3.38	3.35	–2.99	5.73	–0.21	1.37	–5.37

Referencing Wickham's (Wickham et al. 2013) analysis of NLCD accuracy, we see experts performed relatively well, whereas NLCD, Experiment 1, and Experiment 2 did not, in classifying *cultivated crops* (CC). Conversely, the experts match the NLCD and are relatively inaccurate in other grass-dominated classes such as *emergent herbaceous wetlands* (EW) and *grassland* (GS).

When examining the confusion matrix below (Figure 13.8), the following relationships between classes found in Experiments 1 and 2 persist in Experiment 3: *woody wetlands* (WW) is almost exclusively confused as *forest* (FO), the confusion of *barren* (BA) as *shrub/scrub* (SS) continues to increase, the *developed* classes are confused between each other, and the *emergent herbaceous wetlands* (EW), *grassland* (GS), and *pasture/hay* (PH) confused across many classes. Differences between Experiment 3 and the previous experiments are as follows: Experts were more successful in classifying *developed, low intensity* (dL) than *developed, open space* (dO), which is more similar to Experiment 1, and had little confusion when classifying *developed, low intensity* (dL). Although accuracy for *open water* (OW) was high in the previous two experiments, experts were perfect in correctly classifying, and not confusing another class as *open water* (OW). Experts significantly classified *cultivated crops* (CC) correctly more frequently than expected, which was not accomplished in the previous experiments. Experts significantly classified *barren* (BA) and *grassland* (GS) less correctly than expected, which was not accomplished in the previous experiments.

13.3.3.5 Discussion

Referring to the grass-dominated classes that Wickham et al. (2013) note are the cause for most confusion in NLCD 2006, experts are relatively successful in classifying *cultivated crops* (CC), and relatively poor at classifying *grassland* (GS). The *cultivated crops* (CC) success differs from the relative successes of

	BA	CC	dL	dO	EW	FO	GS	OW	PH	SS	WW
BA	14.29	3.57	0	0	14.29	3.57	0	0	0	84.29	0
CC	0	67.86	0	0	0	0	0	0	25	7.14	0
dL	0	3.57	78.57	10.71	0	0	3.57	0	3.57	0	0
dO	0	0	46.43	46.43	0	0	0	0	7.14	0	0
EW	0	17.86	0	0	17.86	42.86	3.57	0	3.57	10.71	3.57
FO	0	0	0	0	0	78.57	0	0	3.57	17.86	0
GS	0	14.29	0	0	14.29	0	21.43	0	21.43	28.57	0
OW	0	0	0	0	0	0	0	100	0	0	0
PH	0	21.43	0	0	0	0	25	0	46.43	7.14	0
SS	17.86	0	0	0	0	3.57	17.86	0	0	60.71	0
WW	0	0	0	0	0	92.86	0	0	0	7.14	0
Total	2.92	11.69	11.36	5.19	4.22	20.13	6.46	9.09	10.06	18.51	0.32

FIGURE 13.8 (See color insert.)
Confusion matrix for Experiment 3 (experts).

NLCD, Experiment 1, and Experiment 2. This could indicate that experts are uniquely capable in recognizing anthropogenically induced patterns relating to crop fields that lay participants are unable to visually recognize, and remote sensors are unable to spectrally identify. Conversely, experts are unsuccessful in recognizing *grassland* (GS), almost equally confusing the class with four other classes. While the previous two experiments struggled with *grassland* (GS), this indicates that experts uniquely conceptualize *grassland* (GS) as a broader class that includes many other land covers that previous experiments do not consider.

The experts' success in recognizing *developed, low intensity* (dL) could indicate their ability to successfully recognize anthropogenic influences in land cover as also shown in the accuracy of *cultivated crops* (CC), and their overall ability to recognize *developed* classes. This is further indicated by confusion of *cultivated crops* (CC) mostly with *pasture/hay* (PH) which is another class that has some degree of anthropogenic influence by definition.

13.4 Conclusions/Outlook

The overall match between participants' classifications and NLCD is rather low (40.19%–48.37%). Accuracy increased statistically significantly using an intervention of providing definitions and prototypical images as examples as mentioned in Section 13.3.2.4. The misclassifications are not random but rather systematic. This is the case on the level of land cover classes as well as on the level of individual images.

Classification accuracy naturally increases the more land cover classes are aggregated. The Anderson Level 1 classification groups *pasture/hay* (PH) and *cultivated crops* (CC) as a single land cover class, all of the *developed* classes as a single land cover class, and *woody wetlands* (WW) and *emergent herbaceous wetlands* (EW) as a single land cover class. Even though other research that analyzes the quality of human classification of land cover (Perger et al. 2012; See et al. 2013) gives the human participants a similar amount of land cover classes to choose from (10 classes compared to our 11), accuracy results are either presented after some level of aggregation to account for potential confusion between similar land cover classes (Perger et al. 2012), or some land cover classes' accuracy results omitted (See et al. 2013). When using humans to classify land cover, the level of aggregation in class representation becomes a heavily influencing factor. As seen in the results above, humans are much more accurate in discerning specific land cover classes, and naturally more accurate overall when distinguishing between fewer land cover classes.

See et al. (2013) show results of *shrub cover, grassland,* and *mosaiced cropland* as having the lowest accuracies. They thus argue that there is a need to provide more examples of how classes that are often confused are represented specifically within Google Earth. When comparing Experiment 3 (experts) results to the land cover classes that were most often confused in See's study (most specifically *shrub cover* and *mosaiced cropland*), human classification accuracy is relatively high in our experiment for those land cover classes when using on the ground photographs. This perhaps indicates the necessity for more contextual information when classifying particular land cover classes, such as shrub and crop type land cover.

When assigning complex tasks to be performed by the crowd, one must ensure that the volunteered data quality is appropriate and sustainable. In the context of land cover validation, humans are very successful in correctly classifying certain land cover via on the ground photographs, and poor in classifying others. Lessons learned from these three experiments are currently integrated in additional experiments that will, among other things, provide additional information about the area to be classified in form of aerial images, ask participants to perform classifications along individual dimensions, and allow for an indication of uncertainty of classifications.

Acknowledgments

This research is funded by the National Science Foundation under grant #0924534. We thank the Degree Confluence Project for permission to use photos from the confluence.org website for our research.

References

Abler, R., J. S. Adams, and P. Gould. 1971. *Spatial Organization: The Geographer's View of the World.* Upper Saddle River, NJ: Prentice-Hall.

Aitkenhead, M. J., and I. H. Aalders. 2011. "Automating land cover mapping of Scotland using expert system and knowledge integration methods." *Remote Sensing of Environment* 115.5: 1285–1295.

Ahlqvist, Ola. 2008. "In search of classification that supports the dynamics of science: the FAO Land Cover Classification System and proposed modifications." *Environment and Planning B* 35.1: 1691–86.

Ahlqvist, Ola. 2012. "Semantic issues in land cover studies: Representation and analysis." *Remote Sensing of Land Use and Land Cover: Principles and Applications.* Edited by Giri, C. Boca Raton, FL: Taylor and Francis: 25–36.

Anderson, J. R. E.E. Hardy, J.T. Roach, and R.E. Witmer, 1976. "A Land Use and Land Cover Classification System for Use with Remote Sensor Data." *Geological Survey Professional Paper* 964.

Arsanjani, Jamal J. M. Helbich, M. Bakillah, J. Hagenauer and A. Zipf 2013. "Toward mapping land-use patterns from volunteered geographic information." *International Journal of Geographical Information Science* 27.12: 2264–2278.

Clery, D. 2011. "Galaxy zoo volunteers share pain and glory of research." *Science* 333: 173–175.

Coeterier, J. F. 1996. "Dominant attributes in the perception and evaluation of the Dutch landscape." *Landscape and Urban Planning* 34.1: 27–44.

Comber, Alexis, Peter Fisher, and Richard Wadsworth. 2004. "Integrating land-cover data with different ontologies: Identifying change from inconsistency." *International Journal of Geographical Information Science* 18.7: 691–708.

Comber, Alexis Peter Fisher, and Richard Wadsworth 2005. "What is land cover?" *Environment and Planning B* 32: 199–209.

Comber, Alexis Peter Fisher, Chris Brunsdon, and Abdulhakim Khmag 2012. "Spatial analysis of remote sensing image classification accuracy." *Remote Sensing of Environment* 127: 237–246.

Comber, Alexis C. Brunsdon, L. See, S. Fritz and I. McCallum 2013. "Comparing expert and non-expert conceptualisations of the land: An analysis of crowd-sourced land cover data." *Spatial Information Theory*. Edited by Thora Tenbrink et al. Springer, Cham, Switzerland: 243–60.

Comber, Alexis, Linda See, and Steffen Fritz. 2014. "The Impact of contributor confidence, expertise and distance on the crowdsourced land cover data quality." *GI_Forum 2014: Geospatial Innovation for Society*.

Foody, G. M. 2002. "Status of land cover classification accuracy assessment." *Remote Sensing of Environment* 80: 185–201.

Foody, G. M. 2008. "Harshness in image classification accuracy assessment." *International Journal of Remote Sensing* 29.11: 3137–3158.

Foody, G. M. L. See, S. Fritz, M. Van der Velde, C. Perger, C. Schill and D. S. Boyd 2013a. "Assessing the accuracy of volunteered geographic information arising from multiple contributors to an internet based collaborative project." *Transactions in GIS* 17.6: 847–860.

Foody, Giles M., and Doreen S. Boyd. 2013b. "Using volunteered data in land cover map validation: Mapping West African forests." *IEEE Journal of Selected Topics in Applied Earth Observations Remote Sensing* 6.3: 1305–1312.

Fritz, Steffen I. McCallum, C. Schill, C. Perger, R. Grillmayer, F. Achard, F. Kraxner and M. Obersteiner 2009. "Geo-Wiki.Org: The use of crowdsourcing to improve global land cover." *Remote Sensing* 1.3: 345–354.

Fry, J. G. Xian, S. Jin, J. Dewitz, C. Homer, L. Yang, C. Barnes, N. Herold and J. Wickham 2011. "Completion of the 2006 national land cover database for the conterminous United States." *Photogrammetric Engineering & Remote Sensing* 77.9: 858–864.

Habron, Dominic. 1998. "Visual perception of wild land in Scotland." *Landscape and Urban Planning* 42.1: 45–56.

Hoffman, Robert, and Richard Pike. 1995. "On the specification of the information available for the perception and description of the natural terrain." *Local Applications of the Ecological Approach to Human Machine Systems*. Edited by Peter Hancock. Hillsdale, NJ: Lawrence Erlbaum Associates.

Iwao, Koki K. Nishida, T. Kinoshita and Y. Yamagata 2006. "Validating land cover maps with Degree Confluence Project information." *Geophysical Research Letters* 33: 1–5.

Jepsen, Martin, and Gregor Levin. 2013. "Semantically based reclassification of Danish land-use and land-cover information." *International Journal of Geographical Information Science* 27.12: 2375–2390.

Khatib, Firas F. DiMaio, S. Cooper, M. Kazmierczyk, M. Gilski, S. Krzywda, H. Zabranska, I. Pichova, J. Thompson, Z. Popovic, M. Jaskolski and D. Baker 2011.

"Crystal structure of a monomeric retroviral protease solved by protein folding game players." *Natural Structural & Molecular Biology* 18.10: 1175–1177.

Kinley, Laura. 2013. "Towards the use of citizen sensor information as an ancillary tool for the thematic classification of ecological phenomena." *Proceedings of the 2nd AGILE (Association of Geographic Information Laboratories for Europe) PhD School 2013*.

Klippel, A., M. Worboys, and M. Duckham. 2008. "Identifying factors of geographic event conceptualisation." *International Journal of Geographical Information Science* 22.2: 183–204.

Malt, B. C. 1995. "Category coherence in cross-cultural perspective." *Cognitive Psychology* 29.2: 85–148.

Mark, D. A. Turk, N. Burenhult, and D. Stea 2011. "Landscape in language: An introduction." *Landscape in Language: An Introduction*. Edited by D. Mark A. Turk, N. Burenhult, and D. Stea. John Benjamins Publishing Company, Amsterdam, The Netherlands.

Pattison, William. 1964. "The four traditions of geography." *Journal of Geography* 63: 211–216.

Perger, Christoph S. Fritz, L. See, C. Schill, M. Van der Velde, I. McCallum, and M. Obersteiner 2012. "A campaign to collect volunteered geographic information on land cover and human impact." *GI_Forum 2012: Geovizualisation, Society and Learning*.

Robbins, Paul. 2003. "Beyond ground truth: GIS and the environmental knowledge of herders, professional foresters, and other traditional communities." *Human Ecology* 31.2: 233–253.

See, Linda A. Comber, C. Salk, S. Fritz, M. Van der Velde, C. Perger, C. Schill, I. McCallum, F. Kraxner, M. Obersteiner, and T. Preis 2013. "Comparing the quality of crowdsourced data contributed by expert and nonexperts." *PLoS ONE* 8.7 (2013): e69958.

Wickham, James D. S. Stehman, L. Gass, J. Dewitz, J. Fry, T. Wade 2013. "Accuracy assessment of NLCD 2006 land cover and impervious surface." *Remote Sensing of Environment* 130: 294–304.

Wilkinson, G. G. 2005. "Results and implications of a study of fifteen years of satellite image classification experiments." *IEEE Transactions on Geoscience and Remote Sensing* 43.3: 433–440.

Index

A

Abiotic surface, 117
Absolute orientation relations, 98
Accumulation process, 60
Active forestry, land with, 180
Activity approach, 66, 71
Africover project, 64
Agricultural land, 16
 categories of, 183
Albanian National Forest Inventory
  (ANFI) project, 69
Amazon Mechanical Turk (AMT),
  302, 303
American Planning Association
  (APA), 273
 Land-Based Classification Standards
  (LBCS), 12
AMT, *see* Amazon Mechanical Turk
Anderson classification system Level I, 14
Anderson LU/LC Classification System
  (LCCS), 90
ANFI project, *see* Albanian National
  Forest Inventory project
Annexation process, 234–235
 identification of, 228
APA, *see* American Planning Association
Aquatic areas, 182
Arable land, 178
ArableNonIrrigated, 236
ArcGIS software, 302
ArcObjects software, ESRI, 278
AreaToBeSubmerged, 161–163
Arrangement, spatial properties, 92
Artificiality/naturality, 92
Artificial non-built up surface, 265
Automatically mapping land use coding
  systems, 278, 279
Automatic classification of quantitative
  data, 149

B

"Baptism of fire," 264
Barren (BA), 304, 307–310

Basic Formal Ontology (BFO), 100, 150
 class definitions, 154
BDOT database, 21
BDOT10k, 28
BFO, *see* Basic Formal Ontology
BioPortal code, 282
Biotic LCC, 123
Biotic surface, 117
Biotic/Vegetation block in the EAGLE
   data model, 124, 125
"Black box" model, 51
Body of water, 148
Bottom-up approach, 87, 194
British Ordnance Survey (OS), 148
Broadleaved forests (FOB), 69–71
Broadleaved wood-lands (WLB), 69, 70
Built-up areas (BU), 71
BULCOVER project, 245

C

CAP, *see* Common Agricultural Policy
Cardinal direction relations, 96
Carpathians, 22
Categorization, 60, 76–78
 defined as, 61
 international efforts at, 62–64
CatScan software, 303, 304
Causal relations, 95
CBC Project, *see* Cross-Border
    Cooperation Project
CC, *see* Crown cover
Central focus of ontology, 150
Change detection, parameterized
    categorizations for, 67–69, 76
 Land-Cover Classification System ,
    69–71
 Land-Use Information System for
    Albania, 71–74
 potential driving factors, at semantic
    levels, 74–75
CIGARS, *see* Common Integrated
    Generalization and
    Aggregation Rules Set